Elementary Particles and Their Currents

A SERIES OF BOOKS IN PHYSICS

Editors:

Henry M. Foley *Malvin A. Ruderman*

Elementary Particles and Their Currents

JEREMY BERNSTEIN *Stevens Institute of Technology*

W. H. FREEMAN AND COMPANY *San Francisco*

Printed in the United States of America (C2)

Library of Congress Catalog Card Number 68-21404

The Hedgehog and the Fox

The Fox knows many things,
but the Hedgehog knows one great thing.

—*Archilochus*

Philosophical Preface

It is probably no exaggeration to say that all of theoretical physics proceeds by analogy. In constructing new theories one takes over those features of the preceding theories that appear "universal" and, in analogy, one constructs the new theories to look as much as possible like the preceding ones. There are two very good reasons for doing this. In the first place the human imagination is, after all, rather limited. The prior theories provide the material on which imagination feeds. Thus even the most "radical" new theory retains some of the flavor of the preceding ones. In the second place the older, accepted theories must have had a domain of validity, and one of the conditions put upon any new theory is that it agree with previous ones in their common domains of validity—the "correspondence principle." The two most successful dynamical theories in physics—Newtonian mechanics and classical electrodynamics—have in common a concern with the way in which certain quantities (the particle coordinates in the one case, and the fields, in the other) change in time. All the theories that have followed after, have, in analogy, dealt with the time development of some object*—a field, a wave function, or an operator. As the theories have become more and more sophisticated these objects have become farther and farther removed from direct physical measurement and, to make matters worse, the equations of motion that govern them have become progressively more difficult to solve. As is well known, there is at present only one theory in elementary particle physics in which one still attempts to solve the equations of motion—quantum electrodynamics. However, most books in elementary particle physics place considerable emphasis on the field equations for the strongly interacting fields and on various, generally unsuccessful, attempts to solve them. It appears to me that this emphasis ignores a change in attitude that has taken place toward these equations on the part of elementary particle physicists, during the last twenty years.

If one reads the issues of the *Physical Review* of the early 1950's and compares them to the contemporary literature, one is immediately struck by the fact that almost no contemporary author is much concerned with the field equa-

* *S*-matrix theory is an exception, but it is not clear, at least to me, that this approach has led to significant new discoveries.

tions, but the literature of the 1950's abounds in papers that start from these equations, especially for the pion-nucleon systems, and which make use of them in an attempt to construct nuclear potentials, phase shifts, and the like, from "first principles." Lack of success is, of course, the mother of moderation, and the present generations of physicists would certainly regard anyone who attempted to start, for example, with the pion-nucleon equations of motion to compute the form factors for electron-proton scattering as being, to say the least, extremely naive. In the 1950's this might have been considered a reasonable thing to try. In the present book the reader will find almost no mention of the field equations. Instead, the Lagrangian is used simply to express symmetries and to generate the currents whose charges obey commutation relations dictated by those symmetries.

It is a remarkable experimental fact that currents play such a universal role in elementary particle physics. If one had asked a physicist in 1950 where currents entered physics he would have answered "electrodynamics," and that would have been that. However, since that time it has become clear that the weak interactions at low energy can be expressed as the coupling of currents to currents and that currents, or vector mesons, play a crucial role in the strong interactions as well. For electrodynamics it is clear why currents are important. The photon exists and, since it is a vector particle, the simplest coupling it can have is to a vector operator—that is, a current. For the weak interactions the role of the currents is still something of a mystery since the "weak photon," which would be exchanged by the weak interactions, has not yet been found. If it is discovered, then the weak interactions will also fit nicely into the pattern of currents and vector mesons. If it does not show up, then the current-current picture must be considered as a convenient phenomenological shorthand.

The discovery of the strongly interacting vector mesons has emphasized another aspect of the role of currents,* the expression of conservation laws. Here, as in most of elementary particle theory, electrodynamics has provided the guide to analogy. In electrodynamics the photon plays a dual role. It is the medium by which charged particles exchange energy and it is also the "gauge field" that expresses the conservation of the electric current. As is well known (we discuss it in detail later in the book), the electromagnetic current can be introduced into the elementary particle Lagrangian in two ways. One may make the replacement $\partial_\mu \longrightarrow \partial_\mu - ieA_\mu$ (that is, A_μ in the "free" Lagrangian, L^0) and then define the current J_μ^γ by the equation

$$eJ_\mu^\gamma = \frac{\delta L}{\delta A_\mu},$$

where L is the Lagrangian obtained after introducing the field A_μ.

However, one may also subject all of the "charged" fields $\psi(x)$ in L^0

---

* J. J. Sakurai, *Ann. Phys.* **11**:1 (1960).

to a space-dependent change in phase, which we can write, symbolically, for $\Lambda(x) \ll 1$,

$$\psi(x) \longrightarrow (1 + i\Lambda(x))\psi(x),$$

and then define

$$J_\mu^\gamma = \frac{\delta L}{\delta \Lambda,_\mu},$$

where L is the Lagrangian obtained after introducing $\Lambda(x)$. From this point of view the conservation of J_μ^γ reflects the fact that L^0 is invariant under phase transformations with constant phase. (See Chapters 2 and 4 for the details.) Hence we can say that the photon is coupled to a current, J_μ^γ, whose conservation expresses the invariance of L^0 under the Abelian gauge group

$$\psi \longrightarrow e^{i\Lambda}\psi.$$

Thus the dual character of the photon.

However, the photon current is not the only exactly conserved current in elementary particle physics. In addition there is the baryon current, whose conservation means that single baryons cannot disappear from the universe. In analogy to electrodynamics one may speculate that this current is also coupled to a vector meson, and, as we shall see later, nature appears to have been kind enough to furnish several strongly interacting vector mesons, one of which may very well couple to the conserved baryon current. In addition to the exactly conserved currents there are also several "approximately" conserved currents. These include the isotopic spin current, the hypercharge or strangeness current, and several additional ones that are conserved in the useful mathematical limit in which certain mass differences among the strongly interacting particles are set equal to zero. These currents may also be coupled to vector mesons, and their approximate conservation again reflects approximate symmetries—however, now more general than Abelian gauge invariance.

In summary: two of the basic interactions in elementary particle physics, the electromagnetic and the weak, take the form of currents coupling to currents or currents coupling to vector mesons. In addition, the third basic interaction, the strong interaction, can be at least partially understood as the coupling of currents to vector mesons. Moreover, these currents and their associated charges express the exact, or approximate, conservation laws obeyed by these interactions. Hence there is good reason for focusing attention on currents as a subject for a book on elementary particles.

The plan of the book is as follows. The first chapter contains a brief review of the elements of field theory, especially of the free fields. Apart from fixing conventions on the Lorentz metric and the like, a sophisticated reader will not learn much by reading it. In Chapter 2 the currents are introduced in a general way and the connection between current conservation and an invariance group is illustrated with the classical case of the group of invariances of a Lagrangian containing two degenerate scalar fields—the group SU_2. Although this chapter deals only with free fields we use throughout the *LSZ* technique for constructing one-particle states via the asymptotic condition. The formulas are then easily generalized to the interacting field case, and, in fact, the third chapter deals with SU_2 in the context of interacting pion-nucleon fields. In the fourth chapter we turn on the electromagnetic field and study SU_2 as a "broken symmetry."

With Chapter 5 the real work of the book gets underway. Here we begin the study of electron-proton scattering. As the reader will discover, there is a detailed analysis of the form factor structure of the matrix elements of the currents as they are limited by Lorentz, parity, time reversal, and charge conjugation invariance. I have chosen to treat the latter symmetries here, rather than in the beginning of the book, since my own experience in reading texts is that it is more convenient to have a general concept and its concrete applications in close juxtaposition, whereas otherwise one tends to forget the former before coming to grip with the latter. In Chapter 6 the experimental electron-proton situation is treated in a preliminary way and it is here that we first meet the vector mesons. Historically the existence of the vector mesons was first conjectured to explain the electron-proton scattering data. Only later were these mesons actually found. In the seventh chapter there is a brief discussion of dispersion relations, which are introduced mainly to give a logical foundation to some of the formulas introduced in the previous chapters. I do not think that dispersion relations "explain" anything in electron-proton scattering, but they do provide a formalism for considering the structure of certain matrix elements and for connecting one matrix element to another.

The next five chapters deal with the weak interactions of nonstrange particles. The discussion divides naturally into the two parts of leptonic interactions and semileptonic interactions. The former provide a perfect example of the present impasse in theoretical elementary particle physics. The theory, as is discussed in detail in the text, involves the coupling of currents to currents and is in splendid agreement with experiment, but it suffers from an almost complete absence of logical foundation. Leptons do not have strong interactions but if we attempt to treat the current-current couplings in anything but lowest-order perturbation theory, we are bedeviled by infinities of the worst sort. Moreover, since the "weak photon" has not yet been observed, the basic role of the currents must be regarded as mysterious.

It should be emphasized that even if the weak vector meson *is* found, this will still leave open the mathematical problem of the structure of its theory, especially in higher orders. In fact, the success of the lowest-order theory is something of a handicap since, so far, experiment has not given any guidance on how to construct the full theory. These matters are discussed, in some detail, in Chapters 7 through 9.

The semileptonic weak interactions, such as ordinary β-decay, involve both lepton and baryons in initial and final states. Because of this, semileptonic decays combine aspects of both the strong and the weak interactions. Chapter 10 treats the vector current part of the semileptonic coupling, which is given, as we shall see, by the isotopic vector part of the electromagnetic current. The vector current is conserved so long as SU_2 is an approximately valid symmetry group. In this chapter I review the consequences of this property of the current. Because the vector current can be identified with the isotopic spin current one is spared the problem of the details of the strong couplings, since one may take the unknown form factors in β-decay matrix elements from electron-proton scattering experiments. However, with the axial vector current there is no such luck, and in Chapters 11 and 12 I review the various attempts at making dynamical models for the axial current. I have included a selection of "proofs" of the Goldberger-Treiman relation and an introduction to the current algebra of $SU_2 \times SU_2$. An exploitation of this algebra leads to the discovery of the celebrated Adler-Weisberger sum role, which is derived in Chapter 12.

The rest of the book deals with the weak and electromagnetic interactions of strange particles. Chapter 13 is a general introduction to SU_3, which now appears to be the correct approximate symmetry group of the elementary particle Lagrangian. Chapter 14 contains a review of some of the calculations using naive quark models. Chapter 15 considers the Cabibbo theory of strange particle decays and some of the results of the $SU_3 \times SU_3$ current algebra. At the very end of the book the reader will find a discussion of CP violation in the K°, $\bar{K}^\circ$ system.

No doubt another physicist writing the same book would have emphasized different aspects of the subject or would have treated the same aspects differently. One of the few pleasures in writing such a book is that the author can present the subject as he would like to see it presented. There are too few books that treat the frontier of elementary particle theory, and if this one encourages someone else to write a better one, then the present author will be among its most enthusiastic readers.

CERN, June 1967

Jeremy Bernstein

Acknowledgments

The work on this book was begun in the summer of 1966 at CERN. I am grateful to the members of the laboratory and, in particular, to J. S. Bell, J. Prentki, and L. Van Hove, both for the hospitality of the laboratory and for numerous discussions. I am also in debt to many physicists who were kind enough to read sections of the manuscript and to point out mistakes and to make suggestions for improvement. Among them I would like to list with special thanks M. A. B. Bég, R. H. Dalitz, G. Feinberg, T. D. Lee, R. H. Marshak, A. Pais, J. J. Sakurai, J. Schwinger, K. Szymanzik, and G. C. Wick. I am also grateful to Andy Cubeta for typing the manuscript, with efficiency and good humor, and to Joel Primack for his help in preparing the manuscript for the printer.

Contents

1
Introduction and Notation

Anyone rash enough to write a book on elementary particles runs the risk of having the material out of date before the book appears in print. He can of course restrict his choice of material to only those items that are really well known, but there are by now several excellent texts that cover the basics of the quantum theory of fields, Feynman diagrams, symmetries, and so on. These books give the student a feeling for the foundations of the subject, but there is still a considerable step between where the books leave off and where the latest papers begin. What this book proposes to do is to help bridge the gap by treating, in great detail, one subject of fundamental importance—currents.

In teaching the theory of elementary particles I find that it is possible to illuminate large areas of the subject by focusing on currents: electromagnetic currents, weak currents, and the currents conserved by the strong interactions. I assume that my students are familiar with the Dirac equation, some aspects of Feynman graphs, and some of the experimental phenomena, such as the existence of mesons, their spins and parities, and the like. I will assume here that the reader knows about what my students do. I will make little or no attempt at mathematical rigor. The derivations given in the text will be complete, but not rigorous. With patience, and mathematical sophistication, they can probably be made rigorous, but it is more important, I think, to know what is going on than to fill in all the epsilons and deltas.

We shall begin with a review of elementary formulas to give the reader an idea of the notation to be used in the text and to indicate what he is expected to know in order to follow the rest of the development.[1]

If A and B are two Lorentz four-vectors, we adopt a metric such that

$$(AB) = \mathbf{A}\cdot\mathbf{B} - A_0B_0 = \boldsymbol{A}\cdot\boldsymbol{B} + A_4B_4, \tag{1.1}$$

with $A_4 = iA_0$. Thus the rest mass of a particle, with $\hbar = c = 1$, is given by the equation

$$m^2 = -(pp) = E_{\mathbf{p}}{}^2 - \mathbf{p}^2 = p_0{}^2 - \mathbf{p}^2. \tag{1.2}$$

1. For an especially careful review of the fundamentals, see J. Hamilton, *The Theory of Elementary Particles*, Oxford (1959).

Therefore the equation of motion for a free neutral scalar or pseudoscalar field is given by

$$(p^2 + m^2)\phi(\mathbf{r}, t) = \left(-\nabla^2 + \frac{\partial^2}{\partial t^2} + m^2\right)\phi(\mathbf{r}, t) = 0. \tag{1.3}$$

The solutions to this equation may be expanded in terms of plane waves and creation and annihilation operators $a(\mathbf{p})$ and $a^\dagger(\mathbf{p})$. [For Φ and Ψ, any two states in Hilbert space, $O^\dagger$, the Hermitian conjugate of O, any operator, is defined by the equations

$$(\Phi, O^\dagger\Psi) = (O\Phi, \Psi) = (\Psi, O\Phi)^*,$$

so that $O^\dagger = O^{\mathsf{T}*}$, where T means transposition and $*$ will always refer to complex conjugation.]

$$\phi(\mathbf{r}, t) = \frac{1}{(2\pi)^{3/2}} \int \frac{d\mathbf{p}}{\sqrt{2p_0}} [a(\mathbf{p})e^{i(px)} + a^\dagger(\mathbf{p})e^{-i(px)}], \tag{1.4}$$

with $(px) = \mathbf{p}\cdot\mathbf{r} - p_0 t$ and

$$[a(\mathbf{p}), a(\mathbf{p}')]_- = [a^\dagger(\mathbf{p}), a^\dagger(\mathbf{p}')]_- = 0 \tag{1.5}$$

and

$$[a(\mathbf{p}), a^\dagger(\mathbf{p}')]_- = \delta^3(\mathbf{p} - \mathbf{p}'). \tag{1.6}$$

For operators A and B we use the notation $[A, B]_\pm = AB \pm BA$. From this commutation relation it follows that

$$[\phi(\mathbf{r}, t), \dot\phi(\mathbf{r}', t')]_{-t=t'} = i\delta^3(\mathbf{r} - \mathbf{r}'). \tag{1.7}$$

We call $|0\rangle$ the vacuum state, and it is defined so that

$$a(\mathbf{p})|0\rangle = 0. \tag{1.8}$$

The state

$$|\mathbf{p}\rangle = a^\dagger(\mathbf{p})|0\rangle \tag{1.9}$$

will be called the one free-particle state.

Then

$$\langle\mathbf{p}'|\mathbf{p}\rangle = \langle 0|a(\mathbf{p}')a^\dagger(\mathbf{p})|0\rangle = \delta^3(\mathbf{p} - \mathbf{p}'). \tag{1.10}$$

The field equations are, in general, derivable from a Lagrangian density $\mathcal{L}(x)$, which yields the Euler-Lagrange equations[2]

$$\partial_\mu\left(\frac{\delta\mathcal{L}}{\delta\phi,_\mu}\right) = \frac{\delta\mathcal{L}}{\delta\phi} \tag{1.11}$$

(Repeated indices, in this case $\mu = 1, 2, 3, 4$, are summed over.) For the neutral spin zero field,

---

2. We use the notation

$$\phi(x)_{,\mu} = \partial_\mu\phi(x) = \frac{\partial}{\partial x_\mu}\phi(x).$$

$$\mathcal{L}(x) = -\frac{1}{2}\left\{\left(\frac{\partial\phi(x)}{\partial x_\mu}\right)^2 + m^2\phi^2\right\}. \tag{1.12}$$

Associated with each field, ϕ, there is a canonical momentum, π_μ, with

$$\pi_\mu = \frac{\delta\mathcal{L}}{\delta\phi,\, \mu} \tag{1.13}$$

and a stress-energy tensor

$$T_{\mu\nu} = \pi_\mu\, \partial_\nu\phi - g_{\mu\nu}\mathcal{L} \tag{1.14}$$

with

$$g_{\mu\nu} = \begin{pmatrix} 1 & 0 & 0 & 0 \\ 0 & 1 & 0 & 0 \\ 0 & 0 & 1 & 0 \\ 0 & 0 & 0 & -1 \end{pmatrix}. \tag{1.15}$$

In terms of $T_{\mu\nu}$ and the four-vector surface element

$$d\sigma_\mu = (dx_2\, dx_3\, dt,\, dx_3\, dt\, dx_1,\, dt\, dx_1\, dx_2,\, -i\, dx_1\, dx_2\, dx_3), \tag{1.16}$$

we may define the energy-momentum operator

$$P_\nu = \int d\sigma_\mu\, T_{\mu\nu}. \tag{1.17}$$

The integral is taken over an arbitrary spacelike surface (a surface in four-space all of whose points are spacelike with respect to each other). If we choose a surface at constant time t, then

$$P_\nu = -i\int d\mathbf{r}\, T_{4\nu}(\mathbf{r}, t). \tag{1.18}$$

As an example we construct P_ν for the scalar field

$$\mathbf{P} = -\tfrac{1}{2}\int d\mathbf{r}\, [\dot{\phi}\, \nabla\phi + \nabla\phi\dot{\phi}], \tag{1.19}$$

with

$$\dot{\phi} = \frac{\partial}{\partial t}\phi. \tag{1.20}$$

($T_{\mu\nu}$ has been symmetrized to make $\mathbf{P}$ manifestly Hermitian.) We use the momentum expansion

$$\begin{aligned} \mathbf{P} &= \tfrac{1}{2}\int d\mathbf{p}\, [a(\mathbf{p})a^\dagger(\mathbf{p}) + a^\dagger(\mathbf{p})a(\mathbf{p})]\mathbf{p} \\ &= \int d\mathbf{p}\, N(\mathbf{p})\mathbf{p}, \end{aligned} \tag{1.21}$$

where we have introduced the number operator $N(\mathbf{p})$:

$$N(\mathbf{p}) = a(\mathbf{p})^\dagger a(\mathbf{p}). \tag{1.22}$$

It follows from the commutation relations that

$$\begin{aligned} N(\mathbf{p})|0\rangle &= 0, \\ N(\mathbf{p})|\mathbf{p}'\rangle &= |\mathbf{p}'\rangle\delta^3(\mathbf{p} - \mathbf{p}'), \end{aligned} \tag{1.23}$$

and

$$\mathbf{P}|\mathbf{p}\rangle = \mathbf{p}|\mathbf{p}\rangle. \tag{1.24}$$

In writing the momentum operator in this way we have dropped a term proportional to $\int d\mathbf{p}\mathbf{p}$ and have used the commutation relations. The Hamiltonian of the system is related to P_4 by the equation

$$\frac{P_4}{i} = H = \int d\mathbf{p}\, p_0 N(\mathbf{p}). \tag{1.25}$$

We have dropped the infinite zero-point energy in arriving at this expression, keeping in mind that adding a constant to a Hamiltonian simply changes the arbitrary reference point with respect to which energies of the system are defined. Only energy differences have physical significance. Clearly

$$H|\mathbf{p}\rangle = p_0|\mathbf{p}\rangle \tag{1.26}$$

and

$$P_\nu|0\rangle = 0. \tag{1.27}$$

The operator P_ν acts like a displacement operator for the space-time points at which the fields are defined. This means that if $O(x)$ is any operator function of the fields,

$$[P_\nu, O(x)]_- = i\,\partial_\nu O(x) \tag{1.28}$$

or[3]

$$O(x) = e^{-i(Px)}O(0)e^{i(Px)}. \tag{1.29}$$

Thus

$$\begin{aligned} \langle 0|\phi(x)|\mathbf{p}\rangle &= e^{i(px)}\langle 0|\phi(0)|\mathbf{p}\rangle \\ &= \frac{e^{i(px)}}{\sqrt{2p_0}(2\pi)^{3/2}}. \end{aligned} \tag{1.30}$$

In other words, $\langle 0|\phi(x)|\mathbf{p}\rangle$ is a complex number function of x that obeys the Klein-Gordon equation. The fact that the field ϕ is free means that it connects the vacuum to the free one-particle state and no others.

We may now repeat this discussion with varying degrees of technical complication for free fields of arbitrary spin. For the moment we shall content ourselves with spin $\frac{1}{2}$ and spin 1.

For spin $\frac{1}{2}$ the fundamental equation is that of Dirac. If $\psi(x)$ is the free Dirac field (ψ is a four-dimensional column vector), we shall write its equation in the form

$$(\gamma_\mu\partial_\mu + m)\psi(x) = 0. \tag{1.31}$$

3. From the definitions of $\mathbf{P}$, H, and $N(\mathbf{p})$, we have $[\mathbf{P}, H]_- = 0$ for free fields. The time dependence in the definition of P_ν is illusory.

The γ's are 4×4 traceless matrices obeying the anticommutation relations

$$\gamma_\mu\gamma_\nu + \gamma_\nu\gamma_\mu = 2\delta_{\mu\nu}. \tag{1.32}$$

We shall always use Hermitian γ's and we shall occasionally use the explicit representation

$$\gamma_4 = \begin{pmatrix} I & 0 \\ 0 & -I \end{pmatrix}, \tag{1.33}$$

where

$$I = \begin{pmatrix} 1 & 0 \\ 0 & 1 \end{pmatrix}, \qquad 0 = \begin{pmatrix} 0 & 0 \\ 0 & 0 \end{pmatrix}, \tag{1.34}$$

and

$$\boldsymbol{\gamma} = \begin{pmatrix} 0 & -i\boldsymbol{\sigma} \\ i\boldsymbol{\sigma} & 0 \end{pmatrix}, \tag{1.35}$$

with

$$\sigma_1 = \begin{pmatrix} 0 & 1 \\ 1 & 0 \end{pmatrix}, \quad \sigma_2 = \begin{pmatrix} 0 & -i \\ i & 0 \end{pmatrix}, \quad \sigma_3 = \begin{pmatrix} 1 & 0 \\ 0 & -1 \end{pmatrix}. \tag{1.36}$$

We can expand $\psi(x)$ in terms of creation and annihilation operators and the Dirac wave functions $u_s(\mathbf{p})$ and $v_s(\mathbf{p})$; $u_s(\mathbf{p})$ is a solution of the momentum space equation

$$(i(\gamma p) + m)u_s(\mathbf{p}) = 0 \tag{1.37}$$

and represents a free spin $\frac{1}{2}$ particle of spin s moving in the direction $\mathbf{p}$, while $v_s(\mathbf{p})$, which satisfies the Dirac equation for negative energy and momentum, is interpreted as representing the antiparticle of mass m moving in the direction $\mathbf{p}$ with spin s. Thus

$$\psi(x) = \frac{1}{(2\pi)^{3/2}} \int d\mathbf{p} \sum_{s=-1/2}^{1/2} [a_s(\mathbf{p})u_s(\mathbf{p})e^{i(px)} + b^\dagger_s(\mathbf{p})v_s(\mathbf{p})e^{-i(px)}]. \tag{1.38}$$

The a's and b's satisfy anticommutation relations

$$\begin{aligned} [a_s(\mathbf{p}), a_{s'}(\mathbf{p}')]_+ &= [a_s(\mathbf{p}), b_{s'}(\mathbf{p}')]_+ = [a_s{}^\dagger(\mathbf{p}), a_s{}^\dagger{}_{'}(\mathbf{p})]_+ \\ &= [a_s(\mathbf{p}), b_s{}^\dagger{}_{'}(\mathbf{p}')]_+ = [b_s{}^\dagger(\mathbf{p}), b_s{}^\dagger{}_{'}(\mathbf{p}')]_+ \\ &= [a_s{}^\dagger(\mathbf{p}), b_{s'}(\mathbf{p}')]_+ = 0 \end{aligned} \tag{1.39}$$

and

$$[a_s(\mathbf{p}), a_s{}^\dagger{}_{'}(\mathbf{p}')]_+ = [b_s(\mathbf{p}), b_s{}^\dagger{}_{'}(\mathbf{p}')]_+ = \delta_{ss'}\,\delta^3(\mathbf{p} - \mathbf{p}'). \tag{1.40}$$

To $\psi(x)$ we may associate $\psi^\dagger(x)$, its Hermitian conjugate field, and

$$\bar{\psi}(x) = \psi^\dagger(x)\gamma_4. \tag{1.41}$$

$\psi^\dagger(x)$ satisfies the equation

$$\nabla\psi^\dagger(x)\cdot\boldsymbol{\gamma} - \partial_4\psi^\dagger(x)\gamma_4 + m\psi^\dagger(x) = 0, \tag{1.42}$$

so that $\bar{\psi}(x)$ satisfies

$$-\partial_\mu \bar{\psi}(x)\gamma_\mu + m\bar{\psi}(x) = 0. \tag{1.43}$$

These equations are derivable from the Lagrange density

$$\mathfrak{L}(x) = -\bar{\psi}(x)(\gamma_\mu \partial_\mu + m)\psi(x). \tag{1.44}$$

$\psi^\dagger(x)$ clearly has the expansion

$$\psi^\dagger(x) = \frac{1}{(2\pi)^{3/2}} \int d\mathbf{p} \sum_{s=-1/2}^{1/2} [a_s^\dagger(\mathbf{p}) u_s^\dagger(\mathbf{p}) e^{-i(px)} + b_s(\mathbf{p}) v_s^\dagger(\mathbf{p}) e^{i(px)}]. \tag{1.45}$$

We have the equal-time anticommutation relation (from the corresponding relations among the a's and b's)

$$[\psi_\rho(\mathbf{r}, t), \psi_{\rho'}^\dagger(\mathbf{r}', t')]_+ = \delta_{\rho\rho'}\, \delta^3(\mathbf{r} - \mathbf{r}'), \tag{1.46}$$

providing that the u's and v's are normalized so that

$$u_s^\dagger(\mathbf{p}) u_{s'}(\mathbf{p}) = v_s^\dagger(\mathbf{p}) v_{s'}(\mathbf{p}) = \delta_{ss'} \tag{1.47}$$

and

$$u_s^\dagger(\mathbf{p}) v_s(\mathbf{p}) = v_s^\dagger(\mathbf{p}) u_s(\mathbf{p}) = 0. \tag{1.48}$$

In general,

$$\begin{aligned} 2E\bar{u}_s(\mathbf{p}) u_s(\mathbf{p}) &= \bar{u}_s(\mathbf{p})[i\gamma \cdot \mathbf{p} + m, \gamma_4]_+ u_s(\mathbf{p}) \\ &= 2m u_s^\dagger(\mathbf{p}) u_s(\mathbf{p}), \end{aligned} \tag{1.49}$$

which fixes the normalization of $\bar{u}$ and $\bar{v}$.

If $|0\rangle$ is again the vacuum state, the one-particle state of momentum $\mathbf{p}$ and spin s is

$$|\mathbf{p}_+\rangle_s = a_s^\dagger(\mathbf{p})|0\rangle \tag{1.50}$$

and the one-antiparticle state of momentum $\mathbf{p}$ and spin s is

$$|\mathbf{p}_-\rangle_s = b_s^\dagger(\mathbf{p})|0\rangle. \tag{1.51}$$

We may use the Lagrangian density $\mathfrak{L}(x)$ to construct the displacement operator P_ν:

$$\begin{aligned} \mathbf{P} &= \frac{1}{2i} \int d\mathbf{r}\, [\psi^\dagger(\mathbf{r}, t) \nabla \psi(\mathbf{r}, t) - \nabla \psi^\dagger(\mathbf{r}, t) \psi(\mathbf{r}, t)] \\ &= \int d\mathbf{p}\, \mathbf{p} \sum_{s=-1/2}^{1/2} [n_+(\mathbf{p})_s + n_-(\mathbf{p})_s], \end{aligned} \tag{1.52}$$

where

$$\begin{aligned} n_+(\mathbf{p})_s &= a_s^\dagger(\mathbf{p}) a_s(\mathbf{p}), \\ n_-(\mathbf{p})_s &= b_s^\dagger(\mathbf{p}) b_s(\mathbf{p}) \end{aligned} \tag{1.53}$$

are to be interpreted as the number operators for particles and antiparticles respectively. Then

$$H = \int d\mathbf{p}\, E(\mathbf{p}) \sum_{s=-1/2}^{1/2} [n_+(\mathbf{p})_s + n_-(\mathbf{p})_s], \tag{1.54}$$

so that **P** is again time-independent. Thus

$$\langle 0|\psi(x)|\mathbf{p}_+\rangle_s = \frac{e^{i(px)}}{(2\pi)^{3/2}} u(\mathbf{p})_s \tag{1.55}$$

is a solution of the Dirac equation with positive energy, and

$$\langle 0|\psi^\dagger(x)|\mathbf{p}_-\rangle_s = \frac{e^{i(px)}}{(2\pi)^{3/2}} v(\mathbf{p})_s \tag{1.56}$$

is a negative-energy solution.

The quantization of the electromagnetic field is a vexing problem; a reader interested in its nuances is advised to consult one of the specialized texts. Since the photon has spin one, it is represented by a vector field $A_\nu(\mathbf{r}, t)$. In the classical theory of fields the condition $\partial_\nu A_\nu = 0$ can be imposed directly and unambiguously on the field A_ν. This condition is necessary in order that the field represent a true spin-one particle rather than a mixture of spin-zero and spin-one particles.[4] In the quantum theory of fields this condition becomes a restriction on the Hilbert space of states. Only those states with the property

$$\partial_\nu A_\nu|\phi\rangle = 0 \tag{1.57}$$

are admitted into the discussions of the electromagnetic field. Furthermore, the massless character of the photon means that the field A_ν has only two independent degrees of polarization. We shall return to this point later when we discuss the gauge invariance of the electromagnetic field. The equation of motion for A_ν is

$$\Box^2 A_\nu = 0, \tag{1.58}$$

and it can be derived from the Lagrange density

$$\mathcal{L}(x) = -\tfrac{1}{2}(\partial_\mu A_\nu)(\partial_\mu A_\nu). \tag{1.59}$$

We can expand A_ν in terms of creation and annihilation operators:

$$A_\nu(\mathbf{r}, t) = \frac{1}{(2\pi)^{3/2}} \int \frac{d\mathbf{p}}{\sqrt{2p_0}} [e^{i(px)} a_\nu(\mathbf{p}) + e^{-i(px)} a_\nu^\dagger(\mathbf{p})]. \tag{1.60}$$

The nonvanishing commutators among the a's can be written for $i = 1, 2, 3$:

$$[a_i(\mathbf{p}), a_j^\dagger(\mathbf{p}')]_- = \delta_{ij}\delta^3(\mathbf{p} - \mathbf{p}') \tag{1.61}$$

---

4. It is also necessary if A_μ is to be coupled to a conserved current that is proportional to $\Box^2 A_\mu$.

and

$$[a_0(\mathbf{p}), a_0{}^\dagger(\mathbf{p})]_- = -\delta^3(\mathbf{p}-\mathbf{p}'), \tag{1.62}$$

with a change in sign for the zeroth component that reflects the Lorentz metric.

The commutation relations imply that, at equal times,

$$[A_\nu(\mathbf{r}, t), \dot{A}_\mu(\mathbf{r}, t)]_- = i\,\delta_{\mu\nu}\,\delta^3(\mathbf{r}-\mathbf{r}'). \tag{1.63}$$

For each degree of freedom associated with one of the $a_\nu(\mathbf{p})$ we may define a number operator

$$n_\nu(\mathbf{p}) = a_\nu{}^\dagger(\mathbf{p})a_\nu(\mathbf{p}). \tag{1.64}$$

However, the supplementary condition $\partial_\mu A_\mu = 0$, applied to the admissible states, means, in terms of $a_\nu(\mathbf{p})$ and $a_\nu{}^\dagger(\mathbf{p})$, that,

$$\langle\psi|(a_0(\mathbf{p}) - \hat{\mathbf{p}}\cdot\mathbf{a}(\mathbf{p}))|\phi\rangle = 0. \tag{1.65}$$

(Recall that for a mass-zero field, $E(\mathbf{p}) = |\mathbf{p}|$.) We can choose[5] $\hat{\mathbf{p}}$ to be in the 3 direction corresponding to photon polarizations in the 1,2 directions, so that, in effect,

$$a_0(\mathbf{p}) - a_3(\mathbf{p}) = 0. \tag{1.66}$$

Thus, with this choice,

$$n_0(\mathbf{p}) - n_3(\mathbf{p}) = \tfrac{1}{2}\{(a_3{}^\dagger(\mathbf{p}) - a_0{}^\dagger(\mathbf{p}))(a_3(\mathbf{p}) + a_0(\mathbf{p}))\} + \tfrac{1}{2}\{(a_3{}^\dagger(\mathbf{p}) + a_0{}^\dagger(\mathbf{p}))(a_3(\mathbf{p}) - a_0(\mathbf{p}))\} = 0, \tag{1.67}$$

which means that only quanta transverse to $\hat{\mathbf{p}}$ are observable. So, in effect,

$$H = \int d\mathbf{p}|\mathbf{p}|\,[n_1(\mathbf{p}) + n_2(\mathbf{p})] \tag{1.68}$$

and

$$\mathbf{P} = \int d\mathbf{p}\,\mathbf{p}[n_1(\mathbf{p}) + n_2(\mathbf{p})], \tag{1.69}$$

while

$$\langle 0|\mathbf{A}_{1,2}(x)|\mathbf{p}\rangle_{1,2} = \frac{e^{i(px)}}{(2\pi)^{3/2}}\,\epsilon_{1,2}, \tag{1.70}$$

with $\epsilon_{1,2}$ being the two polarization directions transverse to $\hat{\mathbf{p}}$.

In concluding this section we note that there is an ambiguity in the Lagrangian method, to which we shall later return. Two Lagrange densities that

---

5. $\hat{\mathbf{A}}$ means a unit vector.
6. The O_μ are restricted if we insist that $\mathcal{L}$ has at most first derivatives in the fields. To see how this happens, consider a typical $\partial_\mu O_\mu$ of the form

$$\partial_\mu O_\mu = (\psi O_{\mu\nu}\psi_{,\nu})_{,\mu}, \tag{1.71}$$

where $O_{\mu\nu}$ is some arbitrary matrix acting on the fields. It must be that the term propor-

differ by a term of the form $\partial_\mu O_\mu(x)$ yield the same equations of motion providing that[6]

$$\int d^4x\, \partial_\mu O_\mu(x) = 0, \tag{1.72}$$

since they both give rise to the same Lagrange function, $L = \int \mathcal{L}(x)\, d^4x$.

---

tional to $(\psi_{,\nu})_{,\mu}$ vanishes, otherwise we would have second derivatives in $\mathcal{L}$. The vanishing can be assured by the antisymmetry of $O_{\mu\nu}$. We can express this condition in a general way by the requirement that

$$\frac{\delta O_\mu}{\delta \psi_{,\nu}} = -\frac{\delta O_\nu}{\delta \psi_{,\mu}}. \tag{1.73}$$

2
Currents: A First Look

A current density $J_\nu(x)$ can be defined as any 4-vector function of the quantized fields ϕ, ψ, etc., with dimension $1/L^3$. Needless to say, there is a vast assortment of operators that fall under the scope of this definition. However, some currents are more interesting than others. The interest in a particular current or set of currents is usually related to three things:

1. Currents can interact with each other.
2. Currents can have very simple divergences; for example, $\partial_\mu J_\mu = 0$.
3. The commutation relations among sets of currents, or integrals of certain current densities, can express symmetries in physical systems.

In the rest of this chapter we shall discuss points 1 and 2, leaving 3 for later.

1. The best-known example of currents that interact is, of course, the mutual coupling of the electromagnetic currents of two charge-bearing particles. In the context of the quantum theory of fields, this coupling is visualized as the exchange of quanta between the charged particles, each of which emits and absorbs quanta. This exchange is pictured by means of a Feynman diagram,

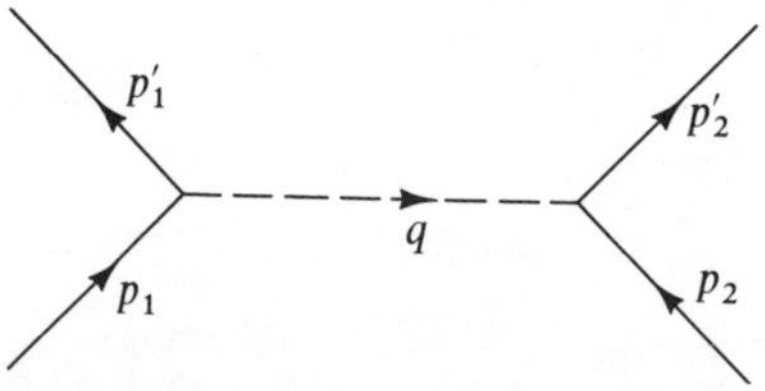

$$q = (p_1 - p_1') = (p_2' - p_2)$$

in which the unbroken lines represent the charged particles and the broken line represents the exchanged quantum. Without, for the moment, exhibiting explicitly the functional form of the currents, the exchange is described mathematically by an integral of the form

$$\int d^4x \int d^4x' J_\mu^{(1)}(x) D_{\mu\nu}(x - x') J_\nu^{(2)}(x'),$$

where the J_μ's are the current densities and $D_{\mu\nu}$ is a function (a propagator or Green's function) that describes the propagation of the quantum. In principle, the electromagnetic currents might have interacted directly at a point, without an exchange of quanta, as shown in the following diagram:

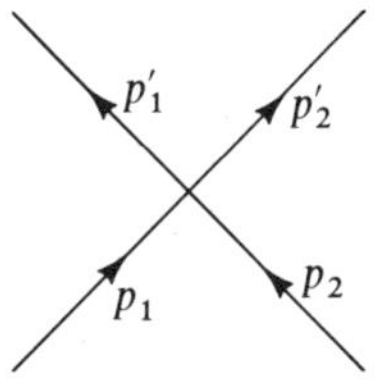

This picture leads to an electromagnetism entirely different from that observed in nature.[1] In particular, the force between two particles would have zero range, whereas the force between charged particles (the Coulomb force) is well known to have infinite range. The range of the force is related to the mass of the exchanged quantum. We can see this in two ways.

a. If a particle emits a quantum of mass m but retains its identity, as in $e \longrightarrow e + \gamma$, it must do so by violating the conservation of energy by an amount

$$\Delta E \sim m.$$

Therefore the quantum must be reabsorbed in a time

$$\Delta t \sim \frac{1}{m}.$$

If it moves with the velocity of light, then the distance it travels from its source is on the average about $1/m$, so that the range of the force is inversely proportional to the mass of the exchanged quantum.

b. We can make this notion more quantitative by considering the equation of the field quanta coupled to their source. To avoid irrelevant complications we can take the quanta to be scalar mesons. If the source is a time-independent function of position, $\rho(\mathbf{r})$, we may write the field equation

$$(-\nabla^2 + m^2)\phi(\mathbf{r}) = 4\pi\rho(\mathbf{r}), \tag{2.1}$$

which has a solution

$$\phi(\mathbf{r}) = \int d\mathbf{r} \frac{e^{-m|\mathbf{r}-\mathbf{r}'|}}{|\mathbf{r}-\mathbf{r}'|} \rho(\mathbf{r}). \tag{2.2}$$

This means that the field falls off exponentially, with a scale $1/m$, away from the source. If two sources exchange quanta there will be an energy of interaction of the form

$$W = -\tfrac{1}{2} \int d\mathbf{r}' \int d\mathbf{r}''[\rho_1(\mathbf{r}')\rho_2(\mathbf{r}'') + \rho_1(\mathbf{r}'')\rho_2(\mathbf{r}')] \frac{e^{-m|\mathbf{r}'-\mathbf{r}''|}}{|\mathbf{r}'-\mathbf{r}''|}. \tag{2.3}$$

If the sources are delta functions, $\delta^3(\mathbf{r} - \mathbf{r}_i)$, then the interaction potential becomes

---

1. The sophisticated reader may want to amuse himself by computing electron-proton scattering using the Feynman diagram shown second, rather than the usual, and physically correct, first diagram.

$$V(|\mathbf{r}_1 - \mathbf{r}_2|) = \frac{e^{-m|\mathbf{r}_1 - \mathbf{r}_2|}}{|\mathbf{r}_1 - \mathbf{r}_2|}, \tag{2.4}$$

which reduces to the Coulomb potential for $m \longrightarrow 0$. In the quantum field theory the D functions (the propagators) have an exponential falloff for spacelike $x - x'$, so that the connection between range and mass holds there as well. Most of the currents that we will encounter in this book appear to be coupled to field quanta with couplings of the general form $J_\mu(x)V_\mu(x)$, implying that these field quanta can have spin 1. As we shall see, some of these quanta have been observed, but the existence of others is still an open experimental question.

2. Many of the interesting currents have very simple divergences: either the currents are exactly conserved or their divergence is a simple enough function of the fields that its properties can be investigated. The advantage of a conserved current is that it leads automatically to a constant of the motion—a good quantum number. Not all of the constants of the motion can be related to currents. (The conservation of parity in strong interactions gives an example of a constant operator not derivable from a current.) Those that can be, we shall call "charges." Among the charges that are conserved by one or more of the known interactions are the electric, isotopic spin, and hyper charges. All of these will be discussed later in the book.

To see how a conserved current leads to a constant charge, consider

$$\int d\mathbf{r}\, \partial_\mu J_\mu = \frac{\partial}{\partial t} \int d\mathbf{r}\, J_0(\mathbf{r}, t) = 0. \tag{2.5}$$

The second step follows if $\mathbf{J}(\mathbf{r}, t)$ satisfies

$$\int d\mathbf{r} \nabla \cdot \mathbf{J}(\mathbf{r}, t) = 0. \tag{2.6}$$

The Lorentz scalar quantity,[2]

$$Q_J = \int d\mathbf{r}\, J_0(\mathbf{r}, t), \tag{2.7}$$

then commutes with H, the Hamiltonian governing the time dependence of the fields out of which J_μ is composed:

---

2. Although Q_J is not *manifestly* a Lorentz scalar, it follows from the current conservation, as in the classical theory of fields, that it is. To see this we may use the four-dimensional Gauss identity. We suppose that $J_\mu(\mathbf{r}, t)$ vanishes outside some finite volume. Thus

$$0 = \int (\partial_\mu J_\mu)\, d^4x = \int J_\mu\, d\sigma_\mu.$$

If the region of integration is taken to be a four volume chosen so that J_μ vanishes on s_1 and s_2, then

$$\int J_0\, d\mathbf{r} = \int J_0'\, d\mathbf{r},$$

$$[H, Q_J]_- = 0. \tag{2.8}$$

If $|\mathbf{p}\rangle$ is any state which is an eigenstate of H with energy $E(\mathbf{p})$—for example, a one-free-particle state—then, for constant Q_J,

$$HQ_J|\mathbf{p}\rangle = E(\mathbf{p})Q_J|\mathbf{p}\rangle, \tag{2.9}$$

which means that either $|\mathbf{p}\rangle$ is an eigenstate of Q_J or there is a degeneracy in the spectrum of H.

With future applications in mind we derive a simple condition on the matrix elements of a conserved current, J_μ, taken between eigenstates of the displacement operator P_μ. From current conservation we have

$$0 = \partial_\mu J_\mu(x) = \frac{1}{i}[P_\mu, J_\mu(x)]_-, \tag{2.10}$$

and hence

$$\begin{aligned} 0 &= \langle \mathbf{p}'|[P_\mu, J_\mu(x)]_-|\mathbf{p}\rangle = (p' - p)_\mu \langle \mathbf{p}'|J_\mu(x)|\mathbf{p}\rangle \\ &= e^{i((p-p')x)}(p' - p)_\mu \langle \mathbf{p}'|J_\mu(0)|\mathbf{p}\rangle. \end{aligned} \tag{2.11}$$

Since the current J_μ is a function of the field operators, finding its divergence is, in general, not possible without knowing the field equations. However, for the currents that appear to play the most important role in elementary particle physics, this problem can be bypassed. These currents reflect symmetry properties of the Lagrangian density $\mathcal{L}$. Suppose that $\mathcal{L}$ is a function of the fields $\psi_1, \ldots, \psi_n$, and, at most, first-order in the derivatives of these fields. (This is really not an extra assumption, since if $\mathcal{L}$ involves only derivatives of finite order it can always be written in terms of first-order derivatives by the introduction of auxiliary fields and Lagrange multipliers.) Let $\Lambda(x)$ be an infinitesimal c-number function of x and let $F_i(\psi_1, \ldots, \psi_n)$ be n essentially arbitrary functions, with $i = 1, \ldots, n$, of the fields $\psi_1, \ldots, \psi_n$. In practice, the F's that we shall consider are usually extremely simple functions of the ψ's.

Consider the transformation

$$\psi_i(x) \longrightarrow \psi_i(x) + \Lambda(x)F_i(\psi_1, \ldots, \psi_n). \tag{2.12}$$

We shall study the change in $\mathcal{L}$ arising from this transformation. From the fact

---

where J_0 and J_0' are related by an arbitrary Lorentz transformation. See, for example, J. Anderson, *Principles of Relativity Physics*, Academic Press, New York (1967).

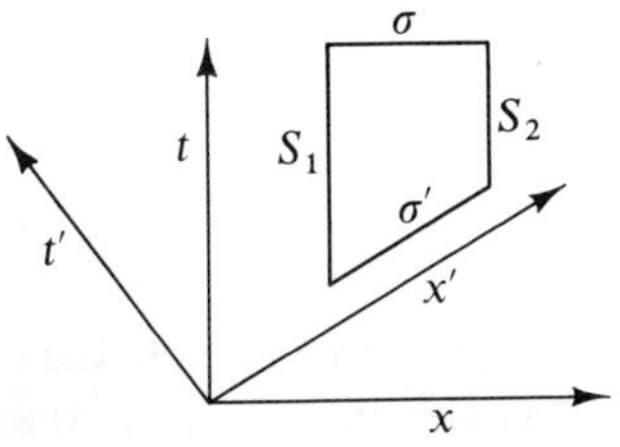

that Λ is infinitesimal, so that $\Lambda^2 \simeq 0$, it follows that both $\delta\mathfrak{L}/\delta\Lambda$ and $\delta\mathfrak{L}/\delta\Lambda_{,\mu}$ are independent of Λ. We will define the current generated by this transformation as

$$J_\mu(x) = \frac{\delta\mathfrak{L}(x)}{\delta\Lambda_{,\mu}}, \tag{2.13}$$

and we shall prove that, for this current,

$$\partial_\mu J_\mu = \frac{\delta\mathfrak{L}}{\delta\Lambda}. \tag{2.14}$$

It then follows that if for *constant* Λ the transformation above is an invariance of $\mathfrak{L}$, then the current generated for nonconstant Λ will be conserved. If the transformation is not an invariance, then the divergence of the associated current is given directly by $\delta\mathfrak{L}/\delta\Lambda$.

To prove this result, consider

$$\frac{\delta\mathfrak{L}}{\delta\Lambda} = \frac{\delta\mathfrak{L}}{\delta\psi_i}\frac{\delta\psi_i}{\delta\Lambda} + \frac{\delta\mathfrak{L}}{\delta\psi_{i,\mu}}\frac{\delta\psi_{i,\mu}}{\delta\Lambda}. \tag{2.15}$$

But

$$\frac{\delta\psi_i}{\delta\Lambda} = F_i \tag{2.16}$$

and

$$\psi_{i,\mu} \longrightarrow \psi_{i,\mu} + \Lambda_{,\mu}F_i + \Lambda F_{i,\mu}, \tag{2.17}$$

or

$$\frac{\delta\psi_{i,\mu}}{\delta\Lambda} = F_{i,\mu}. \tag{2.18}$$

Thus

$$\frac{\delta\mathfrak{L}}{\delta\Lambda} = \frac{\delta\mathfrak{L}}{\delta\psi_i}F_i + \frac{\delta\mathfrak{L}}{\delta\psi_{i,\mu}}F_{i,\mu}, \tag{2.19}$$

which is, as expected, independent of Λ. Moreover,

$$\frac{\delta\mathfrak{L}}{\delta\Lambda_{,\mu}} = \frac{\delta\mathfrak{L}}{\delta\psi_i}\frac{\delta\psi_i}{\delta\Lambda_{,\mu}} + \frac{\delta\mathfrak{L}}{\delta\psi_{i,\mu}}\frac{\delta\psi_{i,\mu}}{\delta\Lambda_{,\mu}} = \frac{\delta\mathfrak{L}}{\delta\psi_{i,\mu}}F_i. \tag{2.20}$$

Therefore

$$\partial_\mu\frac{\delta\mathfrak{L}}{\delta\Lambda_{,\mu}} = \partial_\mu\frac{\delta\mathfrak{L}}{\delta\psi_{i,\mu}}F_i + \frac{\delta\mathfrak{L}}{\delta\psi_{i,\mu}}F_{i,\mu}. \tag{2.21}$$

We may now use the Euler-Lagrange equation,

$$\partial_\mu\frac{\delta\mathfrak{L}}{\delta\psi_{i,\mu}} = \frac{\delta\mathfrak{L}}{\delta\psi_i}, \tag{2.22}$$

to reduce the right side of Eq. (2.21) to $\delta\mathfrak{L}/\delta\Lambda$ and hence to complete the proof. Note that the ambiguity mentioned at the end of Chapter 1 means that two

Lagrange densities differing by a total divergence yield the same equations of motion but different currents.

However, the contribution to the current from a term $\partial_\mu O_\mu$ to the Lagrangian will not change the total charge under the condition

$$\frac{\delta O_\mu}{\delta \psi_{i,\nu}} = -\frac{\delta O_\nu}{\delta \psi_{i,\mu}}, \tag{2.23}$$

discussed at the end of Chapter 1. For the contribution to the current from the added term is

$$J_\nu = \left(\frac{\delta \partial_\mu O_\mu}{\delta \Lambda_{,\nu}}\right) = \partial_\mu \left(\frac{\delta O_\mu}{\delta \Lambda_{,\nu}}\right), \tag{2.24}$$

using the condition above, which guarantees that $\partial_\mu O_\mu$ contains only first derivatives. Thus

$$J_4 = \nabla \cdot \frac{\delta \mathbf{O}}{\delta \Lambda_{,4}} \tag{2.25}$$

which integrates to zero over all space. (See G. Wentzel, *Preludes in Theoretical Physics*, North-Holland Publishing Company, Amsterdam, p. 199 and ff., for an especially nice discussion of this point.)

We may now consider some elementary applications of this method.

For a single, free scalar field we are not led to any interesting currents. But suppose we consider two free scalar fields[3] ϕ_1 and ϕ_2, degenerate in mass. (This problem is related to the isotopic spin formulation of pion physics, as we shall shortly see.) Their Lagrangian will be of the form

$$\mathcal{L} = -\tfrac{1}{2}(\boldsymbol{\phi}_{,\mu}\cdot\boldsymbol{\phi}_{,\mu} + m^2\boldsymbol{\phi}\cdot\boldsymbol{\phi}), \tag{2.26}$$

where

$$\boldsymbol{\phi}\cdot\boldsymbol{\phi} = \phi_1\phi_1 + \phi_2\phi_2. \tag{2.27}$$

Let

$$\begin{aligned} \phi_1 &\longrightarrow \phi_1 + \Lambda\phi_2, \\ \phi_2 &\longrightarrow \phi_2 - \Lambda\phi_1 \end{aligned} \tag{2.28}$$

In the space of 1, 2 this corresponds to the matrix transformation

$$\begin{pmatrix}\phi_1' \\ \phi_2'\end{pmatrix} = \begin{pmatrix}1 & \Lambda \\ -\Lambda & 1\end{pmatrix}\begin{pmatrix}\phi_1 \\ \phi_2\end{pmatrix}, \tag{2.29}$$

which, since Λ is infinitesimal, is unitary and unimodular.[4] This transformation corresponds to a real rotation of the fields. The degeneracy in mass implies that $\mathcal{L}$ is invariant under these unitary, unimodular transformations for constant Λ. For $\Lambda = \Lambda(x)$,

3. We assume that $[\phi_1, \phi_2]_- = 0$.
4. Unit determinant.

$$\phi_{1,\mu}\phi_{1,\mu} + \phi_{2,\mu}\phi_{2,\mu} \longrightarrow \boldsymbol{\phi}_{,\mu}\cdot\boldsymbol{\phi}_{,\mu} + 2\Lambda_{,\mu}(\phi_2\phi_{1,\mu} - \phi_1\phi_{2,\mu}). \tag{2.30}$$

Thus[5]

$$J_{\mu 3} = \phi_2\phi_{1,\mu} - \phi_1\phi_{2,\mu} \tag{2.31}$$

is a conserved current. The corresponding charge is

$$Q_3 = \int d\mathbf{r}(\phi_1\dot{\phi}_2 - \phi_2\dot{\phi}_1). \tag{2.32}$$

Apart from the general argument, it is obvious from the equations of motion that this charge is time-independent. It is interesting to study the effect of this charge on the one-particle states $|\mathbf{p}\rangle_1$ and $|\mathbf{p}\rangle_2$ corresponding to the fields 1 and 2, where, for example,

$$|\mathbf{p}\rangle_1 = a_1{}^\dagger(\mathbf{p})|0\rangle. \tag{2.33}$$

We could apply this definition directly but, in anticipation of later work on interacting fields, it is useful to carry out the whole computation in a way that can be directly applied to the general case as well. We therefore proceed in two steps. First we compute $[Q_3, \phi_i(\mathbf{r}, t)]_-$. Q_3 is independent of time, so that this commutator can be explicitly evaluated using the equal time commutators between ϕ_i and $\dot{\phi}_i$. Since the commutation relations between charges and fields play an essential role in what follows, we shall work this one out in detail:

$$[Q_3, \phi_i(\mathbf{r}, t)]_- = \int d\mathbf{r}'[(\phi_2(\mathbf{r}', t)\dot{\phi}_1(\mathbf{r}', t) - \phi_1(\mathbf{r}', t)\dot{\phi}_2(\mathbf{r}', t)), \phi_i(\mathbf{r}, t)]_- \tag{2.34}$$

Let $i = 1$; then

$$[Q_3, \phi_1(\mathbf{r}, t)]_- = i\int d\mathbf{r}'\, \phi_2(\mathbf{r}', t)\delta^3(\mathbf{r}' - \mathbf{r}) = i\phi_2(\mathbf{r}, t) \tag{2.35}$$

and

$$[Q_3, \phi_2(\mathbf{r}, t)]_- = -i\phi_1(\mathbf{r}, t). \tag{2.36}$$

Thus

$$[Q_3, \phi_i] = i\epsilon_{ij3}\phi_j, \tag{2.37}$$

where ϵ_{ijk} is the totally antisymmetric alternating symbol, with

$$\epsilon_{123} = 1.$$

All other values are fixed by the assumption of antisymmetry in all index pairs.

Next we construct the one-particle state in an equivalent manner to the one used in Chapter 1 but with a technique that generalizes to interacting fields. We shall make use of a limiting process designed to pick out from the Fourier decomposition of a field operator a single Fourier component.

We note that

$$\lim_{t\to\pm\infty} e^{i\alpha t} = \begin{cases} 1 & \text{for } \alpha = 0, \\ 0 & \text{for } \alpha \neq 0, \end{cases} \tag{2.38}$$

5. The reason for the subscript 3 will become clear later.

with the understanding that these limits are to be taken under appropriate integrals. Consider, for example,

$$\lim_{t\to\pm\infty}\left(i\int d\mathbf{r}\left\{\phi_i(\mathbf{r},t)\frac{\partial}{\partial t}\frac{e^{i(px)}}{(2\pi)^{3/2}\sqrt{2p_0}}\right\}\right)\cdot\equiv\cdot\lim_{t\to\pm\infty}(\quad),$$

where p is some specified four-momentum. Using the limiting procedure, expanding ϕ_i in creation and annihilation operators, and taking the time derivative, we have

$$\lim_{t\to\pm\infty}(\quad)|0\rangle=\int d\mathbf{r}\,d\mathbf{p}'\,\frac{a_i(\mathbf{p})^\dagger e^{-i(\mathbf{p}'-\mathbf{p})\cdot\mathbf{r}}p_0|0\rangle}{(2\pi)^3\sqrt{4p_0p_0'}}. \tag{2.39}$$

The space integral gives $(2\pi)^3\delta^3(\mathbf{p}-\mathbf{p}')$, and therefore

$$\lim_{t\to\pm\infty}(\quad)|0\rangle=\tfrac{1}{2}a_i{}^\dagger(\mathbf{p})|0\rangle. \tag{2.40}$$

Likewise,

$$\lim_{t\to\pm\infty}\left(-i\int d\mathbf{r}\,\dot{\phi}_i(\mathbf{r},t)\frac{e^{i(px)}}{(2\pi)^{3/2}\sqrt{2p_0}}\right)|0\rangle=\tfrac{1}{2}a_i{}^\dagger(\mathbf{p})|0\rangle. \tag{2.41}$$

Thus

$$|\mathbf{p}\rangle_i=\lim_{t\to\pm\infty}\left(i\int d\mathbf{r}\left\{\phi_i(\mathbf{r},t)\frac{\partial}{\partial t}\frac{e^{i(px)}}{(2\pi)^{3/2}\sqrt{2p_0}}-\frac{\dot{\phi}_i(\mathbf{r},t)e^{i(px)}}{(2\pi)^{3/2}\sqrt{2p_0}}\right\}\right)|0\rangle. \tag{2.42}$$

This rather complicated looking operator will also serve to manufacture one-particle states from the vacuum for interacting scalar fields. Now, since Q_3 is independent of time we can take it under the limit,

$$Q_3|\mathbf{p}\rangle_i=\lim_{t\to\pm\infty}i\int d\mathbf{r}\,Q_3\left\{\phi_i(\mathbf{r},t)\frac{\partial}{\partial t}\frac{e^{i(px)}}{(2\pi)^{3/2}\sqrt{2p_0}}-\frac{\dot{\phi}_i(\mathbf{r},t)e^{i(px)}}{(2\pi)^{3/2}\sqrt{2p_0}}\right\}|0\rangle. \tag{2.43}$$

To proceed we note that, since $\dot{Q}_3=0$,

$$\frac{\partial}{\partial t}([Q_3,\phi_i]_-)=[Q_3,\dot{\phi}_i]_-=i\epsilon_{ij3}\dot{\phi}_j. \tag{2.44}$$

Moreover,

$$Q_3|0\rangle=0. \tag{2.45}$$

This follows explicitly from the fact that, in its representation in terms of creation and annihilation operators,

$$Q_3=i\int d\mathbf{p}(a_1(\mathbf{p})a_2{}^\dagger(\mathbf{p})-a_1{}^\dagger(\mathbf{p})a_2(\mathbf{p})). \tag{2.46}$$

But the a_i's annihilate the vacuum.

We can also argue in a somewhat different way and in a way that also applies to interacting fields, that $Q_3|0\rangle=0$. We assume that the Hamiltonian is such that the vacuum state is unique; that is, we assume that there exists one, and only one, normalized state such that $H|0\rangle = 0$. We will now show that if $Q_3|0\rangle \neq 0$, there must be a degeneracy of the vacuum. Since

$$[Q_3, H]_- = 0, \tag{2.47}$$

we can always choose $|0\rangle$ so that it is an eigenstate of Q_3. Therefore

$$Q_3|0\rangle = q|0\rangle. \tag{2.48}$$

But consider the transformation defined by the equations

$$\begin{aligned} c\phi_1 c^{-1} &= \phi_1, \\ c\phi_2 c^{-1} &= -\phi_2. \end{aligned} \tag{2.49}$$

Clearly for the Hamiltonian corresponding to the Lagrangian of Eq. (2.26)

$$cHc^{-1} = H \tag{2.50}$$

and

$$c^2 = 1, \tag{2.51}$$

and thus

$$\begin{aligned} c\dot{\phi}_1 c^{-1} &= \dot{\phi}_1, \\ c\dot{\phi}_2 c^{-1} &= -\dot{\phi}_2. \end{aligned} \tag{2.52}$$

Therefore

$$cQ_3c^{-1} = -Q_3. \tag{2.53}$$

Now we can see that for $q \neq 0$, $c|0\rangle$ is another vacuum state orthogonal to $|0\rangle$. First,

$$Hc|0\rangle = 0, \tag{2.54}$$

since c is time-independent. Therefore $c|0\rangle$ is a vacuum state. Second,

$$Q_3c|0\rangle = -qc|0\rangle. \tag{2.55}$$

Therefore $|0\rangle$ and $c|0\rangle$ are two states with the same energy but with different eigenvalues of the Hermitian operator Q_3. Thus

$$\langle 0|cQ_3|0\rangle = q\langle 0|c|0\rangle = -q\langle 0|c|0\rangle. \tag{2.56}$$

Therefore, if $q \neq 0$, $|0\rangle$ and $c|0\rangle$ are orthogonal, which means that the vacuum is degenerate. In all of our later work we shall explicitly assume that the vacuum is nondegenerate.[6]

It is now simple to finish computing $Q_3|\mathbf{p}\rangle_i$. Making use of the commutation relations and the fact that $Q_3|0\rangle = 0$, it follows at once that

$$Q_3|\mathbf{p}\rangle_i = i\epsilon_{ij3}|\mathbf{p}\rangle_j. \tag{2.57}$$

6. We may note that this argument fails if, instead of using Q_3, the charge, we were to use the charge density $J_0(x)_3$. Indeed, $[H, J_0(x)_3]_- \neq 0$. Only $[H, Q_3]_- = 0$. This is fortunate, since it can be shown that if $J_0(x)_3|0\rangle = 0$, then $J_0(x)_3 = 0$, a result that does *not* hold for

Thus the states $|\mathbf{p}\rangle_i$ are not eigenstates of Q_3.[7] Since $[Q_3, H]_- = 0$ we can find states that are normalized eigenstates of Q_3 and H.

Let

$$|\mathbf{p}\rangle_+ = \frac{|\mathbf{p}_1\rangle + i|\mathbf{p}_2\rangle}{\sqrt{2}} \tag{2.58}$$

and

$$|\mathbf{p}\rangle_- = \frac{|\mathbf{p}_1\rangle - i|\mathbf{p}_2\rangle}{\sqrt{2}}. \tag{2.59}$$

We see at once that

$$\begin{aligned} Q_3|\mathbf{p}\rangle_+ &= |\mathbf{p}\rangle_+, \\ Q_3|\mathbf{p}\rangle_- &= -|\mathbf{p}\rangle_-. \end{aligned} \tag{2.60}$$

These states have "charge" ± 1.

If we define

$$\phi^\pm = \frac{\phi_1 \pm i\phi_2}{\sqrt{2}}, \tag{2.61}$$

then $\phi^\pm$ can be identified as the fields that create the states of $\pm$ charge. Using the c defined above, we note that

$$c\phi^\pm c^{-1} = \phi^\mp, \tag{2.62}$$

so that this c can be identified with the operation of "charge conjugation," the operation that takes particle into antiparticle. For a scalar field the antiparticle to ϕ^+ is ϕ^-. It is useful to write $\mathcal{L}$ and Q_3 directly in terms of $\phi^\pm$. Note that at equal times

$$[\phi^+, \dot{\phi}^-]_- = [\phi^-, \dot{\phi}^+]_- = i\delta^3(\mathbf{r} - \mathbf{r}'). \tag{2.63}$$

Thus

$$\mathcal{L} = -\tfrac{1}{2}\{\partial_\mu\phi^+\partial_\mu\phi^- + \partial_\mu\phi^-\partial_\mu\phi^+ + m^2[\phi^+\phi^- + \phi^-\phi^+]\}. \tag{2.64}$$

The transformation

$$\begin{pmatrix}\phi_1' \\ \phi_2'\end{pmatrix} = \begin{pmatrix}1 & \Lambda \\ -\Lambda & 1\end{pmatrix}\begin{pmatrix}\phi_1 \\ \phi_2\end{pmatrix}, \tag{2.65}$$

which left $\mathcal{L}$ invariant, becomes in the $\pm$ language

$$\begin{aligned} \phi^{+\prime} &= (1 - i\Lambda)\phi^+, \\ \phi^{-\prime} &= (1 + i\Lambda)\phi^-. \end{aligned} \tag{2.66}$$

---

the nonlocal operator Q_3 (see P. Federbush and K. A. Johnson, *Phys. Rev.* **120**:1926 (1960)).

7. As mentioned above, this implies that $|\mathbf{p}\rangle_1$ and $|\mathbf{p}\rangle_2$ are degenerate in energy, as is clear anyway from the symmetry of the Lagrangian.

Thus the rotational invariance in the 1, 2 space corresponds to a phase invariance in the $\pm$ space, since

$$e^{i\Lambda} = (1 + i\Lambda) \qquad \text{for } \Lambda^2 \ll 1. \tag{2.67}$$

Indeed, we can use the phase transformation to generate J_μ and Q_3 directly as a function of the $\pm$ fields, since the theorem proved above applies to these fields as well as to the fields 1, 2.

Thus

$$J_\mu = \frac{i}{2}\{[\phi^+\partial_\mu\phi^- - \partial_\mu\phi^+\phi^-] - [+ \leftrightarrow -]\} \tag{2.68}$$

and

$$Q_3 = \frac{i}{2}\int d\mathbf{r}\,\{[\dot\phi^+\phi^- - \phi^+\dot\phi^-] - [+ \leftrightarrow -]\}, \tag{2.69}$$

with the definitions

$$\begin{aligned} n_+ &= a_+{}^\dagger a_+, \\ n_- &= a_-{}^\dagger a_-, \\ Q_3 &= \int d\mathbf{p}\,[n_+(\mathbf{p}) - n_-(\mathbf{p})]. \end{aligned} \tag{2.70}$$

From the commutation relation

$$[Q_3, \phi_i]_- = i\epsilon_{ij3}\phi_j, \tag{2.71}$$

it follows that

$$[Q_3, \phi^\pm]_- = \pm\phi^\pm. \tag{2.72}$$

This commutation relation implies that Q_3 is the *generator* of phase transformations on the $\phi^\pm$. By *generator* we mean the following.

Consider the unitary operators (Q is Hermitian and Λ is real):

$$U(\Lambda) = e^{i\Lambda Q}. \tag{2.73}$$

These $U(\Lambda)$ form a group in the obvious sense that

$$\begin{aligned} U(\Lambda_1)U(\Lambda_2) &= U(\Lambda_1 + \Lambda_2), \\ U(\Lambda)^{-1} &= U(-\Lambda) = U(\Lambda)^\dagger, \end{aligned} \tag{2.74}$$

and all of the elements of this group commute with each other. The group of the $U(\Lambda)$ is an example of a so-called Abelian gauge group. We may now study the action of this group on the fields $\phi^\pm$. Let A be any Hermitian operator and O be *any* operator and consider $F(\Lambda)$, where

$$F(\Lambda) = e^{i\Lambda A}Oe^{-i\Lambda A}. \tag{2.75}$$

Then

$$F'(\Lambda) = i(AF(\Lambda) - F(\Lambda)A) = i[A, F(\Lambda)]_- \tag{2.76}$$

and

$$F''(\Lambda) = -[A, [A, F(\Lambda)]_-]_-. \tag{2.77}$$

Expanding $F(\Lambda)$ in a Taylor series in Λ, we have

$$F(\Lambda) = \mathrm{F}(0) + \Lambda F'(0) + \Lambda^2 \frac{F''(0)}{2!} + \cdots = O + i\Lambda[A, O]_- - \frac{\Lambda^2}{2!}[A, [A, O]_-]_- + \cdots. \tag{2.78}$$

Letting $O = \phi^\pm$ and $A = Q_3$, we have

$$e^{i\Lambda Q_3}\phi^\pm e^{-i\Lambda Q_3} = \phi^\pm \pm i\Lambda\phi^\pm - \Lambda^2\phi^\pm + \cdots = e^{\pm i\Lambda}\phi^\pm. \tag{2.79}$$

Thus, when applied to the fields, the operator $U(\Lambda) = e^{i\Lambda Q_3}$ generates a change in phase by the amount $\pm\Lambda$. The charge Q_3 generates infinitesimal changes in phase in the sense that for $\Lambda \ll 1$

$$\phi^{\pm\prime} = \phi^\pm + i\Lambda[Q_3, \phi^\pm]_-. \tag{2.80}$$

It is characteristic of the differentiable groups (Lie groups) such as an Abelian gauge group that the finite transformations can be determined from the infinitesimal ones by making use of the commutation relations between the generators and the fields.

We may make two further remarks before passing on to the Fermion case.

First, we note that in deriving Q_3 by transforming the Lagrangian under $\phi^\pm \longrightarrow (1 \mp i\Lambda)\phi^\pm$, the Q_3 that we have derived generates in turn the infinitesimal transformation $(1 + i\Lambda Q_3)\phi^\pm(1 - i\Lambda Q_3) = (1 \pm i\Lambda)\phi^\pm$ on the fields. The sign of Λ is opposite in the two cases. This is to be expected, since in the method that we use to generate currents we transform the fields as if they were state vectors. This is an unorthodox but perfectly well-defined procedure. But when we make unitary transformations on the fields, in the usual sense, using the generators we obtain from the conserved currents, the fields transform contragradiently to the states. This means that if we transform the fields in the Lagrangian to generate Q_3 according to $\phi^{\pm\prime} = e^{\mp i\Lambda}\phi^\pm$, then $e^{i\Lambda Q_3}\phi^\pm e^{-i\Lambda Q_3} = e^{\pm i\Lambda}\phi^\pm$.

Second, it is clear that, for constant Λ,

$$e^{i\Lambda Q_3}\mathcal{L}e^{-i\Lambda Q_3} = \mathcal{L}, \tag{2.81}$$

where $\mathcal{L}$ is the Lagrange density above. Therefore, from the connection between $\mathcal{L}$ and H, clearly

$$e^{i\Lambda Q_3}He^{-i\Lambda Q_3} = H. \tag{2.82}$$

Therefore for $\Lambda \ll 1$ we have, expanding the exponentials,

$$[H, Q_3]_- = 0. \tag{2.83}$$

Hence the fact that the charge is a constant of the motion can also be viewed as the statement that the Hamiltonian is invariant under the constant changes of phase that Q_3 generates.

The free Fermion fields are characterized by the Lagrangian density,

$$\mathcal{L} = -\bar{\psi}(\gamma_\mu\partial_\mu + m)\psi. \tag{2.84}$$

This Lagrangian is clearly invariant under the transformation

$$\begin{aligned}\psi' &= e^{i\Lambda}\psi,\\ \bar{\psi}' &= e^{-i\Lambda}\bar{\psi}\end{aligned} \tag{2.85}$$

for constant real Λ. Thus, letting $\Lambda(x) \ll 1$ and applying the gauge transformation Eq. (2.85) to $\mathcal{L}$, we are led at once to the conserved current

$$J_\mu = i\bar{\psi}\gamma_\mu\psi \tag{2.86}$$

and the constant charge

$$Q = \int d^3\mathbf{r}\, \psi^\dagger\psi. \tag{2.87}$$

The Hermiticity of this current and its charge is somewhat less transparent than in the scalar case. It is, however, easy to see that the condition for the Hermiticity of a bilinear form such as $\bar{\psi}O\psi$, where O is some 4×4 matrix; that is,

$$(\bar{\psi}O\psi)^\dagger = \bar{\psi}O\psi \tag{2.88}$$

is equivalent to having

$$O = \gamma_4 O^\dagger \gamma_4. \tag{2.89}$$

Thus, taking into account the i in the definition of J_μ, we see that, with all of our γ's Hermitian,

$$\mathbf{J}^\dagger = \mathbf{J} \tag{2.90}$$

and

$$J_4^\dagger = -J_4. \tag{2.91}$$

Thus

$$Q^\dagger = Q. \tag{2.92}$$

It is obvious that if J_μ is a conserved current, then so is αJ_μ, where α is any number. There is nothing in the method that we have used to generate currents that sets the scale of the charge Q. This can only be done by bringing to bear additional physical assumptions. In anticipation of future work we shall define B, the baryon number, as

$$B = \int d\mathbf{r}\, \psi^\dagger\psi. \tag{2.93}$$

In this connection we shall introduce a slight technical refinement that will enable us to eliminate the infinite zero-point energy from the free Hamiltonian and an infinite constant from B from the beginning. Whenever, in $\mathcal{L}$ or J_μ, or such like we encounter a term of the form $\bar{\psi}O\psi$, where O is a 4×4 matrix, we shall replace it by

$$:\bar{\psi}O\psi: = \tfrac{1}{2}\sum_{\alpha,\beta}(\bar{\psi}_\alpha O_{\alpha\beta}\psi_\beta - \psi_\beta O_{\alpha\beta}\bar{\psi}_\alpha). \tag{2.94}$$

This replacement does not change the Hermiticity of $\vdots J_\mu \vdots$, and if we apply the phase transformation to $\vdots \mathcal{L} \vdots$ we generate $\vdots J_\mu \vdots$ directly. Unless it is of special significance we shall not, in the future, write the $\vdots\ \vdots$ explicitly, but we shall understand that this replacement has been made in all the bilinear forms in ψ and $\bar{\psi}$.

We see at once that B ($\vdots\ \vdots$ understood) can be written in the form

$$B = \int d\mathbf{p}\,(n_+(\mathbf{p}) - n_-(\mathbf{p})). \tag{2.95}$$

Thus the baryon number measures the number of baryons minus the number of antibaryons. B is one of the few quantum numbers in elementary particle physics that is conserved in each and every interaction. All Lagrangians in elementary particle physics are constructed to be invariant against the phase transformations that generate a conserved B.

We may construct the one-particle fermion states, with spins, by making use of a limiting process similar to the one used above for constructing the scalar states. For free fields, ψ and $\bar{\psi}$, it is easy to see, if p is a four-vector energy-momentum with $p_0 > 0$, that

$$\lim_{t\to\pm\infty} \left(\int d\mathbf{r}\,\bar{\psi}(\mathbf{r}, t)\,\gamma_4 \frac{e^{i(px)}}{(2\pi)^{3/2}}\,u_s(\mathbf{p})\right)|0\rangle = a(\mathbf{p})_s{}^\dagger|0\rangle. \tag{2.96}$$

For the interacting ψ we shall use the same limiting process to construct the states. In the literature the state

$$\lim_{t\to-\infty} (\quad)|0\rangle = |\mathbf{p}\rangle_s \tag{2.97}$$

is often referred to as $|\mathbf{p}\rangle_{s\,\text{in}}$, the "in" reflecting the idea that this is a state in the remote past, before the interaction, if there is one, has been turned on. Likewise,

$$\lim_{t\to+\infty} (\quad)|0\rangle = |\mathbf{p}\rangle_{s\,\text{out}} \tag{2.98}$$

again with the idea that in the remote future the particles will again be free of interaction. For one particle states, in general, $|\mathbf{p}\rangle_{s\,\text{in}} = |\mathbf{p}\rangle_{s\,\text{out}}$. We have not discussed the construction of the many particle states; we will have some occasion to do this later, but it is clear that for *free* particles

$$|\mathbf{p}, \mathbf{p}'\rangle_{s,s'} = a^\dagger(\mathbf{p})_s a^\dagger(\mathbf{p}')_{s'}|0\rangle, \tag{2.99}$$

so that for these *free* states,

$$|\mathbf{p}, \mathbf{p}'\rangle_{ss'\,\text{in}} = |\mathbf{p}, \mathbf{p}'\rangle_{ss'\,\text{out}}. \tag{2.100}$$

For interacting fields this result will no longer hold and the difference between $|\rangle_{\text{in}}$ and $|\rangle_{\text{out}}$ reflects the possibility that interacting particles can collide and scatter.

For the fermion fields we can also define a charge conjugation or particle antiparticle transformation C.

If we let

$$\bar{\psi}(\mathbf{r}, t)' \cdot\!\equiv\!\cdot\ C\bar{\psi}(\mathbf{r}, t)C^{-1} = -c^{-1}\psi(\mathbf{r}, t)\eta^*, \tag{2.101}$$

with n^* an arbitrary phase,[8] $n^*n = 1$, and demand that the c matrix obeys

$$\begin{aligned} c^\dagger &= -c, \\ c^\dagger c &= 1, \\ c\gamma_\mu{}^\dagger c^{-1} &= -\gamma_\mu, \end{aligned} \tag{2.102}$$

then we verify, with a little algebra ($:\ :$ understood) that for the Dirac Lagrangian

$$C\mathcal{L}C^{-1} = \mathcal{L}, \tag{2.103}$$

so that

$$[H, C]_- = 0 \tag{2.104}$$

and

$$C\mathbf{J}C^{-1} = -\mathbf{J}, \tag{2.105}$$

and finally,

$$CBC^{-1} = -B. \tag{2.106}$$

The antiparticle state of momentum $\mathbf{p}$ and spin s is constructed from

$$\lim_{t\to\pm\infty} \int \bar{\psi}'(\mathbf{r}, t)\gamma_4 v_s(\mathbf{p}) \frac{e^{i(px)}}{(2\pi)^{3/2}} |0\rangle \, d\mathbf{r}. \tag{2.107}$$

Just as in the scalar case, it is very useful to study the commutation relations between B and $\psi(\mathbf{r}, t)$ and $\psi^\dagger(\mathbf{r}, t)$. These are fixed by the equal time anticommutation relations between ψ and $\psi^\dagger$, since B is time-independent. Thus we find that[9]

$$\begin{aligned} [B, \psi(\mathbf{r}, t)]_- &= -\psi(\mathbf{r}, t), \\ [B, \psi^\dagger(\mathbf{r}, t)]_- &= \psi^\dagger(\mathbf{r}, t). \end{aligned} \tag{2.108}$$

Where we have made use of the operator indentity

$$[AB, C]_- = A[B, C]_+ - [C, A]_+ B, \tag{2.109}$$

this means that B generates the transformation

$$\begin{aligned} \psi' &= e^{-i\Lambda}\psi, \\ \psi^{\dagger\prime} &= e^{i\Lambda}\psi^\dagger, \end{aligned} \tag{2.110}$$

again with the sign change in Λ, as we expect.

---

8. The appearance of this arbitrary phase is related to the phase invariance of $\mathcal{L}$ and hence, ultimately, to the conservation of baryon number. If C is any charge conjugation operator that leaves $\mathcal{L}$ invariant, then $e^{i\Lambda B}C$ for any constant, real Λ, also leaves $\mathcal{L}$ invariant, since we must have, to conserve baryons,

$$e^{i\Lambda B}\mathcal{L}e^{-i\Lambda B} = \mathcal{L}.$$

Hence C is only defined up to an arbitrary phase.

9. Note that J_0 and $:J_0:$ have the same commutation relation with ψ, since they differ from each other by a constant, albeit an infinite one.

From the nondegeneracy of the vacuum and the properties of C, the charge conjugation operator, it follows that

$$B|0\rangle = 0, \tag{2.111}$$

and so, using the creation operator above to make the one-particle states,

$$B|\mathbf{p}\rangle_s = |\mathbf{p}\rangle_s, \tag{2.112}$$

and, for antiparticles,

$$B|\bar{\mathbf{p}}\rangle_s = -|\bar{\mathbf{p}}\rangle_s. \tag{2.113}$$

To derive these results we have, as in the scalar case, written

$$\begin{aligned} B|\mathbf{p}\rangle_s &= \lim_{t\to\pm\infty} \int d\mathbf{r}\, B\psi^\dagger(\mathbf{r}, t) \frac{e^{i(px)}}{(2\pi)^{3/2}} u_s(\mathbf{p})|0\rangle \\ &= \lim_{t\to\pm\infty} \int d\mathbf{r}\, ([B, \psi^\dagger(\mathbf{r}, t)]_- + \psi^\dagger(\mathbf{r}, t)B) \frac{e^{i(px)}}{(2\pi)^{3/2}} u_s(\mathbf{p})|0\rangle \end{aligned}$$

and have applied the B commutation relations and the fact that $B|0\rangle = 0$.

It is also interesting to discuss $\langle\mathbf{p}'|_{s'}B|\mathbf{p}\rangle_s$. It clearly satisfies

$$\langle\mathbf{p}'|_{s'}B|\mathbf{p}\rangle_s = \langle\mathbf{p}'|_{s'}|\mathbf{p}\rangle_s = \delta_{ss'}\delta^3(\mathbf{p} - \mathbf{p}'). \tag{2.114}$$

We can understand the occurrence of this δ function as follows. For the free fields,

$$B = \int d\mathbf{r}\, \psi^\dagger(\mathbf{r}, t)\psi(\mathbf{r}, t) = \int d\mathbf{r}\, e^{-i\mathbf{P}\cdot\mathbf{r}}\psi^\dagger(0, t)\psi(0, t)e^{i\mathbf{P}\cdot\mathbf{r}}. \tag{2.115}$$

Thus ($\colon\ \colon$ understood),

$$\begin{aligned} \langle\mathbf{p}'|_{s'}B|\mathbf{p}\rangle_s &= \int d\mathbf{r}\, e^{i(\mathbf{p}-\mathbf{p}')\cdot\mathbf{r}}\langle\mathbf{p}'|\psi^\dagger(0)\psi(0)|\mathbf{p}\rangle \\ &= (2\pi)^3\delta^3(\mathbf{p} - \mathbf{p}')\langle\mathbf{p}'|_{s'}\psi^\dagger(0)\psi(0)|\mathbf{p}\rangle_s = \delta^3(\mathbf{p}' - p)\delta_{ss'}, \end{aligned} \tag{2.116}$$

where we have used the expansion of $\psi(0)$ in terms of creation operators. We see clearly from this derivation, that any operator, such as B, which can be written as the integral over all space of a density function, that is,

$$O = \int d\mathbf{r}\, \mathcal{O}(\mathbf{r}), \tag{2.117}$$

can have, as a consequence of the translation invariance of the theory, only matrix elements between states of the same momentum. The derivation we have given earlier for the matrix elements of B between one-particle states, based on the commutation relations between B and ψ, will also work for the interacting fields even though, in this case, B has no simple expansion in terms of creation and annihilation operators. We begin, in the next chapter, the discussion of the interacting fields.

3
Interacting Fields

In general, if two or more quantized fields are coupled together this results in a nonlinear mathematical problem that is essentially impossible to solve exactly. We shall however assume the following.[1]

1. The field equations are derivable from a Lagrange density $\mathcal{L}$, so that using the canonical formalism we can construct $\mathbf{P}_\nu$, the space-time displacement operator.

2. This operator, $\mathbf{P}_\nu$, has a complete set of eigenstates $|\mathbf{p}\rangle$ and a unique vacuum state $|0\rangle$.

The exact solutions to the field equations $\boldsymbol{\psi}$, $\boldsymbol{\phi}$, and so on, assuming that they exist, may now connect the vacuum to states of several particles, as well as to the one-particle state. As we shall see, the symmetry properties of $\boldsymbol{\phi}$ and $\boldsymbol{\psi}$ restrict the connections that can be made between states. For example, if $\boldsymbol{\phi}^+$ has charge $+1$ and if this charge is conserved by the dynamics, then $\boldsymbol{\phi}^+$ can only connect states whose charges differ by one unit, no matter how strong the interaction is.

To fix ideas it is helpful to study, in detail, one fairly realistic situation. We choose, to begin with, the theory of pions and nucleons in interaction. In the first instance we shall neglect their electromagnetic interactions. In the absence of electromagnetic interactions it is generally assumed that these particles collapse into "multiplets" as illustrated in the following diagram. (N means neutron; P, proton; and $\pi_0^{\pm}$, the three known pi mesons. All masses are given in units of MeV, millions of electron volts.)

Observed mass spectrum			Conjectured mass spectrum in the absence of e. m. forces		
N 939.550 ±0.005	P 938.256 ±0.005		N		P
π^+ 139.579 ±0.014	π^0 134.975 ±0.014	π^- 139.579 ±0.014	π^+	π^0	π^-

1. When there is a possibility of confusion, we shall denote quantities referring to interacting fields by boldface letters.

This view is encouraged by the observation that the mass differences between charged and neutral particles are of order $e^2 \times$ mass of the charged particle where, in the usual dimensionless units, $e^2 \simeq 1/137$. This would indicate that the mass splittings are some sort of second-order perturbation in the electromagnetic couplings. The fact that the neutron is heavier than the proton means, presumably, that this mass difference is not a simple Coulomb phenomenon but involves the magnetic couplings as well. As we shall later see, the pions, having spin zero, have no magnetic couplings.

Thus, in the absence of electromagnetism, any realistic pion nucleon Lagrangian should have this multiplet symmetry between the nucleons and among the pions built in *ab initio*. For this purpose we introduce a "nucleon" field ψ:

$$\psi = \begin{pmatrix} \psi_P \\ \psi_N \end{pmatrix}, \tag{3.1}$$

which is clearly an eight-component spinor object. We represent, to begin with, the three pions by three Hermitian fields ϕ_1, ϕ_2, and ϕ_3. After the work of the last chapter it will hardly be a mystery how to make $\phi^{\pm}$ from ϕ_1 and ϕ_2; ϕ_3 will turn out to represent the neutral pi-meson. The noninteracting part of the Lagrangian can be written

$$\mathfrak{L}_0 = \mathfrak{L}_N + \mathfrak{L}_\pi = -\{\bar{\psi}(\gamma_\mu \partial_\mu + m)\psi\} - \frac{1}{2}\left\{\frac{\partial \boldsymbol{\phi}}{\partial x_\nu} \cdot \frac{\partial \boldsymbol{\phi}}{\partial x_\nu} + m_\pi{}^2 \boldsymbol{\phi} \cdot \boldsymbol{\phi}\right\}, \tag{3.2}$$

where m is the common nucleon mass and m_π is the common meson mass:

$$\begin{aligned} \boldsymbol{\phi} \cdot \boldsymbol{\phi} &= \phi_1\phi_1 + \phi_2\phi_2 + \phi_3\phi_3, \\ \bar{\psi}\psi &= \bar{\psi}_P\psi_P + \bar{\psi}_N\psi_N. \end{aligned} \tag{3.3}$$

We may ask for the conserved currents associated with this Lagrangian.

Clearly,

$$\begin{aligned} \psi &\longrightarrow e^{i\Lambda}\psi, \\ \boldsymbol{\phi} &\longrightarrow \boldsymbol{\phi}, \end{aligned} \tag{3.4}$$

for constant real Λ is an invariance of $\mathfrak{L}_0$. Thus, letting $\Lambda = \Lambda(x)$, we generate the conserved current

$$\mathbf{J}_\mu = i\bar{\psi}\gamma_\mu\psi \tag{3.5}$$

and the conserved baryon number

$$\mathbf{B} = \int d\mathbf{r}\, \psi^\dagger(x)\psi(x) = \mathbf{N}_P + \mathbf{N}_N - \mathbf{N}_{\bar{P}} - \mathbf{N}_{\bar{N}}. \tag{3.6}$$

So long as we do not include interactions, **B** can be expanded in terms of simple creation and annihilation operators. However, when we do take interactions into account, the expansions of the fields become very complicated, reflecting the fact that an interacting field can connect a variety of states. We will shortly see, however, that even in this case

$$\mathbf{B} = \int d\mathbf{r}\, \psi^\dagger(x)\psi(x) \tag{3.7}$$

still has the simple properties of the baryon number operator.

The degeneracy between N and P introduces a new symmetry—the so-called isotopic spin invariance. The quantity

$$\bar{\psi}\psi = \bar{\psi}_1\psi_1 + \bar{\psi}_2\psi_2, \tag{3.8}$$

where we replace N and P by 1 and 2 to emphasize that these fields are completely equivalent in the absence of electromagnetism, is invariant under all 2×2 unitary transformations in the complex two-dimensional 1, 2 space. If U is a 2×2 matrix such that $U^{-1} = U^\dagger$, then, letting

$$\psi' = U\psi, \tag{3.9}$$

we have

$$\bar{\psi}'\psi' = \bar{\psi}\psi. \tag{3.10}$$

As has become customary in discussing isotopic spin, let us call the three Pauli matrices

$$\boldsymbol{\sigma} \equiv \boldsymbol{\tau} \tag{3.11}$$

Any Hermitian 2×2 matrix A can be written in the form

$$A = aI + \boldsymbol{\Lambda}\cdot\boldsymbol{\tau}, \tag{3.12}$$

where

$$I = \begin{pmatrix} 1 & 0 \\ 0 & 1 \end{pmatrix},$$

since these four matrices are independent and Hermitian; a and $\boldsymbol{\Lambda}$ are four arbitrary real numbers. Thus the U above can be written[2]

$$U = e^{iaI}e^{i\boldsymbol{\Lambda}\cdot\boldsymbol{\tau}} \tag{3.13}$$

We know that the invariance of $\mathfrak{L}_N$ under e^{iaI} simply leads back to baryon conservation. The interesting new symmetry is contained in the factor $e^{i\boldsymbol{\Lambda}\cdot\boldsymbol{\tau}}$. From the identity for finite matrices,

$$\log(\det A) = \operatorname{Tr}(\log A), \tag{3.14}$$

and the fact that

$$\operatorname{Tr}(\boldsymbol{\tau}) = 0, \tag{3.15}$$

it follows that

$$\det(e^{i\boldsymbol{\Lambda}\cdot\boldsymbol{\tau}}) = 1. \tag{3.16}$$

Thus the set of unitary transformations that are given by the $e^{i\boldsymbol{\Lambda}\cdot\boldsymbol{\tau}}$ is restricted to the unimodular ones—the ones with unit determinant. In general, if we have a complex quadratic form in n variables,

$$x_1^*x_1 + \cdots + x_n^*x_n, \tag{3.17}$$

---

2. Since I commutes with all the $\boldsymbol{\tau}$, it can be factored from the sum.

then the set of all unitary unimodular transformations that leave this form invariant is a group; it is, in fact, a group that mathematicians have given the name SU_n. Thus the symmetry of $\mathcal{L}_N$ under the $e^{i\boldsymbol{\Lambda}\cdot\boldsymbol{\tau}}$ is an example of an SU_2 symmetry.

That the $e^{i\boldsymbol{\Lambda}\cdot\boldsymbol{\tau}}$ form a group is not difficult to see. Consider

$$U = e^{i\boldsymbol{\Lambda}\cdot\boldsymbol{\tau}}e^{i\boldsymbol{\Lambda}'\cdot\boldsymbol{\tau}}. \tag{3.18}$$

Since

$$UU^\dagger = e^{i\boldsymbol{\Lambda}\cdot\boldsymbol{\tau}}e^{i\boldsymbol{\Lambda}'\cdot\boldsymbol{\tau}}e^{-i\boldsymbol{\Lambda}'\cdot\boldsymbol{\tau}}e^{-i\boldsymbol{\Lambda}\cdot\boldsymbol{\tau}} = 1 \tag{3.19}$$

and

$$\det U = \det U(\boldsymbol{\Lambda}) \det U(\boldsymbol{\Lambda}') = 1, \tag{3.20}$$

it must be possible to write U in the form

$$U = e^{i\boldsymbol{\Lambda}''\cdot\boldsymbol{\tau}}. \tag{3.21}$$

However, the relation between $\boldsymbol{\Lambda}$, $\boldsymbol{\Lambda}'$, and $\boldsymbol{\Lambda}''$ is not entirely trivial and, as we shall see, depends in the last analysis on the commutation relations among the $\boldsymbol{\tau}$:

$$[\tau_i, \tau_j]_- = 2i\epsilon_{ijk}\tau_k. \tag{3.22}$$

The SU_2 group is *not* Abelian; in general, $e^A e^B \neq e^{A+B}$. (For example, a rotation of $\pi/2$ about $\hat{\mathbf{x}}$, followed by a rotation of $\pi/2$ around $\hat{\mathbf{y}}$ corresponds to a rotation about $\hat{\mathbf{z}}$.) To simplify the arithmetic and in anticipation of work to come, we call

$$T_i = \frac{\tau_i}{2}, \tag{3.23}$$

so that

$$[T_i, T_j]_- = i\epsilon_{ijk}T_k. \tag{3.24}$$

We may then study the group given by the $e^{i\boldsymbol{\Lambda}\cdot\mathbf{T}}$. For $\boldsymbol{\Lambda} \ll 1$,

$$U(\boldsymbol{\Lambda}) \simeq (1 + i\boldsymbol{\Lambda}\cdot\mathbf{T}), \tag{3.25}$$

which is unitary to order Λ. To this order

$$U(\boldsymbol{\Lambda})U(\boldsymbol{\Lambda}') = (1 + i(\boldsymbol{\Lambda} + \boldsymbol{\Lambda}')\cdot\mathbf{T}), \tag{3.26}$$

so the infinitesimal part of the group is Abelian. We can then proceed to construct the finite transformations by stringing together infinite products of infinitesimal ones. In textbooks on group theory,[3] it is shown that the commutation relations given above correspond to the condition that the finite group can be recovered from the infinitesimal one by iteration or integration. We can get a feeling for how this comes about by considering the next terms in the expansion of $U(\boldsymbol{\Lambda})U(\boldsymbol{\Lambda}')$:

---

3. See, in this connection, *Group Theory and Spectroscopy*, by G. Racah, reprinted in Ergebwisse der Exakten Naturwissenschaften, Vol. 37, Springer, Berlin, 1965.

$$U(\Lambda)U(\Lambda') = 1 + i(\Lambda + \Lambda' \cdot \mathbf{T}) + \frac{((\Lambda + \Lambda') \cdot \mathbf{T})^2}{2}$$

$$- \frac{1}{4}(\Lambda_i \Lambda_j' - \Lambda_i' \Lambda_j)[T_i, T_j]_- \cdots . \qquad (3.27)$$

We see at this stage that the commutators among the T_i begin to enter. In fact, if we use them,

$$U(\Lambda)U(\Lambda') = 1 + i(\Lambda + \Lambda') \cdot \mathbf{T} - \frac{((\Lambda + \Lambda') \cdot \mathbf{T})^2}{2}$$

$$- \frac{1}{2} i\mathbf{T} \cdot (\Lambda \times \Lambda') \cdots \simeq e^{i\left(\mathbf{T} \cdot \left(\Lambda + \Lambda' - \frac{\Lambda \times \Lambda'}{2}\right)\right)}, \qquad (3.28)$$

where it is understood that the exponential is to be expanded only up to second order. Thus, to this order,

$$\Lambda'' = \Lambda + \Lambda' - \frac{\Lambda \times \Lambda'}{2}. \qquad (3.29)$$

In general, Λ'' is given in terms of an infinite series involving Λ and Λ'.

However, following the method of Chapter 2, we can use the transformations for $\Lambda \ll 1$ to generate the conserved current corresponding to the invariance of $\mathfrak{L}_N$ under the $U(\Lambda)$. This conserved current is

$$\mathbf{j}_\mu^V = i\bar{\psi}\gamma_\mu \frac{\boldsymbol{\tau}}{2}\psi. \qquad (3.30)$$

(The significance of the superscript ^{V} will become clear in what follows.)

Thus the three isotopic charges

$$\mathbf{T} = \int \psi^\dagger \frac{\boldsymbol{\tau}}{2} \psi \, d\mathbf{r} \qquad (3.31)$$

are Lorentz scalar constants of the motion. Moreover, the **T** charges have the same commutation relations as do the $\boldsymbol{\tau}/2$. This can be seen by direct computation, making frequent use of the identity

$$[AB, CD]_- = [AB, C]_+ D - C[AB, D]_+. \qquad (3.32)$$

It is crucial in this computation to use the fact that **T** is time-independent, so that in doing the commutation T_i and T_j can be taken at the same times, which means that the equal-time anticommutators among the ψ can be employed in the form

$$[\psi_\tau^\dagger(\mathbf{r}, t), \psi_{\tau'}(\mathbf{r}, t)]_+ = \delta_{\tau\tau'}\, \delta^3(\mathbf{r} - \mathbf{r}'), \qquad (3.33)$$

where τ and τ' are isotopic indices that indicate whether the four-component field ψ_τ is 1 or 2.

Thus[4]

$$[T_i, T_j]_- = i\epsilon_{ijk}T_k. \tag{3.34}$$

In addition we can easily establish the relations

$$[\mathbf{T}, \psi]_- = -\frac{\boldsymbol{\tau}}{2}\psi, \tag{3.35}$$

$$[\mathbf{T}, \psi^\dagger]_- = \psi^\dagger\frac{\boldsymbol{\tau}}{2},$$

where

$$\psi = \alpha\begin{pmatrix}\psi_1\\0\end{pmatrix} + \beta\begin{pmatrix}0\\\psi_2\end{pmatrix}$$

is a general nucleon field.

These relations taken together imply that the unitary transformations U, which operate in the Hilbert space of fields, rather than on the 2×2 finite-dimensional vector space 1, 2; that is,

$$U = e^{i\boldsymbol{\Lambda}\cdot\mathbf{T}}, \tag{3.36}$$

with

$$\mathbf{T} = \int d\mathbf{r}\,\psi^\dagger\frac{\boldsymbol{\tau}}{2}\psi,$$

generate phase transformations on the ψ:

$$\begin{aligned}\psi' &= e^{i\boldsymbol{\Lambda}\cdot\mathbf{T}}\psi e^{-i\boldsymbol{\Lambda}\cdot\mathbf{T}} = e^{-i\boldsymbol{\Lambda}\cdot\boldsymbol{\tau}/2}\psi,\\ \psi^{\dagger\prime} &= e^{i\boldsymbol{\Lambda}\cdot\mathbf{T}}\psi^\dagger e^{-i\boldsymbol{\Lambda}\cdot\mathbf{T}} = \psi^\dagger e^{i\boldsymbol{\Lambda}\cdot(\boldsymbol{\tau}/2)}.\end{aligned} \tag{3.37}$$

An object that transforms like ψ under the U's we call an isotopic spinor. The quantity **B**, under the U's, has the transformation property

---

4. In some of the most recent work in elementary particle physics there has been interest in the equal time commutation relations among different components of the isotopic current densities $\mathbf{j}_\mu{}^V$. For example, a naive application of the equal-time commutation relations, in which i and j are isotopic indices, gives

$$[\mathbf{j}_i(\mathbf{r}, t), j_{4j}(\mathbf{r}', t)]_- = -i\epsilon_{ijk}\mathbf{j}_k(\mathbf{r}, t)\,\delta^3(\mathbf{r} - \mathbf{r}').$$

However, these equations have been questioned especially by J. Schwinger (*Phys. Rev. Letters*, **3**: 296 (1959)), who points out that the right side of the equation above may well contain more singular terms of the form $\nabla\delta^3(\mathbf{r} - \mathbf{r}')$. Ultimately, the origin of these terms is traced back to the singular character of products of field operators such as those that enter the currents. If the currents are more carefully defined than we have done, the Schwinger terms arise in the commutation relations. However, these singularities do not appear to affect integrated quantities such as the charges—the singular parts integrate away. Hence we seem to be on safe ground so long as we commute charges. Later we shall have occasion to come back to problems connected with commuting nonconserved charges, but we shall avoid, as long as we can, commuting the current components.

$$\mathbf{B}' = e^{i\boldsymbol{\Lambda}\cdot\mathbf{T}}\left(\int d\mathbf{r}\,\psi^\dagger\psi\right)e^{-i\boldsymbol{\Lambda}\cdot\mathbf{T}}$$
$$= \int d\mathbf{r}\,\psi^{\dagger\prime}\psi' = \int d\mathbf{r}\,\psi^\dagger\psi = \mathbf{B}. \tag{3.38}$$

B is an example of an isotopic scalar. On the other hand, the transformation of

$$\mathbf{j}_\mu{}^V = i\bar{\psi}\frac{\boldsymbol{\tau}}{2}\gamma_\mu\psi \tag{3.39}$$

is much more complex. It is sufficient to study it for $\Lambda \ll 1$. Then

$$\mathbf{j}_{\mu i}{}^{V\prime} = i\bar{\psi}\gamma_\mu\left(1 + i\frac{\boldsymbol{\Lambda}\cdot\boldsymbol{\tau}}{2}\right)\frac{\tau_i}{2}\left(1 - i\boldsymbol{\Lambda}\cdot\frac{\boldsymbol{\tau}}{2}\right)\psi$$
$$= i\bar{\psi}\frac{\tau_i}{2}\gamma_\mu\psi - \bar{\psi}\gamma_\mu\left[\boldsymbol{\Lambda}\cdot\frac{\boldsymbol{\tau}}{2}, \frac{\tau_i}{2}\right]_-\psi$$
$$= i\bar{\psi}\gamma_\mu\left(\frac{\tau_i}{2} + \left(\boldsymbol{\Lambda}\times\frac{\boldsymbol{\tau}}{2}\right)_i\right)\psi \tag{3.40}$$

or

$$\mathbf{j}_\mu{}^{V\prime} = \mathbf{j}_\mu{}^V + \boldsymbol{\Lambda}\times\mathbf{j}_\mu{}^V. \tag{3.41}$$

This is the transformation equation for an ordinary vector under an infinitesimal rotation of the coordinate system. Hence objects that transform like $\mathbf{j}_\mu{}^V$ are called isotopic vectors.[5]

We can use the field ψ to create the nucleon state (a two-dimensional vector in the iso-spin space),

$$|\mathbf{p}\rangle = \alpha\begin{pmatrix}|\mathbf{p}\rangle_1\\ 0\end{pmatrix} + \beta\begin{pmatrix}0\\ |\mathbf{p}\rangle_2\end{pmatrix}, \tag{3.42}$$

and use the commutation relations between **T** and ψ to find $\mathbf{T}|\mathbf{p}\rangle$. We shall assume that the vacuum is an eigenstate of **T** with eigenvalue zero,[6]

$$\mathbf{T}|0\rangle = 0. \tag{3.43}$$

In fact, since

$$[H, \mathbf{T}]_- = 0, \tag{3.44}$$

we can always diagonalize H with one of the T_i (not *more* than one, since they do not commute). If we choose T_3, then the vacuum becomes an eigenstate of T_3. But, expanding for free nucleons $\int :\psi^\dagger\tau_3\psi: d\mathbf{r}$,

$$T_3 = \frac{N_P - N_{\bar{P}}}{2} - \frac{(N_N - N_{\bar{N}})}{2}. \tag{3.45}$$

5. In particular, the T_i are components of an isotopic vector.
6. This comes down to the assumption that the vacuum is an isotopic scalar, since it implies that $e^{i\boldsymbol{\Lambda}\cdot\mathbf{T}}|0\rangle = |0\rangle$, and vice versa.

(Incidentally, this equation clearly implies that the charge-conjugation operator C *anticommutes* with T_3.) Hence the condition

$$T_3|0\rangle = 0 \tag{3.46}$$

is consistent with the fact that the vacuum is empty.

We then find for the two-dimensional states $|\mathbf{p}\rangle$ that

$$\mathbf{T}|\mathbf{p}\rangle = \frac{\boldsymbol{\tau}}{2}|\mathbf{p}\rangle. \tag{3.47}$$

The states

$$\begin{pmatrix} |\mathbf{p}\rangle \\ 0 \end{pmatrix}, \qquad \begin{pmatrix} 0 \\ |\mathbf{p}\rangle \end{pmatrix}$$

diagonalize T_3, since they are diagonal in $\tau_3/2$ with eigenvalues $\pm\frac{1}{2}$. There is clearly a one-to-one correspondence between the transformations $e^{i\boldsymbol{\Lambda}\cdot\mathbf{T}}$ on the Hilbert space of fields and the transformation $e^{i\boldsymbol{\Lambda}\cdot\boldsymbol{\tau}/2}$ on the two-dimensional isotopic state space. We can speak of $e^{i\boldsymbol{\Lambda}\cdot\boldsymbol{\tau}/2}$ as the 2×2 dimensional representation of the group SU_2 in the sense that it is a group of 2×2 matrices generated by three generators, the $\boldsymbol{\tau}/2$, which have the same commutation relations as the $\boldsymbol{\tau}$. Often physicists call the states themselves, rather than the matrices, the representation. Thus, in this language, the nucleon is the doublet representation of SU_2 while the vacuum, which is assumed to be invariant under the group, is the singlet or "trivial" representation.

Before considering the interaction Lagrangian, we must discuss the invariances of

$$\mathfrak{L}_\pi = -\frac{1}{2}\left\{\frac{\partial\boldsymbol{\phi}}{\partial x_\nu}\cdot\frac{\partial\boldsymbol{\phi}}{\partial x_\nu} + m_\pi^2\boldsymbol{\phi}\cdot\boldsymbol{\phi}\right\}. \tag{3.48}$$

Clearly from the way we have written the quantity $\boldsymbol{\phi}\cdot\boldsymbol{\phi}$ we anticipate that it will be invariant under all real rotations in the three dimensional space defined by ϕ_1, ϕ_2, ϕ_3. In the first example discussed in Chapter 2, in which we had two Hermitian fields ϕ_1 and ϕ_2, we studied a special case of this invariance which, from our present standpoint, corresponded to rotations in the 1–2 plane, leaving the 3 direction fixed. The general infinitesimal rotation in 3-space can be written in terms of three real infinitesimal parameters $\boldsymbol{\Lambda}$ as

$$\boldsymbol{\phi} \longrightarrow \boldsymbol{\phi} - \boldsymbol{\Lambda} \times \boldsymbol{\phi}. \tag{3.49}$$

one may verify explicitly that this transformation leaves $\mathfrak{L}_\pi$ invariant for constant $\boldsymbol{\Lambda}$. Letting $\boldsymbol{\Lambda} = \boldsymbol{\Lambda}(x)$, we generate the conserved current

$$\mathbf{J}_\mu = -\boldsymbol{\phi} \times \partial_\mu\boldsymbol{\phi}. \tag{3.50}$$

The third component of this current was discussed in Chapter 2.[7] This current leads to three conserved "isotopic" charges,

---

7. In dealing with scalar fields we must remember that $[\phi_i, \phi_i]_- \neq 0$. In the current we always have ϕ_i and ϕ_j, with $i \neq j$, so the vector product obeys the same rules as if the $\boldsymbol{\phi}$ were c-numbers.

$$\mathbf{T} = \int d\mathbf{r}\,(\boldsymbol{\phi} \times \dot{\boldsymbol{\phi}}). \tag{3.51}$$

We may use the equal time commutation relations to verify that

$$[T_i, \phi_l]_- = i\epsilon_{ilj}\phi_j \tag{3.52}$$

and that

$$[T_i, T_j] = i\epsilon_{ijk}T_k \tag{3.53}$$

so that these charges obey the same characteristic SU_2 algebraic relations that the nucleon charges do. We can construct the unitary Hilbert space operators,

$$U = e^{i\boldsymbol{\Lambda}\cdot\mathbf{T}}, \tag{3.54}$$

and for $\Lambda \ll 1$ verify that

$$U\boldsymbol{\phi}U^{-1} = \boldsymbol{\phi} + \boldsymbol{\Lambda} \times \boldsymbol{\phi}, \tag{3.55}$$

which is the isotopic rotation in the opposite sense.[8]

We may construct the pion states (using the creation operators) and, again assuming that the vacuum is an isotopic scalar, we find that

$$T_i|\mathbf{p}\rangle_j = i\epsilon_{ijl}|\mathbf{p}\rangle_l \tag{3.56}$$

(A special case of this formula, for T_3, was given in Chapter 2.) The states $|\mathbf{p}\rangle_\pm$ can be defined as in Chapter 2 and we call $|\mathbf{p}\rangle_0$ the state created by ϕ_3. These states diagonalize T_3 and have eigenvalues $+$, $-$, 0. They constitute the three-dimensional representation space of the SU_2 group. The three dimensional matrix representations of this group are the set of all 3×3 matrices of the form $e^{i\boldsymbol{\Lambda}\cdot\mathbf{t}/2}$ where, in a specific coordinate system, in 1, 2, 3 space,

$$t_1 = \begin{pmatrix} 0 & 0 & 0 \\ 0 & 0 & -i \\ 0 & i & 0 \end{pmatrix}, \quad t_2 = \begin{pmatrix} 0 & 0 & i \\ 0 & 0 & 0 \\ -i & 0 & 0 \end{pmatrix}, \quad t_3 = \begin{pmatrix} 0 & -i & 0 \\ i & 0 & 0 \\ 0 & 0 & 0 \end{pmatrix}. \tag{3.57}$$

A student of quantum mechanics will, of course, have recognized that the discussion of isotopic spin mimics the discussion of angular momentum, which is in turn based on the theory of the three-dimensional rotations of ordinary space. Indeed, the commutation relations of the **T** are just the angular momentum commutation relations. Hence we know without doing any further work that

$$\mathbf{T}^2 = T_1^2 + T_2^2 + T_3^2 \tag{3.58}$$

commutes with H and **T**. (It can be shown in the context of local field theory

8. The $\boldsymbol{\phi}$ thus form an isotopic vector.
9. For a detailed discussion of the consequences of *i*-spin invariance see, for example, *Invariance Principles and Elementary Particles*, by J. J. Sakarai, Princeton University Press, 1964.

that $\mathbf{T}$ is in fact the only isovector that commutes with $\mathbf{T}^2$.) As in the angular momentum case, all of the representations of SU_2 can be labeled by the values of $\mathbf{T}^2$. In particular, the vacuum state has $\mathbf{T}^2 = 0$, the nucleon state has $\mathbf{T}^2 = \frac{1}{2}(\frac{1}{2} + 1)$, and the pions have $\mathbf{T}^2 = 1(1 + 1)$. T_3 labels the different particles within a given isomultiplet.

With these preliminaries in mind we may now discuss pions and nucleons in interaction. The experimental basis for isotopic spin conservation in pion-nucleon physics is now very firm. In particular, there are all sorts of selection rules[9] dictated by the conservation of *i*-spin (we shall study some later) that are observed to hold within a few percent. (As we shall see, the electromagnetic interactions violate *i*-spin conservation. These violations are characterized by $e^2 \simeq 1/137$ and we should not expect that *i*-spin invariance will hold to better accuracy than this.) Apart from everything else, the observed multiplet structure of pions and nucleons indicates how well iso-spin invariance holds. Thus we must be sure that the interaction Lagrangian $\mathfrak{L}_{\pi N}$ is an isotopic scalar like $\mathfrak{L}_N$ and $\mathfrak{L}_\pi$.

In addition, we would like to make the interaction between pions and nucleons look as much like the interaction between electrons and photons as possible. This is at least a good first guess, since that theory, when quantized, is known to give brilliant agreement with experiment in those situations where it is applicable. The salient feature of this theory is that it generally deals with the emission and absorption of photons one at a time. The interaction Lagrangian at least for spin-$\frac{1}{2}$ particles interacting with photons (for spin zero, which we study in the next chapter, there are additional complications) contains the photon field A_μ once. This field is then coupled to a current J_μ to produce the interaction. Thus, mimicking this procedure, we suppose that $\mathfrak{L}_{\pi N}$ contains the pion field once. Thus the Lagrangian will have some general form like $\mathfrak{L}_{\pi N} = \bar{\psi}\mathbf{O}\psi \cdot \boldsymbol{\phi}$ summing over 1, 2, 3 for the pions, or perhaps something like $\bar{\psi}\mathbf{O}_\mu\psi \cdot \partial_\mu\boldsymbol{\phi}$. The O's are 4×4 matrices in Dirac space. The π's are well known to be pseudoscalar particles;[10] thus O must also be a pseudoscalar or O_μ must be a pseudovector. It is well known[11] that $\bar{\psi}\gamma_5\psi$ is a pseudoscalar combination of Dirac particles and $\bar{\psi}\gamma_5\gamma_\mu\psi$ is a pseudovector. Hence we can take $\mathbf{O}$ to be $\gamma_5\mathbf{T}$, where $\mathbf{T}$ carries the isotopic spin dependence. We might guess that a combination like

$$\bar{\psi}\gamma_5 \frac{\boldsymbol{\tau}}{2} \psi \cdot \boldsymbol{\phi} \tag{3.59}$$

is an isotopic scalar. But we must remember that up to now we have only studied the transformation properties of free pions and nucleons, whose isotopic spin

---

10. If P is the parity operation then for pi-mesons

$$P\boldsymbol{\phi}(\mathbf{r}, t)P^{-1} = -\boldsymbol{\phi}(-\mathbf{r}, t).$$

11. For the Dirac field the parity operation is defined by

$$P\psi(\mathbf{r}, t)P^{-1} = \gamma_4\psi(-\mathbf{r}, t).$$

character can in this case be studied separately. Let us study the transformation invariances of

$$\mathfrak{L}_{\pi N} = -i\bar{\psi}\gamma_5 \frac{\boldsymbol{\tau}}{2} \psi \cdot \boldsymbol{\phi} g_{\pi N}. \tag{3.60}$$

We have added the i to make it Hermitian and $g_{\pi N}$ is a dimensionless pion-nucleon coupling constant. In the first place the transformation $\boldsymbol{\phi} \longrightarrow \boldsymbol{\phi}$, $\psi \longrightarrow (1 + i\Lambda)\psi$, with constant Λ, clearly leaves $\mathfrak{L}_{\pi \mathrm{N}}$ invariant. Thus, from our previous work, it also leaves

$$\mathfrak{L} = \mathfrak{L}_N + \mathfrak{L}_\pi + \mathfrak{L}_{\mathrm{N}\pi} \tag{3.61}$$

invariant. Hence

$$\mathbf{j}_\mu = i\bar{\psi}\gamma_\mu\psi \tag{3.62}$$

is a conserved current even in the presence of interactions. (We have written ψ for the nucleon fields to emphasize that these are the interacting fields.) Viewed in this way the conservation of $\mathbf{j}_\mu$ is an obvious consequence of the phase invariance of $\mathfrak{L}$. The ψ and $\bar{\psi}$ obey Dirac equations of the form

$$\begin{aligned} \gamma_\mu \partial_\mu \psi &= -m\psi + \mathbf{O}\psi, \\ -\partial_\mu \bar{\psi}\gamma_\mu &= m\bar{\psi} + \bar{\psi}\gamma_4 O^\dagger \gamma_4, \end{aligned} \tag{3.63}$$

where O involves the coupling of pions to nucleons. Thus

$$\partial_\mu(\bar{\psi}\gamma_\mu\psi) = \bar{\psi}(O - \gamma_4 O^\dagger \gamma_4)\psi. \tag{3.64}$$

Viewed this way, the conservation of $\mathbf{j}_\mu$ looks a little less obvious. The phase-invariant construction of $\mathfrak{L}_{\pi \mathrm{N}}$ guarantees that

$$\bar{\psi}(O - \gamma_4 O^\dagger \gamma_4)\psi = 0. \tag{3.65}$$

The quantity ($: \; :$ understood),

$$\mathbf{B} = \int d\mathbf{r}\, \psi^\dagger \psi, \tag{3.66}$$

is the constant of the motion that we expect to represent the conservation of baryons—that is, nucleons and antinucleons. To see this we must consider afresh the problem of constructing the one-particle states. Again we let

$$|\mathbf{p}\rangle_s = \lim_{t \to \pm\infty} \int d\mathbf{r}\, \bar{\psi}(\mathbf{r}, t)\, \gamma_4 \frac{e^{i(px)}}{(2\pi)^{3/2}} u_s(p)|0\rangle. \tag{3.67}$$

Here $|0\rangle$ is the vacuum state of the coupled system which we assume to be a nondegenerate state with the property that

$$\mathbf{P}_\nu|0\rangle = 0.$$

(This follows from the assumption that the vacuum is invariant under Lorentz transformations.) Furthermore, we assume that ψ is such that in its Fourier decomposition there is a nonvanishing component with the property that $(p^2) = -m^2$, where m is the observed mass of the nucleon. There is no reason why this mass should be the same as the mass parameter that appears in the nucleonic part of the Lagrangian. To be clear about this matter we should label this latter mass m_0. If we could compute out the dynamics of the coupled fields correctly, we would presumably find that m is some specific function of m_0. As it is, we shall simply assume that ψ does contain such a Fourier component. The limiting process then picks out this component. Furthermore, we assume (this actually follows from the usual Lagrangian field theory formalism[12]) that at *equal times* all the fields ψ, ϕ, etc. have the same commutation relations with each other as if they were free:

$$[\psi(\mathbf{r}, t), \psi^\dagger(\mathbf{r}', t)]_+ = \delta^3(\mathbf{r} - \mathbf{r}'). \tag{3.68}$$

Up to this point we have not had occasion to need the commutation relations among the ϕ and ψ. For free fields it is clear that

$$[\phi, \psi]_- = 0, \tag{3.69}$$

and we shall assume that all boson and fermion fields commute at equal times in the interacting case.[13] This assumption is vital to much of the work that follows. Knowing these equal-time commutation relations, we can work out quan-

---

12. In the Lagrangian formalism one defines the momentum π canonical to the field ϕ to be

$$\pi = -\frac{\delta \mathfrak{L}}{\delta \dot{\phi}},$$

and then shows that for boson fields

$$[\phi(\mathbf{r}, t), \pi(\mathbf{r}', t)]_- = i\,\delta^3(\mathbf{r} - \mathbf{r}').$$

For the pseudoscalar pion-nucleon coupling

$$\pi = \frac{\partial \phi}{\partial t},$$

which leads to the commutation relation given earlier in the text. For the pseudovector coupling

$$\mathfrak{L}_{\pi \mathrm{N}} = -\frac{if}{m_\pi} \frac{\partial \phi}{\partial x_\mu} \cdot \bar{\psi} \gamma_5 \gamma_\mu \mathbf{t} \psi$$

so that

$$\pi = \frac{\partial \phi}{\partial t} + \frac{f}{m_\pi} \bar{\psi} \gamma_5 \gamma_4 \mathbf{t} \psi.$$

If at equal times $[\phi, \psi]_- = 0$, we again have, for equal times,

$$[\phi, \dot{\phi}]_- = i\,\delta^3(\mathbf{r} - \mathbf{r}').$$

But

$$[\dot{\phi}, \psi] \neq 0.$$

13. We always assume that distinct fields within an isotopic multiplet commute or anticommute, depending on whether they are bosons or fermions. Thus

$$[\psi_N, \psi_P{}^\dagger]_+ = 0.$$

tities like $\mathbf{B}|\mathbf{p}\rangle_s$, using exactly the same algebraic considerations that we have already applied for the free fields in Chapter 2. The essential point is that we can take the fields that define **B** at any arbitrary time, since **B** does not depend on time. We find at once the relations

$$\begin{aligned} \mathbf{B}|\mathbf{p}\rangle_s &= |\mathbf{p}\rangle_s, \\ \mathbf{B}|\bar{\mathbf{p}}\rangle_s &= -|\bar{\mathbf{p}}\rangle_s. \end{aligned} \tag{3.70}$$

Hence, although **B** does not have a simple expansion in terms of creation and annihilation operators, so far as it acts on one-particle states,

$$\mathbf{B} = \mathbf{N}_P + \mathbf{N}_N - \mathbf{N}_{\bar{N}} - \mathbf{N}_{\bar{P}}. \tag{3.71}$$

We may next apply the nucleon isotopic spin transformation, to

$$\mathfrak{L}_{\pi N} = -g_{N\pi} i\bar{\psi}(\gamma_5 \boldsymbol{\phi}\cdot\boldsymbol{\tau})\psi. \tag{3.72}$$

If

$$\begin{aligned} \boldsymbol{\phi} &\longrightarrow \boldsymbol{\phi}, \\ \psi &\longrightarrow \left(1 + i\boldsymbol{\Lambda}\cdot\frac{\boldsymbol{\tau}}{2}\right)\psi, \qquad \Lambda^2 \ll 1, \end{aligned} \tag{3.73}$$

we find that

$$\mathfrak{L}'_{N\pi} = \mathfrak{L}_{N\pi} + \boldsymbol{\Lambda}\times\boldsymbol{\phi}\cdot i g_{N\pi}(\bar{\psi}\gamma_5\boldsymbol{\tau}\psi). \tag{3.74}$$

Thus the interacting fields do *not* conserve

$$\mathrm{j}_\mu{}^V = i\bar{\psi}\gamma_\mu\frac{\boldsymbol{\tau}}{2}\psi. \tag{3.75}$$

In other words, the isotopic spin of the interacting nucleons *by themselves* is not conserved. This is completely reasonable since pions also carry isospin and, in reactions involving pions and nucleons, pions can carry away some of this spin. In any reaction, for example, in which $p \longrightarrow N + \pi^+$ such as $p + p \longrightarrow n + p + \pi^+$, the nucleons lose one unit of i-spin, which is carried away by the meson. However, the combined transformation

$$\begin{aligned} \psi &\longrightarrow \left(1 + i\boldsymbol{\Lambda}\cdot\frac{\boldsymbol{\tau}}{2}\right)\psi, \\ \boldsymbol{\phi} &\longrightarrow \boldsymbol{\phi} + \boldsymbol{\phi}\times\boldsymbol{\Lambda}, \end{aligned} \tag{3.76}$$

does leave the total $\mathfrak{L}$ invariant. Thus the current

$$\mathbf{J}_\mu = i\bar{\psi}\gamma_\mu\frac{\boldsymbol{\tau}}{2}\psi - (\boldsymbol{\phi}\times\partial_\mu\boldsymbol{\phi}) \tag{3.77}$$

is conserved for the interacting case, and hence

$$\mathbf{T} = \int d\mathbf{r}\left(\psi^\dagger\frac{\boldsymbol{\tau}}{2}\psi + (\boldsymbol{\phi}\times\dot{\boldsymbol{\phi}})\right) \tag{3.78}$$

is a constant of the motion. Moreover, from the equal-time commutation relations it follows that

$$[\mathbf{T}_i, \mathbf{T}_j] = i\epsilon_{ijk}\mathbf{T}_k, \tag{3.79}$$

since all cross-terms between pions and nucleons commute away to zero. From the equal-time commutation relations we again show that $\boldsymbol{\phi}$ is an isotopic vector, as is $\bar{\psi}\gamma_5\boldsymbol{\tau}\psi$, so that

$$e^{i\boldsymbol{\Lambda}\cdot\mathbf{T}}\mathfrak{L}e^{-i\boldsymbol{\Lambda}\cdot\mathbf{T}} = \mathfrak{L}. \tag{3.80}$$

Finally, it follows from the construction of the states and the equal-time commutators that $\mathbf{T}$ acting on the states $|\mathbf{p}\rangle_i$ has all of the properties derived earlier in this chapter for the noninteracting fields. Having constructed an invariant $\mathfrak{L}$ we now promptly proceed in the next chapter to break the invariance by introducing the electromagnetic field.

4

Turning on the Electromagnetic Field

Before we couple the electrically charged particles to the photon field we shall give a brief discussion of the photon field Lagrangian itself. As we have chosen to write it

$$\mathcal{L}_\gamma = -\tfrac{1}{2}(\partial_\mu A_\nu\, \partial_\mu A_\nu) \tag{4.1}$$

The most celebrated invariance of this system is gauge invariance. This is the freedom to add to any vector potential $A_\mu(x)$ a quantity of the form $\partial_\mu \Lambda(\mathbf{r}, t)$ where Λ is a c-number function of space-time with the property that $\Box^2\Lambda(\mathbf{r}, t) = 0$. This transformation clearly leaves the equation $\Box^2 A_\mu(\mathbf{r}, t) = 0$, the equation of motion of the photon field, invariant. It is, however, easy to see that the Lagrangian *density* $\mathcal{L}_\gamma$ is *not* invariant under this transformation. A little algebra convinces us that the extra piece added to the Lagrangian after the transformation—call it $\mathcal{L}_\gamma'$—can be written as

$$\mathcal{L}_\gamma' = -\tfrac{1}{2}\, \partial_\mu\{[2A_\nu + \partial_\nu\Lambda]\, \partial_\nu\, \partial_\mu\Lambda\}, \tag{4.2}$$

providing that

$$\Box^2\Lambda(x) = 0. \tag{4.3}$$

Thus the Lagrangian $L = \int d^4x\mathcal{L}(x)$ *is* invariant, since the total 4-divergence can be integrated away to zero. Hence there is invariance of the equations of motion. It is this freedom of gauge that guarantees that the photon has only two independent degrees of polarization even though it has spin 1. For, given any A_μ, we can always add a suitable gauge vector defined to insure that $\mathbf{k}\cdot\mathbf{A} = 0$, where $\mathbf{k}$ is the photon momentum vector. Hence $\mathbf{A}$ is transverse to $\mathbf{k}$, which means that there are only two polarization directions.

There is a time-honored prescription for grafting electromagnetic interactions onto the strong interactions. We begin with the Lagrangian density $\mathcal{L} = \mathcal{L}_0 + \mathcal{L}_s$, where $\mathcal{L}_0$ is the free and $\mathcal{L}_s$ the strong Lagrangian. We then isolate from the Lagrangian those particles that are electrically charged, and so write $\mathcal{L} = \mathcal{L}_{\text{neut}} + \mathcal{L}_{\text{ch}}$. In $\mathcal{L}_{\text{ch}}$, the Lagrangian of the charged particles, we make the replacement $\partial_\mu \longrightarrow (\partial_\mu - ieA_\mu)$ for each derivative, and define $\mathcal{L}_{\text{em}}$, the electromagnetic coupling, after the replacement, as that part of the new Lagrangian that is proportional to e. Needless to say, there is a certain circularity in this procedure. To apply it we must know which particles are electrically charged before turning on the electromagnetic field. However, in practice it seems to work quite well. Experimentally, there is no problem in discovering which particles carry electrical charge and which do not. For example, the electron that is negatively charged has no other interaction than the electromagnetic (and weak), so for it the procedure can be applied directly to the free Lagrangian. Thus, starting with

$$\mathcal{L}_e = -\bar{\psi}_e(\gamma_\mu\partial_\mu + m_e)\psi_e, \tag{4.4}$$

if we let $\partial_\mu \longrightarrow (\partial_\mu - ieA_\mu)$, we find that

$$\mathcal{L}_{\text{em}} = -ie\bar{\psi}_e\gamma_\mu\psi_e A_\mu. \tag{4.5}$$

This coupling gives splendid agreement with electromagnetic experiments when computed in the standard Feynman-Dyson perturbation theory. However, even here, there is some ambiguity in the method of deriving $\mathcal{L}_{\text{em}}$. This has to do with the ambiguity in $\mathcal{L}_e$ discussed in general at the end of Chapter 1: we can add total divergences to $\mathcal{L}_e$ without altering the free equations of motion. These total divergences will, however, generate new terms in $\mathcal{L}_{\text{em}}$ proportional to e and hence they lead to an entirely different electromagnetic interaction than the one written above. To take a specific example, let us write

$$\mathcal{L}_e = -\bar{\psi}_e(\gamma_\mu\partial_\mu + m_e)\psi_e - \lambda\partial_\mu(\bar{\psi}_e\sigma_{\mu\nu}\,\partial_\nu\psi_e), \tag{4.6}$$

where λ is a totally arbitrary constant with dimensions of $1/m$. This Lagrangian will generate an additional electromagnetic interaction of the form

$$i\lambda e\bar{\psi}\sigma_{\mu\nu}\,\partial_\nu\psi A_\mu.$$

As we shall later discuss in more detail, such a coupling looks exactly like the coupling of a magnetic moment to the electric field. This added coupling is often called an *intrinsic* Pauli-moment coupling, "intrinsic" since when it is put into the theory in this way λ is a fundamental constant *intrinsic* to the electron and not connected in any specific way with the electric charge e. There is something unattractive about adding such terms, since one would like to keep the number of fundamental constants in the theory down to an absolute minimum. Furthermore, for the electron certainly, and for the rest of the elementary particles most probably, such terms seem to be entirely irrelevant, since the magnetic properties of these particles can be computed or estimated without them. Thus we would like to have a principle that eliminates these terms from the beginning. This principle usually goes under the name of "minimal electromagnetic coupling." Boiled down to its essentials, this principle states that all Lagrangian densities should contain a minimum number of derivatives. Even stated this way, the principle is not simple to interpret. For the electron and the muon, whose interactions are relatively well understood, we may use the principle to argue that the Lagrangian should never contain more than one derivative, since all the properties of these particles can be correctly computed with such Lagrangians. For spin-zero mesons the free Lagrangian already contains two derivatives. We might try to use the principle to argue against the pseudovector coupling of pions to nucleons,

$$\mathfrak{L}_{\pi N} = -\frac{if}{m_\pi}\frac{\partial\phi}{\partial x_\mu}\cdot\bar{\psi}\gamma_5\gamma_\mu\tau\psi, \tag{4.7}$$

in favor of the pseudoscalar coupling exhibited in the last chapter. Since it is not easy to calculate anything with either coupling, it is essentially impossible to rule out one coupling as opposed to another. The intrinsic Pauli-moment coupling contains two derivatives in the Lagrangian, so we might employ the principle to rule out such couplings for spin-$\frac{1}{2}$ particles. Spin-zero particles, as we shall prove, have no magnetic couplings, hence the question of Pauli terms does not arise for them. Particles of higher spin, such as a charged spin-one particle with mass, automatically exhibit at least two derivatives in their Lagrangian densities. Hence there is no reason to *exclude* Pauli moments for them.[1] Not enough is known, either theoretically or experimentally, to say whether such terms are present for the known vector particles. Thus, at best, the minimal principle is a kind of heuristic guide, based on the success of electrodynamics, to choosing the simplest Lagrangians.

It is a remarkable fact, not deeply understood, that in applying the above prescription for generating the electromagnetic currents belonging to any of the known elementary particles, the only charges that are needed can be expressed as $\pm e$ where e is the electric charge of the electron. In this respect we must distinguish between two questions.

1. Why should the coupling to the photon of *all* the known elementary particles (whatever their other interactions) have the same strength as the electron's coupling to the photon?

2. Given that these coupling constants are all the same, why should the actual *measured* charges of the particles be the same as this coupling constant?

Though the answer to the first question is not known, we can discuss the second one. We must first understand what is meant by the electric charge of a particle. Clearly, if we apply our prescription, the electromagnetic Lagrangian density can always be written in the form

---

1. A massively charged spin-1 field can be characterized by the Lagrangian

$$\mathfrak{L} = -\tfrac{1}{2}\sum_{\mu\nu}\left(\frac{\partial\psi_\nu^\dagger}{\partial x_\mu} - \frac{\partial\psi_\mu^\dagger}{\partial x_\nu}\right)\left(\frac{\partial\psi_\nu}{\partial x_\mu} - \frac{\partial\psi_\mu}{\partial x_\nu}\right) - m^2\sum_\nu \psi_\nu^\dagger\psi_\nu.$$

To this Lagrangian one may add $\partial_\nu\Lambda_\nu$ with

$$\Lambda_\nu = \tfrac{1}{2}\lambda\sum_\mu\left(\frac{\partial\psi_\mu^\dagger}{\partial x_\mu}\psi_\nu - \frac{\partial\psi_\nu^\dagger}{\partial x_\mu}\psi_\mu\right) + \text{Hermitian conjugate.}$$

This new term contains exactly the same number of derivatives as the "original" Lagrangian but it is easy to see that it gives rise to a conserved magnetic current of the form $j_\nu = \partial_\mu P_{\mu\nu}/\partial x_\mu$ with

$$P_{\mu\nu} = i\lambda(\psi_\nu^\dagger\psi_\mu - \psi_\mu^\dagger\psi_\nu),$$

and hence the Pauli ambiguity.

$$\mathcal{L}_{\text{em}} = -\frac{\delta \mathcal{L}}{\delta A_\mu} A_\mu \cdot\equiv\cdot -eJ_\mu{}^\gamma A_\mu. \tag{4.8}$$

We will shortly show that this $J_\mu{}^\gamma$ is conserved; that is,

$$\partial_\mu J_\mu{}^\gamma = 0, \tag{4.9}$$

and thus

$$Q = \int d\mathbf{r} J_0{}^\gamma(\mathbf{r}, t) \tag{4.10}$$

is a Lorentz scalar constant of the motion. Let us confine our attention to the electron for the moment. Its current, with no Pauli terms, is simply

$$J_\mu{}^\gamma = i\bar{\psi}_e \gamma_\mu \psi_e \tag{4.11}$$

and its charge operator is

$$Q = \int d\mathbf{r}\, \psi_e{}^\dagger(\mathbf{r}, t)\psi_e(\mathbf{r}, t). \tag{4.12}$$

These ψ_e are the fully interacting Heisenberg fields. For a given one-particle state, the charge of the state is the matrix element of eQ taken between the state vectors, $|\mathbf{p}\rangle$, constructed from the vacuum using the exact Heisenberg fields. The question that we are raising is the way in which the interactions succeed in making this charge different in value from e itself. Let us call the measured charge e_{obs}. Hence

$$e_{\text{obs}} = e\langle \mathbf{p}|Q|\mathbf{p}\rangle, \tag{4.13}$$

where $|\mathbf{p}\rangle$ is the exact one-particle state.

If we apply the arguments of Chapter 2 naively, first constructing the states $|\mathbf{p}\rangle$ by using the limiting process given there and then noting that from the non-degeneracy of the vacuum and the fact that $[Q, C]_+ = 0$, where C is the charge conjugation operator, defined earlier, with the property

$$CJ_\mu{}^\gamma C^{-1} = -J_\mu{}^\gamma, \tag{4.14}$$

we have

$$Q|0\rangle = 0. \tag{4.15}$$

We would conclude from this that

$$\langle \mathbf{p}|Q|\mathbf{p}\rangle = 1, \tag{4.16}$$

in just the same way we showed that

$$\langle \mathbf{p}|B|\mathbf{p}\rangle = 1 \tag{4.17}$$

for the baryon number.

However, there is reason to believe that this argument is somewhat too simpleminded for the case of the electromagnetic field. The question that arises has to do with the meaning of the limit $t \longrightarrow \pm\infty$ used in constructing the states and its connection with the phenomenon of "vacuum polarization." To see what is involved, a useful analogy is that of measuring the charge of a particle in a polarizable crystal. We consider two situations illustrated by the figures below.

Let us assume that in the figure at left the charge external to the crystal has a measured value $+e$. If we bring the charge inside the crystal and measure the charge in any finite volume around it, we will find that the charge in this volume is no longer e but $e' < e$, since negative "screening" charges will be attracted to e (or positive charges pushed to the surface). In quantum electrodynamics a somewhat similar phenomenon occurs. If at $t = -\infty$ a charge e_0 is introduced into the vacuum and the electromagnetic interaction is turned on, this charge begins creating virtual electron-positron pairs, a phenomenon known as the polarization of the vacuum. If we then measure the electric charge in a small

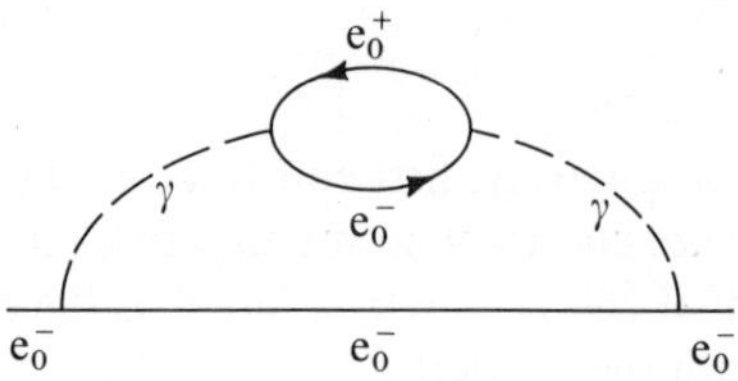

finite volume around, say, an electron, it will no longer be e_0 but rather $e_{obs} = \sqrt{Z}\, e_0$, where $\sqrt{Z}$ is a "renormalization" constant. The difference between e_0 and e_{obs} occurs because the electrons in the virtual pairs are repelled away from the original charge so that the charge that remains in the finite volume is changed. $\sqrt{Z}$ can be computed in perturbation theory.[2] (It turns out to be infinite but this is presumably a result of a breakdown in the perturbation method and not a physical effect.) We must therefore be clear as to whether in our limiting process at $t = -\infty$ we create a particle with charge e_0 or whether at $t = -\infty$ we include the vacuum polarization effects and so create particles with the charge e_{obs}, the measured charge. There would be no problem in identifying the fields at $t = -\infty$ with the ones that bear the physical charge e_{obs}. However,

2. See, for example, *Principles of Quantum Electrodynamics*, by W. E. Thirring, Academic Press, New York, 1958.

what then does the parameter e mean that occurs in the Lagrangian? We choose to identify this e with e_0 and thus write

$$\langle \mathbf{p}|Q|\mathbf{p}\rangle = \sqrt{Z}, \tag{4.18}$$

exhibiting the renormalized charge explicitly. This means that $Q(-\infty) \neq Q(t)$ for finite t, which might appear to violate charge conservation. In particular, we suppose that

$$\lim_{t\to-\infty} \{[Q(t), \psi(t)]_-\} = \sqrt{Z}\psi(-\infty). \tag{4.19}$$

However, before we give up charge conservation we recall that

$$\partial_\mu J_\mu{}^\gamma = 0 \tag{4.20}$$

only implies

$$\frac{\partial}{\partial t} Q(t) = 0 \tag{4.21}$$

if

$$\int (\nabla \cdot \mathbf{J}^\gamma)\, d\mathbf{r} = 0, \tag{4.22}$$

which means only if the contributions to the current from infinitely remote surfaces can be neglected. Because the photon field has infinite range these terms contribute at $t = -\infty$ when the field is turned on. The strong interactions have finite range so currents like $J_\mu{}^B$ the baryon current, will not lead to surface terms of this kind. Indeed, once these surface terms are taken into account for the photon field turning on additional short range strong interactions will not further modify the total electric charge providing that these interactions are charge conserving. In other words the proton, in the presence of electromagnetism, will also have

$$\langle \mathbf{p}|Q|\mathbf{p}\rangle = \sqrt{Z}, \tag{4.23}$$

with the same $\sqrt{Z}$. The strong interactions modify many of the electromagnetic properties of these particles but not their total charge. We shall not have occasion to discuss higher-order electromagnetic corrections to any extent in this book; the reader who would like a more complete discussion would be advised to consult one of the specialized texts.

It is crucial that the prescription given above for constructing $\mathcal{L}_{\text{em}}$ always lead to a conserved electric current. For spin-$\frac{1}{2}$ particles interacting only with the electromagnetic field, this is an immediate application of the work of Chapter 2. The free Lagrange function is simply

$$\mathcal{L}_0 = -\bar{\psi}(\gamma_\mu\partial_\mu + m)\psi, \tag{4.24}$$

and it is clear that the current obtained by letting $\partial_\mu \longrightarrow (\partial_\mu - ieA_\mu)$ is identical to the current obtained by letting

$$\psi \longrightarrow (1 - ie\Lambda(x))\psi \tag{4.25}$$

in $\mathcal{L}_0$ for all the charged ψ's. This current is guaranteed to be conserved because of the phase invariance of $\mathcal{L}_0 + \mathcal{L}_{em}$. In other words, the Lagrangian

$$\mathcal{L} = -\bar{\psi}(\gamma_\mu(\partial_\mu - ieA_\mu) + m)\psi \tag{4.26}$$

is invariant under constant phase transformations of ψ, and the current generated for nonconstant $\Lambda(x)$ is identical to the coefficient of A_μ.

Even for noninteracting spin-zero particles (say pions), the situation is more complex than the spin-$\frac{1}{2}$ case because of the occurrence of the double derivatives in $\mathcal{L}_\pi$. We must begin by identifying the charged particles. Using the three fields ϕ_1, ϕ_2, ϕ_3, we define the charged fields, π^+, π^-, π^0. We give the fields that *create* the corresponding charged particle:[3]

$$\begin{aligned} \pi^+ &= \frac{\phi_1 + i\phi_2}{\sqrt{2}}, \\ \pi^- &= \frac{\phi_1 - i\phi_2}{\sqrt{2}} = (\pi^+)^\dagger, \\ \pi^0 &= \phi_3. \end{aligned} \tag{4.27}$$

We shall see shortly that these definitions fulfill all the characteristics that we would expect for electrically charged pions.

Thus

$$\boldsymbol{\phi}\cdot\boldsymbol{\phi} = \pi^+\pi^- + \pi^-\pi^+ + \pi^0\pi^0 \tag{4.28}$$

and

$$\begin{aligned} \mathcal{L}_\pi &= -\tfrac{1}{2}[\partial_\mu\boldsymbol{\phi}\cdot\partial_\mu\boldsymbol{\phi} + m_\pi^2\boldsymbol{\phi}\cdot\boldsymbol{\phi}] \\ &= -\tfrac{1}{2}[\partial_\mu\pi^+\,\partial_\mu\pi^- + \partial_\mu\pi^-\,\partial_\mu\pi^+ + \partial_\mu\pi^0\,\partial_\mu\pi^0 \\ &\qquad + m_\pi^2(\pi^+\pi^- + \pi^-\pi^+ + \pi^0\pi^0)]. \end{aligned} \tag{4.29}$$

We may now turn on the electric field by making the replacements

$$\partial_\mu \longrightarrow (\partial_\mu \pm ieA_\mu) \tag{4.30}$$

for the $\pi^\pm$. This yields

$$\mathcal{L}_{em} = -ie\left[\left\{\frac{(\pi^+\,\partial_\mu\pi^- - \partial_\mu\pi^+\pi^-)}{2} + \frac{(\partial_\mu\pi^-\pi^+ - \pi^-\,\partial_\mu\pi^+)}{2}\right\} - ieA_\mu\pi^+\pi^-\right]A_\mu \tag{4.31}$$

We see that, for spin-zero particles, the coefficient of A_μ involves the electric charge explicitly, but for spin-$\frac{1}{2}$ particles it does not.

In the absence of electromagnetism we wrote the mass term in the Lagrangian in the form

$$m_\pi^2[\phi_1\phi_1 + \phi_2\phi_2 + \phi_3\phi_3],$$

3. Note that at equal times

$$[\pi^+(\mathbf{r}, t), \dot{\pi}^-(\mathbf{r}', t)]_- = i\,\delta^3(\mathbf{r} - \mathbf{r}').$$

thus assigning to all three pions a common mass parameter. However, most probably it is the electromagnetic interactions that split the pion masses. Thus we might think that we should assign different mass parameters m_1, m_2, and m_3 to ϕ_1, ϕ_2, and ϕ_3 in the presence of the electromagnetic field. There is, however, an additional constraint imposed by current conservation. The conservation of the pionic electromagnetic current is a consequence of the phase invariance of $\mathcal{L}_\pi$. In the language of the Hermitian fields ϕ_1 and ϕ_2 this means (as we shall see in more detail below) that the Lagrangian must be invariant under linear transformations involving these fields (rotations around the 3 directions in isospace) and in particular under the interchange of ϕ_1 and ϕ_2. This is only possible if ϕ_1 and ϕ_2 are assigned the same mass parameter. The parameter assigned to ϕ_3 (that is, m_{π^0}) is arbitrary.

Hence, to take account of electromagnetic mass splittings, we should have written

$$\mathcal{L}_\pi = -\tfrac{1}{2}[\partial_\mu\pi^+\,\partial_\mu\pi^- + \partial_\mu\pi^-\,\partial_\mu\pi^+ + m_{\mathrm{ch}}{}^2(\pi^+\pi^- + \pi^-\pi^+)] - \tfrac{1}{2}[\partial_\mu\pi^0\,\partial_\mu\pi^0 + m_0{}^2\pi^0\pi^0]. \tag{4.32}$$

But it is clear that this Lagrangian is no longer invariant under the full SU_2 group. Those transformations that mix the 3 direction with the others obviously fail to leave $\mathcal{L}_\pi$ invariant. Only transformations around the 3 direction in the 1-2 plane leave $\mathcal{L}_\pi$ invariant. Thus only the transformation that generates rotations around the 3 direction is a conserved quantity. It is clear from examining[4] $\mathcal{L}_{\mathrm{em}}$ that it too is invariant under rotations about the 3-direction in isospace. But a general infinitesimal rotation about the 3-direction can be written

$$\begin{aligned}\phi_1' &= \phi_1 + \Lambda\phi_2,\\ \phi_2' &= \phi_2 - \Lambda\phi_1,\\ \phi_3' &= \phi_3,\end{aligned} \tag{4.33}$$

or equivalently, as in Chapter 2,

$$\begin{aligned}\pi^{+\prime} &= \pi^+(1 - i\Lambda),\\ \pi^{-\prime} &= \pi^-(1 + i\Lambda),\\ \pi^{0\prime} &= \pi^0.\end{aligned} \tag{4.34}$$

Thus the rotation invariance of $\mathcal{L}_0 + \mathcal{L}_{\mathrm{em}}$ is equivalent to its phase invariance for constant phase. But this guarantees the conservation of the electric current, since the current generated by letting

$$\partial_\mu\pi^{\pm\prime} = (1 \mp i\Lambda(x))\,\partial_\mu\pi^\pm \mp i\,\partial_\mu\Lambda(x)\pi^\pm \tag{4.35}$$

in $\mathcal{L}_0 + \mathcal{L}_{\mathrm{em}}$ has exactly the same functional form as the current obtained by allowing $\partial_\mu \longrightarrow \partial_\mu \pm ieA_\mu$ in $\mathcal{L}_\pi$. Thus the conserved pionic electric current

4. The photon has no isospin.

$$J_\mu^\gamma = \tfrac{1}{2}[i(\pi^+ \partial_\mu\pi^- - \partial_\mu\pi^+\pi^-) + i(\partial_\mu\pi^-\pi^+ - \pi^- \partial_\mu\pi^+)] \tag{4.36}$$

yields a constant of the motion

$$Q = \int d\mathbf{r}\, J_0^\gamma(\mathbf{r}, t), \tag{4.37}$$

which generates phase transformations on $\pi^\pm$. This current does not have simple transformation properties under SU_2. In the literature one often finds the equation relating the pion electric charge to the third component of the isotopic spin

$$Q = T_3, \tag{4.38}$$

which might make it appear as if the electric charge were simply the third component of an isovector. In fact this formula should be understood in a perturbation theory sense. To lowest order in e the charge Q is simply T_3, as defined in the last chapter. If this charge is applied to the pion states constructed from the field operators of the strong interactions alone, then, since these states have a well-defined isotopic spin—they belong to the triplet representation of SU_2—the states will have the eigenvalues $+$, $-$, 0 of T_3. However, if the electric field is turned on, then SU_2 is no longer an invariance of the theory. The charged and neutral pions as observed in nature have slightly different masses.[5] The pion states in the presence of electromagnetism no longer belong to simple representations of SU_2 and the classification in terms of T^2, T_3 does not retain an exact validity.[6] However, Q, the electric charge, is still a good quantum number. The fact that it is conserved means that we can use the equal-time commutation relations among the pions which, even in the presence of the electromagnetic interactions, are the same as those of the free fields at any finite time t.[7]

Thus

$$\begin{aligned} [Q, \pi^\pm]_- &= \pm\pi^\pm, \\ [Q, \pi^0]_- &= 0. \end{aligned} \tag{4.39}$$

Hence if we construct the states $|\pi^{\mp}_{0}\rangle$, using for the electromagnetic interactions

---

5. All the members of an SU_2 multiplet must have the same mass. This follows from the assumed invariance of H_{st} under SU_2. The operators $T^\pm = T_1 \pm iT_2$ raise and lower the states within a multiplet just as $J^\pm$ raise and lower angular momentum states. It follows from

$$T^+|TT_3\rangle = \sqrt{(T - T_3)(T + T_3 + 1)}\,|TT_3 + 1\rangle,$$

$$T^-|TT_3\rangle = \sqrt{(T + T_3)(T - T_3 + 1)}\,|TT_3 - 1\rangle$$

that

$$\begin{aligned} \langle TT_3|T^-H_{st}T^+|TT_3\rangle &= \langle TT_3 + 1|H_{st}|TT_3 + 1\rangle(T^2 + T_3^2 - T_3) \\ &= \langle TT_3|H_{st}T^-T^+|TT_3\rangle = \langle TT_3|H_{st}|TT_3\rangle(T^2 + T_3^2 - T_3), \end{aligned}$$

so that all the diagonal elements of H_{st}—that is, the masses—are the same throughout a given multiplet.

the same techniques we used for the strong interactions, and again noting the vacuum polarization effect at $t = -\infty$, we have for these states

$$Q|\pi^{\pm}_{0}\rangle = {}^{\pm}_{0}\sqrt{Z}|\pi^{\pm}_{0}\rangle, \tag{4.40}$$

which means, in turn, for example, that

$$e\langle\pi^+|Q|\pi^+\rangle = e\sqrt{Z} = e_{\text{obs}}. \tag{4.41}$$

We may now add to $\mathfrak{L}_\pi$ and $\mathfrak{L}_{\text{em}}$ the nucleons and their pion couplings. The free nucleon Lagrangian is simply

$$\mathfrak{L}_N = -\bar{P}(\gamma_\mu\partial_\mu + m)P - \bar{N}(\gamma_\mu\partial_\mu + m)N. \tag{4.42}$$

In writing $\mathfrak{L}_N$ this way we adopt the point of view that the mass parameters of neutron and proton in the Lagrangian are identical, but that the observed masses, as a result of the electromagnetic interaction, are distinct. We could also assume that $m_P \neq m_N$ in $\mathfrak{L}_N$, but this would introduce extra parameters into the theory. Allowing $\partial_\mu \longrightarrow \partial_\mu + ieA_\mu$ yields for the proton electromagnetic current simply

$$J_\mu{}^\gamma = ie\bar{P}\gamma_\mu P. \tag{4.43}$$

This current is conserved as a consequence of the invariance of $\mathfrak{L}_N$ under the phase transformations

$$\begin{aligned} P &\longrightarrow (1 + i\Lambda)P, \\ N &\longrightarrow N. \end{aligned} \tag{4.44}$$

The baryon current is generated by the transformations

$$\begin{aligned} P &\longrightarrow (1 + i\Lambda)P, \\ N &\longrightarrow (1 + i\Lambda)N, \end{aligned} \tag{4.45}$$

and it, too, is conserved, but clearly it is not the same as the electric current.

Following the analogy with electromagnetism it is tempting to speculate that all conserved currents are coupled to vector mesons. This speculation is

---

6. There is nothing to stop us from making transformations on $\mathfrak{L}_\pi$ that mix ϕ_1 and ϕ_3. However, these transformations cannot be generated, in the presence of the electromagnetic field, by conserved charges. There is only one conserved charge—Q—and it generates phase transformations. Hence, since the generators are defined in terms of field operators, we have here an example of how certain combinations of field operators, the charges, simply cease to be symmetry generators in the presence of new interactions. In a sense, in the presence of the electromagnetic field, the SU_2 group simply disappears, since its generators can no longer be defined in terms of field operators. The charges T_1, T_2, etc. of the last chapter are not constants in time and do not have simple commutation relations with the fields except at equal times where they remain the same as in the free-field case. The fact that isotopic spin is not conserved in the presence of electromagnetic interactions manifests itself in processes such as $\pi^0 \longrightarrow \gamma + \gamma$, which are well known to occur experimentally but which clearly violate isotopic spin conservation.
7. Note especially that, at equal times, the term $eA_\mu\pi^+\pi^-$ in $J_\mu{}^\gamma$ commutes with π^+ and π^-.

encouraged in the first instance by the fact that such mesons are known to exist in nature. In addition to the photon there exist several mesons of spin one that appear as resonances in processes that involve pions and nucleons. For example, a proton and an antiproton annihilate rapidly ($\sim 10^{-23}$ sec) into several π mesons. Instead of emerging as free uncorrelated particles the pions often appear to have correlated energies and momenta as if they were being emitted as the result of the decay of an intermediate short-lived particle produced in the chain $P + \bar{P} \longrightarrow x \longrightarrow n\pi$. By studying the energies and momenta of the pions the properties of these particles may be inferred. Later we will explore the properties of many of these mesons in detail. Here we note that, among others, two neutral vector mesons, known as ρ^0 and ω^0, have been identified with the following properties.[8]

1. $\rho^0 \longrightarrow \pi^+ + \pi^-$,

   $\Gamma_{\rho^0} = 140$ MeV,

   $m_{\rho^0} = 770$ MeV.

2. $\omega^0 \longrightarrow 3\pi$,

   $\Gamma_{\omega^0} = 11.9 \pm 1.5$ MeV,

   $m_{\omega^0} = 783.4 \pm 0.7$ MeV.

The broad neutral ρ^0 resonance has two charged counterparts $\rho^{\pm}$, at essentially the same central mass, and the narrow neutral ω^0 resonance has no known charged counterparts. This has led to an assignment of $T = 1$ to the $\boldsymbol{\rho}$ multiplet and $T = 0$ to the ω^0. The spin of these resonances has been fixed to be 1. It is easy to see that the parity of the two pion state produced in ρ decay is linked to its angular momentum. Consider the diagram

$$\underset{-\mathbf{r}/2}{\pi^+ \longleftarrow} \cdot \underset{\mathbf{r}/2}{\longrightarrow \pi^-}.$$

Clearly, letting $\mathbf{r} \longrightarrow -\mathbf{r}$ introduces a factor $(-1)^l$, with l the orbital angular momentum, into the 2π wave function. The pions each have an intrinsically odd parity that cancels out here, since there are two of them. Thus, given the spin of the ρ^0, and parity conservation in the decay (which must be caused by the strong parity conserving interactions, since it is so fast), the parity of the ρ^0 is negative so that it is a true vector and not an axial or pseudovector. It cannot decay into two neutral pions, since these must always have even relative angular momentum as they are identical Bose-Einstein particles. The analysis of the ω^0 spin and parity is considerably more complicated and falls outside the scope of this book;[9] however, it too is known to be a 1^- particle. It has been speculated[10]

8. See, for example, A. H. Rosenfeld et al., *Rev. Mod. Phys.*, **39**:1 (1967). The numbers given here are averaged over several experiments. The mass and width of the ρ^0 are quite uncertain and these values are an estimated mean among the experiments.

that the ω^0, since it is an isoscalar, might be coupled to the baryon current with a coupling of the form

$$H_B = g_B \omega_\mu{}^0 J_\mu{}^B. \tag{4.46}$$

This coupling is certainly possible, and elegant, but direct experimental confirmation of its existence is hard to come by, since the ordinary pion nucleon processes can, because the interactions are strong, produce effects that look like those of such a direct coupling. Of one thing we *can* be sure: if there is a vector meson coupled to the baryon current, it cannot have mass zero like the photon, or even a very light mass. To see this, we may compare the potential energy produced by such a coupling, due (say) to the nucleons that compose the Earth, with the ordinary gravitational potential. Suppose we have a particle of mass m at a distance r above the surface of the Earth, with mass M_E. Then

$$V_{\text{grav}} = -\frac{GmM_E}{r}. \tag{4.47}$$

Assume that the test particle p is a point particle of baryon number N_p and mass m_p. We neglect antibaryons in the Earth, and take m_N as the mass of a nucleon and m_B as the mass of the vector particle coupled to $J_\mu{}^B$, and assume that the density of baryons, ρ, is a constant given by[11]

$$\rho = \frac{M_E}{m_N \frac{4}{3}\pi R_E{}^3}. \tag{4.48}$$

Then the potential V_B due to the B force is

$$V_B = \frac{g_B{}^2 M_E N_p}{\frac{4}{3}\pi R_E{}^3 m_N} \int d\mathbf{r}\, \frac{e^{-m_B|\mathbf{r}-\mathbf{r}_p|}}{|\mathbf{r}-\mathbf{r}_p|}, \tag{4.49}$$

where the integral extends over the Earth. For $m_B = 0$,

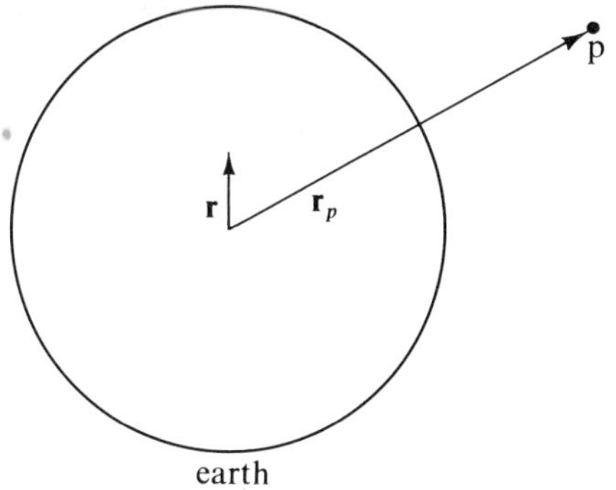

9. See, for example, G. Puppi, *Annual Review of Nuclear Science*, Vol. 13, 1963, page 287 and numerous references found in this review.
10. See especially, J. J. Sakurai, *Ann. Phys.*, **11**:1 (1960). The vector mesons are discussed fully in Chapter 14.
11. It is convenient to measure the gravitational constant in units of $1/m_N{}^2$. Thus $Gm_N{}^2 \simeq 10^{-39}$.

$$V_B = \frac{g_B{}^2 M_E N_p}{m_N r_p}. \tag{4.50}$$

This is identical in appearance to the gravitational potential, so that the potential acting on p would become, in total,

$$V = -G\frac{m_p M_E}{r_p} + g_B{}^2\frac{M_E N_p}{m_N r_p} = -G\frac{m_p M_E}{r_p}\left[1 - \frac{g_B{}^2}{G}\frac{N_p}{m_N m_p}\right]. \tag{4.51}$$

Thus we would be forced to draw the conclusion, if V_B exists, that $V \neq \bar{V}$, where $\bar{V}$ would be the potential acting on $\bar{p}$, the antiparticle to p, with

$$N_{\bar{p}} = -N_p. \tag{4.52}$$

This effect would be observable, in principle, if

$$\frac{g_B{}^2}{m_N{}^2} \sim G. \tag{4.53}$$

That is, if $g_B{}^2 \sim 10^{-39}$.

However, we can see that even a g_B this small can be ruled out experimentally (always assuming that $m_B = 0$).

In a uniform gravitational field, with g the gravitational acceleration, we have

$$m_p g = -\frac{G m_p M_E}{r_p{}^2}, \tag{4.54}$$

so that g is independent of m_p.

Suppose that m_p consists of N_p nucleons; then, neglecting the masses of the electrons,

$$m_p = m_N N_p - \epsilon, \tag{4.55}$$

where ϵ is the binding energy. Thus, taking into account the binding energy, we see that the nucleon number may be expressed as

$$N_p = \frac{m_p}{m_N} + \frac{\epsilon}{m_N}. \tag{4.56}$$

With the B force turned on we may write Newton's equation in the form

$$m_p g = -\frac{m_p M_E}{r_p{}^2} C + \frac{g_B{}^2}{r_p{}^2}\frac{M_E \epsilon}{m_N{}^2}, \tag{4.57}$$

where C is a constant containing G and $g_B{}^2$; that is, $C = G - (g_B{}^2/m_N{}^2)$. The second term in Equation (4.57) violates the equivalence between inertial and gravitational mass. The first term looks just like gravity, so that we would identify this C with the experimentally measured gravitational constant G_{obs}. The Eötvös experiment, especially in its modern versions, sets a limit on this violation, and in fact from such experiments we may deduce that

$$\frac{g_B{}^2}{G m_N{}^2} \lesssim 10^{-5}. \tag{4.58}$$

This means that if the B quantum had zero mass it would have to produce a force that was very much weaker than gravitation. A force this weak might be interesting as a theoretical speculation but it would not effect elementary particle processes in a way that is observable. On the other hand, if $m_B \sim m_{\omega^0}$, then, even if $g_B{}^2 \simeq 1$, there will be no violation of the Eötvös limit. The force is then so short-ranged that the macroscopic effects of the Earth become completely irrelevant, and the B potential (now of Yukawa form) makes an absolutely negligible contribution to anything but nuclear processes. We shall return in later chapters to these vector couplings.

This chapter concludes by adding to $\mathcal{L}_\pi + \mathcal{L}_N + \mathcal{L}_\gamma$ the pion-nucleon coupling $\mathcal{L}_{\pi N}$ and any new terms in the electromagnetic currents that arise from it. We may write, in the good i-spin limit, the pseudoscalar coupling in the form

$$\bar{\psi}\gamma_5\boldsymbol{\tau}\cdot\boldsymbol{\phi}\psi = (\bar{P}\gamma_5P - \bar{N}\gamma_5N)\pi^0 + \sqrt{2}\bar{P}\gamma_5N\pi^- + \sqrt{2}\bar{N}\gamma_5P\pi^+. \tag{4.59}$$

Thus, SU_2 invariance in the coupling term requires that the coupling constants of the charged and neutral pions be related in the ratio $g_{\text{ch}}/g_{\text{neut}} = \sqrt{2}$. Even in the presence of the electromagnetic couplings we shall continue to write $\mathcal{L}_{\pi N}$ in this way, honoring the fact that the electromagnetic corrections are relatively small. This term is clearly phase-invariant and so the electromagnetic current remains conserved. Hence, in this theory, we have two conserved Hermitian currents,

$$\begin{aligned} J_\mu{}^B &= i[\bar{P}\gamma_\mu P + \bar{N}\gamma_\mu N], \\ J_\mu{}^\gamma &= \tfrac{1}{2}[i(\pi^+\,\partial_\mu\pi^- - \partial_\mu\pi^+\pi^-) + i(\partial_\mu\pi^-\pi^+ - \pi^-\,\partial_\mu\pi^+)], \end{aligned} \tag{4.60}$$

and two charges, Q_B and Q_e, that are constants of the motion.[12] Moreover, it is clear that

$$[Q_B, Q_e]_- = 0, \tag{4.61}$$

so that the eigenstates of the energy can also be taken to have definite electric charge and baryon number. In the perturbation theory sense discussed above, to the lowest order in e, we may write

$$Q_e = \frac{Q_B}{2} + T_3, \tag{4.62}$$

which gives the charge of all the strongly interacting states of pions and nucleons in terms of their baryon number and isotopic spin.

In the pseudovector theory the coupling term is

$$\mathcal{L}_{\pi N} = -\frac{if}{m\pi}\,\partial_\mu\boldsymbol{\phi}\cdot\bar{\psi}\gamma_5\gamma_\mu\boldsymbol{\tau}\psi. \tag{4.63}$$

---

12. There is no conserved quantity that measures particle number (as opposed to electric charge or baryon number) in the theory of interacting fields. This is as it should be, since particles can be created and annihilated in many processes and hence their number changes.

This means that the conserved electromagnetic current in this theory will have a different functional form than the current in the pseudoscalar theory. In principle it should be possible to test this difference but, in practice, calculations done with either theory are not reliable since the coupling constants are large. As we shall see in the next chapters, the modern trend is to avoid such calculations (certainly those done in perturbation expansions) in favor of methods that seek to relate different processes to each other rather than to make an absolute computation of a given process. As a first example we turn to electron-proton scattering.

5

Electron-Nucleon Scattering I: Form Factors C, P, T, and CPT

One of the most carefully studied processes in elementary particle physics is the elastic scattering reaction $e + N \longrightarrow e' + N'$, where e is an electron and N a nucleon. In considering this process, both theoretically and experimentally, it is important to state what is meant by *elastic* scattering. Elastic scattering, $a + b \longrightarrow a' + b'$, is a process in which the initial and final energies and momenta are related by the equations

$$\begin{aligned} E(\mathbf{p}_a) + E(\mathbf{p}_b) &= E(\mathbf{p}_a') + E(\mathbf{p}_b'), \\ \mathbf{p}_a + \mathbf{p}_b &= \mathbf{p}_a' + \mathbf{p}_b'. \end{aligned} \tag{5.1}$$

In particular in the barycentric system (often called the center of mass) frame,

$$\mathbf{p}_a + \mathbf{p}_b = \mathbf{p}_a' + \mathbf{p}_b' = 0, \tag{5.2}$$

implying that

$$\sqrt{\mathbf{p}^2 + m_a^2} + \sqrt{\mathbf{p}^2 + m_b^2} = \sqrt{\mathbf{p}'^2 + m_a^2} + \sqrt{\mathbf{p}'^2 + m_b^2}. \tag{5.3}$$

Therefore, in this frame,

$$|\mathbf{p}| = |\mathbf{p}'| \tag{5.4}$$

and

$$\begin{aligned} E(\mathbf{p}_a) &= E(\mathbf{p}_a'), \\ E(\mathbf{p}_b) &= E(\mathbf{p}_b'). \end{aligned} \tag{5.5}$$

Thus to identify a scattering as elastic we must verify that the energies of the initial and final particles in the center of mass are unchanged. In practice this is only possible to within some fixed accuracy, ΔE, since no experimental instrument has perfect energy resolution. However, electrically charged particles emit photons (*bremsstrahlung*) so that along with the elastic scattering there is an inevitable contribution from processes of the type $e + N \longrightarrow e' + N' + \gamma$. Fortunately these "radiative corrections" (and others) can be computed accurately so that we may correct for them and, to lowest order in α, we may confront the experimental results with calculations based simply on the graph shown. This graph consists of three parts, proceeding from right to left.

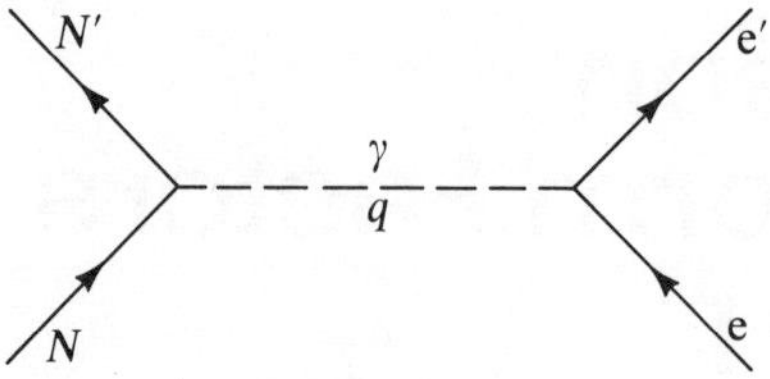

$q = e - e' = N' - N$

The electron-photon vertex:

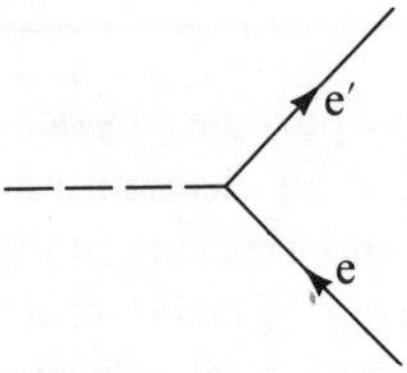

In field theory (we shall come back to this point later) the vertex function $\Gamma_\mu^e(e, e')$ can be given a very general definition in terms of the field operators $\psi_e(x)$, $\psi_e^\dagger(x)$, and $J_\mu^\gamma(x)$, where $J_\mu^\gamma(x)$ is the conserved electric current. So long as we work to order α, thus ignoring corrections like

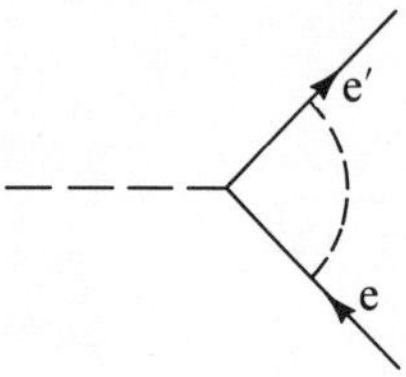

this vertex can be simply written as

$$\Gamma_\mu^e(\mathrm{e}, \mathrm{e}') = e\langle \mathrm{e}'|J_\mu^\gamma(0)|\mathrm{e}\rangle(2\pi)^3, \tag{5.6}$$

where e is the electromagnetic coupling constant in $\mathcal{L}_{\mathrm{em}}$. Here $|\mathrm{e}\rangle$ and $\langle \mathrm{e}'|$ are the one particle electron states which, since we are working to order α, can be taken as states created out of the vacuum by the *free* electron creation operators. $J_\mu^\gamma(0)$ is the electron current evaluated at the origin in space-time. In choosing the origin we make use of the fact that $J_\mu^\gamma(x) = e^{-i(Px)}J_\mu^\gamma(0)e^{i(Px)}$, so that we can always translate the space time point x in $J_\mu(x)$ to the origin—an especially convenient point. For the electron current we write

$$J_\mu^\gamma(0) = i\bar{\psi}_e(0)\gamma_\mu\psi_e(0). \tag{5.7}$$

(We assume *minimality* so that this current contains no intrinsic Pauli terms.) Furthermore, q is the four-momentum transferred to the electron so that, in our metric,

$$q^2 = -2m_e^2 - 2(\mathbf{e}\cdot\mathbf{e}' - E(\mathbf{e})E(\mathbf{e}')) = (p - p')^2 \tag{5.8}$$

Clearly, q^2 is a Lorentz scalar. For the applications of interest the electron energies are several MeV so that we can always set $\frac{m_e}{|\mathbf{e}|} \simeq 0$. With, or without, this approximation it is easy to see that for electron scattering (in our metric)

$$q^2 \geq 0 \tag{5.9}$$

The quantity

$$\hat{\mathbf{e}} \cdot \hat{\mathbf{e}}' \cdot\!\equiv\!\cdot \cos\theta \tag{5.10}$$

is the scattering angle. In the center of mass ($m_e = 0$)

$$q^2 = 4\mathbf{e}^2 \sin^2 \frac{\theta}{2}, \tag{5.11}$$

where $\mathbf{e}^2$ is the electron energy; $\sqrt{\mathbf{e}^2 + m_e^2} \simeq \sqrt{\mathbf{e}^2}$. Since the ψ_e, to the order in which we need them, can be taken as free, we find, using the expansions of Chapter 1, that

$$\langle \mathbf{e}'|_{s'} J_\mu{}^\gamma(0)|\mathbf{e}\rangle_s = \frac{i\bar{u}(\mathbf{e}')_{s'}\gamma_\mu u(\mathbf{e})_s}{(2\pi)^3}. \tag{5.12}$$

There is another way of viewing this result that anticipates the discussion we will shortly give for the proton current matrix element.[1] We may desire the most general form that the matrix element

$$\langle a|J_\mu{}^\gamma(0)|b\rangle$$

can have. Here $\langle a|$ and $|b\rangle$ are two states with the same spin but different four-momenta q_a and q_b. This matrix element must be a Lorentz four-vector,[2] assuming that $J_\mu{}^\gamma(0)$ is a four-vector. The form of this vector matrix element will depend on the spin of the states $|a\rangle$ and $|b\rangle$. We will work it out in detail here for spin-zero and spin-$\frac{1}{2}$.

There are two independent four-vectors that can be associated with two spin-zero particles

$$\begin{aligned} q_+ &= q_a + q_b, \\ q_- &= q_a - q_b, \end{aligned} \tag{5.13}$$

and one[3] Lorentz scalar, say q_-^2. Thus

$$\langle a|J_\mu{}^\gamma(0)|b\rangle = q_{+\mu}F_+(q_-^2) + q_{-\mu}F_-(q_-^2), \tag{5.14}$$

where F_+ and F_- are two arbitrary functions of q_-^2.

---

1. While the work that follows is carried out explicitly for the electromagnetic current $J_\mu{}^\gamma$, the reader should have no difficulty in picking out those features that apply to arbitrary currents.
2. Strictly speaking, this needs proof. In essence the proof is identical to the proof of the Wigner-Eckart theorem in ordinary quantum mechanics, in which we deal with angular momentum states and tensor operators under the group of real rotations.
3. q_+^2 and q_-^2 are not independent since both are functions of the same independent variable (q_+q_-).

In the electromagnetic case J_μ^γ is a Hermitian current, which means, more precisely, that

$$\begin{aligned} \mathbf{J}^{\gamma\dagger}(x) &= \mathbf{J}^\gamma(x), \\ J_4^{\gamma\dagger}(x) &= -J_4^\gamma(x). \end{aligned} \tag{5.15}$$

Thus

$$\begin{aligned} \langle a|\mathbf{J}^\gamma(0)|b\rangle &= \langle b|\mathbf{J}^\gamma(0)|a\rangle,^* \\ \langle a|J_4^\gamma(0)|b\rangle &= -\langle b|J_4^\gamma(0)|a\rangle^*, \end{aligned} \tag{5.16}$$

or, in terms of the representation above,[4]

$$\mathbf{q}_+F_+(q_-^2)_{b\to a} + \mathbf{q}_-F_-(q_-^2)_{b\to a} = \mathbf{q}_+F_+^*(q_-^2)_{a\to b} - \mathbf{q}_-F_-^*(q_-^2)_{a\to b}. \tag{5.17}$$

The change in sign in q_- comes from the fact that it is antisymmetric under $a \longleftrightarrow b$. The matrix elements for $a \longrightarrow b$ and $b \longrightarrow a$ are not, in general, simply related unless a and b are the same particle. In that case the Hermiticity condition implies that

$$\begin{aligned} F_+ &= F_+^*, \\ F_- &= -F_-^*. \end{aligned} \tag{5.18}$$

If J_μ is conserved, we may multiply Eq. (5.14) by $q_{-\mu}$ and we obtain

$$(m_b^2 - m_a^2)F_+(q_-^2) + q_-^2F_-(q_-^2) = 0. \tag{5.19}$$

Thus, in this case there is only one independent form factor, say F_+, so that

$$\langle a|J_\mu^\gamma(0)|b\rangle = \left(q_{+\mu} + q_{-\mu}\left(\frac{m_a^2 - m_b^2}{q_-^2}\right)\right) F_+(q_-^2). \tag{5.20}$$

Since J_μ^γ is the electric current, then this matrix element, multiplied by A_μ, the photon field, would be the matrix element for the transition $b \longrightarrow a + \gamma$. However, in this process (real photon emission), only the space parts of J_μ^γ contribute, since we can always work in the radiation gauge with $A_4 = 0$. We can also choose a frame in which b is at rest so that $\mathbf{q}_+$ and $\mathbf{q}_-$ are essentially the photon's momentum $\mathbf{k}$. Since $\mathbf{k}\cdot\mathbf{A} = 0$, we have shown that a $0 \longrightarrow 0$ transition with the omission of a real photon is forbidden—a well-known selection rule. If, on the other hand, we consider electron-pion scattering,

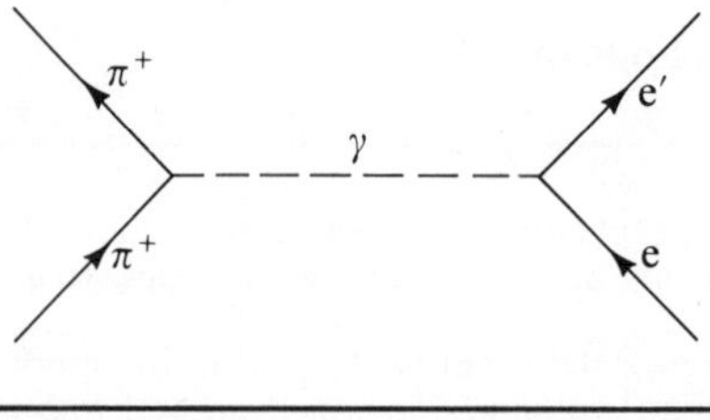

4. The same condition holds for q_4. There are two sign changes that compensate, since $q_4^* = -q_4$.

where the photon is virtual, we cannot invoke the radiation gauge condition and the $\pi\pi\gamma$ vertex is of the form

$$\langle\pi^{\pm}|J_{\mu}{}^{\gamma}(0)|\pi^{\pm}\rangle = q_{+\mu}F_{+}(q_{-}{}^{2}). \tag{5.21}$$

We may ask if there are additional symmetries, or selection rules imposed on the matrix elements of $J_{\mu}{}^{\gamma}$ in processes not involving the emission of real photons. This question becomes especially important, further on, when we study the weak interactions. In particular we may ask for the consequences of the C, P, T, and CPT properties of the currents, and states, on these matrix elements. In general, for any spin, from the fact that C is the charge conjugation operator,

$$CJ_{\mu}{}^{\gamma}(0)C^{-1} = -J_{\mu}{}^{\gamma}(0) \tag{5.22}$$

and from the fact that[5] the antiparticle is obtained from the particle by applying C, that is,

$$C|a\rangle = |\bar{a}\rangle, \tag{5.23}$$

we have

$$\langle a|J_{\mu}{}^{\gamma}(0)|b\rangle = -\langle\bar{a}|J_{\mu}{}^{\gamma}(0)|\bar{b}\rangle. \tag{5.24}$$

In particular, states that are eigenfunctions of C have vanishing matrix elements of $J_{\mu}{}^{\gamma}$. By the way, from the fact that, for the baryon number B,

$$[J_{\mu}{}^{\gamma}, B]_{-} = 0, \tag{5.26}$$

it follows that $J_{\mu}{}^{\gamma}$ cannot change the baryon number.

We have not yet discussed the parity operator, P, in detail. Parity characterizes the transformation property of a field $O(\mathbf{r}, t)$, or a function of the fields, when the underlying space-time coordinate system with respect to which these operators are evaluated is subjected to the transformation $\mathbf{r} \longrightarrow -\mathbf{r}$, $t \longrightarrow t$. As usual in quantum mechanics, we suppose that this reflection is generated by a unitary transformation P on the fields. For example, a *vector* operator V_{α} is, by definition, an operator with the property that letting $\mathbf{r} \longrightarrow -\mathbf{r}$, $t \longrightarrow t$ induces the transformation

$$\begin{aligned} P\mathbf{V}(\mathbf{r}, t)P^{-1} &= -\mathbf{V}(-\mathbf{r}, t) \\ PV_0(\mathbf{r}, t)P^{-1} &= V_0(-\mathbf{r}, t), \end{aligned} \tag{5.27}$$

and a *pseudovector* operator is defined so that

$$\begin{aligned} P\mathbf{A}(\mathbf{r}, t)P^{-1} &= \mathbf{A}(-\mathbf{r}, t) \\ PA_0(\mathbf{r}, t)P^{-1} &= -A_0(-\mathbf{r}, t). \end{aligned} \tag{5.28}$$

Unless we want the electric charge to change sign under space reflection, the $J_{\mu}{}^{\gamma}$, and hence A_{μ} the photon field, must be ordinary Lorentz vectors. If A_{μ} were an improper (or axial) vector, the term $J_{\mu}{}^{\gamma}A_{\mu}$ in H_{em} would not be a scalar. But

5. This is the definition of $|\bar{a}\rangle$, the antiparticle state to $|a\rangle$. Later, when we discuss the weak interactions, which do not conserve C, we will discuss how this definition must be modified.

the free Lagrangian is a scalar under P—namely, $PH_0(\mathbf{r})P^{-1} = H_0(-\mathbf{r})$. Thus H_{em} must also be a scalar if P is to be a good quantum number. Hence

$$\begin{aligned} P\mathbf{J}^\gamma(\mathbf{r}, t)P^{-1} &= -\mathbf{J}^\gamma(-\mathbf{r}, t), \\ PJ_0{}^\gamma(\mathbf{r}, t)P^{-1} &= J_0{}^\gamma(-\mathbf{r}, t). \end{aligned} \tag{5.29}$$

We must define transformations on the fields that guarantee these results. We must also define the parities of the states $|a\rangle$ and $|b\rangle$. This means that here also we must know the parity transformation properties of the fields ϕ_a and ϕ_b that create these states out of the vacuum. For the strong and electromagnetic interactions parity is conserved. The unitary operator P must be such that it commutes with $H_0 + H_{em} + H_{st}$, the strong, electromagnetic, and free Hamiltonians. Of course, the Hamiltonian *densities*, even in a theory in which P is conserved, are *not* invariant; that is,

$$PH(\mathbf{r})P^{-1} = H(-\mathbf{r}). \tag{5.30}$$

This condition is, however, sufficient to guarantee that $\int d\mathbf{r}\, H(\mathbf{r})$, the Hamiltonian, *is* invariant. We may divide spin-zero fields into two classes according to how P transforms them.

1. *Scalars*, with

$$P\phi(\mathbf{r}, t)P^{-1} = \eta_P\phi(-\mathbf{r}, t). \tag{5.31}$$

2. *Pseudoscalars*, with

$$P\phi(\mathbf{r}, t)P^{-1} = -\eta_P\phi(-\mathbf{r}, t). \tag{5.32}$$

We introduce the arbitrary phase η_P to denote the fact that, if P is conserved by H, then so is $P' = e^{i\theta B}P$, where θ is any real angle and B is the baryon number. P' is just as good a definition of parity as is P and we are at liberty to choose any member from this infinite class of operators as the parity operator. So long as the fields are not in interaction, attributing to them one parity, or the other, is a pure convention. The physically measurable quantities are the relative parities of particles in interaction. We may always define the parity operator so that the members of any given isotopic multiplet are all assigned the *same* parity. Thus the electromagnetic current of the pions

$$J_\mu{}^\gamma = \tfrac{1}{2}i[[\pi^+ \partial_\mu\pi^- - \partial_\mu\pi^+\pi^-] + [\partial_\mu\pi^-\pi^+ - \pi^- \partial_\mu\pi^+]] + e\pi^+\pi^-A_\mu \tag{5.33}$$

will transform under parity like a vector, providing that A_μ transforms like a vector, as it must.

All of the eigenstates of H can also be chosen to be eigenstates of parity but not simultaneously of the momentum operator $\mathbf{P}$ except in the special case in which $\mathbf{p}$, its eigenvalue, is zero,[6] since $[\mathbf{P}, P]_- \neq 0$. This means that we can always choose the vacuum state to have even parity,

---

6. In ordinary quantum mechanics, $e^{i(\mathbf{p}\cdot\mathbf{r})}$, the eigenfunction of $\mathbf{P} = (1/i)\nabla$, clearly has no definite parity.

$$P|0\rangle = |0\rangle. \tag{5.34}$$

The one-particle states are not, in general, eigenfunctions of parity. For example, consider a pion state of isotopic spin i,

$$P|\mathbf{p}\rangle_i = \lim_{t\to\pm\infty} iP \int d\mathbf{r} \left\{ \phi_i(\mathbf{r}, t) \frac{\partial}{\partial t} \left(\frac{e^{i(px)}}{(2\pi)^{3/2}\sqrt{2P_0}} \right) - \frac{\partial \phi_i(\mathbf{r}, t)}{\partial t} \frac{e^{i(px)}}{(2\pi)^{3/2}\sqrt{2P_0}} \right\} |0\rangle. \tag{5.35}$$

We bring P under the limit, since it is time-independent, and fix, once and for all, η_P for all pions to be $+1$. Thus

$$\begin{aligned} P|\mathbf{p}\rangle_i &= \lim_{t\to\pm\infty} -i \int d\mathbf{r} \left\{ \phi_i(\mathbf{r}, t) \frac{\partial}{\partial t} \left(\frac{e^{i(-\mathbf{p}\cdot\mathbf{r}-E(\mathbf{p})t)}}{(2\pi)^{3/2}\sqrt{2p_0}} \right) - \frac{\partial}{\partial t} \phi_i(\mathbf{r}, t) \frac{e^{i(-\mathbf{p}\cdot\mathbf{r}-E(\mathbf{p})t)}}{(2\pi)^{3/2}\sqrt{2p_0}} \right\} |0\rangle \\ &= -|-\mathbf{p}\rangle_i. \end{aligned} \tag{5.36}$$

We have made the change $\mathbf{r} \longrightarrow -\mathbf{r}$ in the integration variable. Thus we have the condition[7] for any vector current J_μ, with $|a\rangle$ and $|b\rangle$ any two spin-zero states,

$$\begin{aligned} \langle \mathbf{p}_a|\mathbf{J}(0)|\mathbf{p}_b\rangle &= -P_aP_b\langle -\mathbf{p}_a|\mathbf{J}(0)|-\mathbf{p}_b\rangle, \\ \langle \mathbf{p}_a|J_0(0)|\mathbf{p}_b\rangle &= P_aP_b\langle -\mathbf{p}_a|J_0(0)|-\mathbf{p}_b\rangle, \end{aligned} \tag{5.37}$$

where P_aP_b is the product of the intrinsic parities of a and b.

If a and b have the same parities, this condition is automatically fulfilled; if they have opposite parities, the matrix element must vanish. We will make use of these conditions, and similar ones, extensively when we discuss the weak interactions.

The time-reversal condition is considerably more complicated. It is well known, from nonrelativistic quantum mechanics, that the Schrödinger equation

$$i\frac{\partial}{\partial t}|t\rangle = H|t\rangle \tag{5.38}$$

is not invariant under $t \longrightarrow -t$, which we might, naively, believe is the operation of time reversal. Instead, one can show that if the theory is "time-reversal" invariant, this means that there must exist an operator

$$T = UK,$$

where U is a unitary operator whose specific form depends on H—for example,

---

7. The choice of the space-time origin makes a great simplification in the argument, since there is no need to transform the coordinate at which $J_\mu(0)$ is taken. Thus the translation invariance of the theory plays a crucial underlying role in its symmetry structure.

U will involve the spin if H depends on spin—and K is the operation of complex conjugation. T has the property that

$$TH(\mathbf{r}, t)T^{-1} = H(\mathbf{r}, -t), \tag{5.39}$$

from which it follows that

$$i\frac{\partial}{\partial t}T|-t\rangle = H(t)T|-t\rangle. \tag{5.40}$$

Thus $T|-t\rangle$ obeys the same equation as $|t\rangle$. This T is an "antiunitary" operator, meaning that

$$\langle T\phi|\psi\rangle = \langle\phi|T^{\dagger}\psi\rangle^*, \tag{5.41}$$

so that

$$\langle T\phi|T\psi\rangle = \langle\phi|\psi\rangle^*. \tag{5.42}$$

In the field theory we may also define the T operator as a transformation on the fields. For a scalar (or pseudoscalar) field,

$$\begin{aligned} T\phi(\mathbf{r}, t)T^{-1} &= \eta_T\phi(\mathbf{r}, -t), \\ T^2 &= 1, \end{aligned} \tag{5.43}$$

where η_T is an arbitrary phase. We may write the one-particle spin-zero state symbolically, as

$$|\mathbf{p}E(\mathbf{p})\rangle = O_{\mathbf{p}}|0\rangle, \tag{5.44}$$

where $O_{\mathbf{p}}$ stands for the complicated creation operator given above. We readily show (using the same techniques as in the parity case) that

$$TO_{\mathbf{p}}T^{-1} = \eta_T O_{-\mathbf{p}}. \tag{5.45}$$

If[8]

$$[T, H]_- = 0, \tag{5.46}$$

we can again choose the vacuum, so that

$$T|0\rangle = |0\rangle. \tag{5.47}$$

Thus for the one-particle matrix element of an operator O we can derive the condition

$$\langle\mathbf{p}_a E(\mathbf{p}_a)|O|\mathbf{p}_b E(\mathbf{p}_b)\rangle = \eta^*_{Ta}\eta_{Tb}\langle-\mathbf{p}_a E(\mathbf{p}_a)|TOT^{-1}|-\mathbf{p}_b E(\mathbf{p}_b)\rangle^*. \tag{5.48}$$

To apply this condition to the matrix element of the current we must know how it transforms under T. The answer is suggested by a correspondence principle argument based on the Maxwell equations. We again suppose that the charge does not change sign under time reversal. Thus from the Poisson equation

$$\Box^2\phi(\mathbf{r}, t) = -4\pi\rho(\mathbf{r}, t), \tag{5.49}$$

it follows that

8. H is the Hamiltonian.

$$T\phi(\mathbf{r}, t)T^{-1} = \phi(\mathbf{r}, -t). \tag{5.50}$$

But if **E**, the electric field, is to have a well-defined T invariance we see, from the equation

$$\mathbf{E}(\mathbf{r}, t) = -\nabla\phi(\mathbf{r}, t) - \frac{\partial\mathbf{A}(\mathbf{r}, t)}{\partial t}, \tag{5.51}$$

that

$$T\mathbf{A}(\mathbf{r}, t)T^{-1} = -\mathbf{A}(\mathbf{r}, -t). \tag{5.52}$$

Thus

$$\mathbf{J}^{\gamma}(\mathbf{r}, t) = \Box^2\mathbf{A}(\mathbf{r}, t) \tag{5.53}$$

must have the same property.[9] To assure that the pion current has this property we must define T suitably for the three ϕ_i. We remember that the isotopic spin current whose third component is the electric current to lowest order in e has the form

$$\mathbf{J}_i^{\gamma} = -(\boldsymbol{\phi} \times \nabla\boldsymbol{\phi})_i. \tag{5.54}$$

A choice of phases that assures the correct transformation property is

$$\begin{aligned} T\phi_1(\mathbf{r}, t)T^{-1} &= -\phi_1(\mathbf{r}, -t), \\ T\phi_2(\mathbf{r}, t)T^{-1} &= \phi_2(\mathbf{r}, -t), \end{aligned} \tag{5.55}$$

or, in the language of the charged pions,

$$\begin{aligned} T\pi^+(\mathbf{r}, t)T^{-1} &= -\pi^+(\mathbf{r}, -t), \\ T\pi^-(\mathbf{r}, t)T^{-1} &= -\pi^-(\mathbf{r}, -t). \end{aligned}$$

Since the π^0 does not enter the electromagnetic Hamiltonian explicitly, its phase under time reversal must be fixed by the strong interactions. This we shall do below. Thus for matrix elements of the electromagnetic current we have

$$\langle\mathbf{p}_a|\mathbf{J}^{\gamma}(0)|\mathbf{p}_b\rangle = -\eta_{Ta}{}^*\eta_{Tb}\langle-\mathbf{p}_a|\mathbf{J}^{\gamma}(0)|-\mathbf{p}_b\rangle^*. \tag{5.56}$$

If a and b are the same particle, we have the condition

$$p_{+\mu}F_+{}^* + p_{-\mu}F_-{}^* = p_{+\mu}F_+ + p_{-\mu}F_-, \tag{5.57}$$

so that F_+ and F_- must be real. This contradicts the Hermiticity condition according to which F_+ is real but F_- is pure imaginary. Thus the two conditions combined, in this case, give

$$F_-(q_-{}^2) = 0, \tag{5.58}$$

which is identical to the restriction imposed by current conservation.

Under CPT we have the combined condition

$$\langle\mathbf{a}|O|\mathbf{b}\rangle = P_aP_b\eta_{Ta}{}^*\eta_{Tb}\langle+\bar{\mathbf{a}}|(CPT)O(CPT)^{-1}|+\bar{\mathbf{b}}\rangle^*, \tag{5.59}$$

9. Since $J_0{}^{\gamma} = \rho$, it must transform as $TJ_0{}^{\gamma}(\mathbf{r}, t)T^{-1} = J_0{}^{\gamma}(\mathbf{r}, -t)$.

which serves to connect a particle process $b \longrightarrow a$ with the antiparticle process $\bar{a} \longrightarrow \bar{b}$.

We must now repeat this discussion for the spin-$\frac{1}{2}$ case, which is, physically, one of the most interesting. Here the considerations are complicated by the Dirac algebra. We may always write the matrix element as

$$\langle \mathbf{p}'|_a J_\mu{}^\gamma(0)|\mathbf{p}\rangle_b = \bar{u}(\mathbf{p}')_a \mathcal{J}_\mu(\mathbf{p}', \mathbf{p})u(\mathbf{p})_b, \tag{5.60}$$

where the $u(\mathbf{p})$ are free particle Dirac spinors that satisfy the equation

$$(i(\gamma p) + m)u(\mathbf{p}) = 0 \tag{5.61}$$

and $\mathcal{J}_\mu(\mathbf{p}', \mathbf{p})$ is the most general Lorentz vector operator that can be constructed from the momenta and the γ_μ's. For elastic electron-proton scattering, $m_a = m_b$, but anticipating the later work on the weak interactions we will treat the general case. In finding $\mathcal{J}_\mu(\mathbf{p}', \mathbf{p})$ we begin by listing all of the players in the game.

Numerical vectors:

$p_{-\mu} = (p' - p)_\mu$ $\qquad\qquad$ $p_{+\mu} = (p' + p)_\mu$

Numerical tensors:

$\delta_{\mu\nu}$ $\qquad\qquad$ $\epsilon_{\mu\nu\lambda\sigma}$

$\delta_{\mu\nu}$ is the Kronecker δ, with $\delta_{\mu\nu} = 1$ for $\mu = \nu$, and $\delta_{\mu\nu} = 0$ otherwise, and $\epsilon_{\mu\nu\lambda\sigma}$ is the totally antisymmetric pseudotensor, with $\epsilon_{1234} = 1$.
Dirac matrices

Vectors:	*Pseudovectors:*
γ_μ	$\gamma_5\gamma_\mu$
Tensors:	*Pseudotensors:*
$\sigma_{\mu\nu} = \dfrac{1}{2i}(\gamma_\mu\gamma_\nu - \gamma_\nu\gamma_\mu)$	$\gamma_5\sigma_{\mu\nu}$
$\gamma_\alpha\gamma_\beta\gamma_\delta$	$\gamma_5\gamma_\alpha\gamma_\beta\gamma_\delta$

From these quantities we can construct twelve Lorentz vectors.[10] (We can also construct twelve independent pseudovectors by multiplying each Lorentz vector by γ_5. The pseudovectors will enter in the weak interactions.)

1. $p_{+\mu}$
2. $p_{-\mu}$
3. γ_μ
4. $\sigma_{\mu\nu}p_{+\nu}$
5. $\sigma_{\mu\nu}p_{-\nu}$
6. $(\gamma p_+)p_{-\mu}$
7. $(\gamma p_-)p_{+\mu}$
8. $(\gamma p_+)p_{+\mu}$
9. $(\gamma p_-)p_{-\mu}$
10. $\gamma_5\epsilon_{\mu\nu\lambda\sigma}\gamma_\nu p_{+\lambda}p_{-\sigma}$
11. $(\gamma p_+)(\gamma p_-)p_{+\mu}$
12. $(\gamma p_+)(\gamma p_-)p_{-\mu}$

10. It can be shown by group-theoretical methods that there are only twelve. Here we use the "method" of exhaustion.

Thus introducing twelve invariant functions of $q^2 = p_-^2$, we can write

$$\mathcal{J}_\mu(\mathbf{p}', \mathbf{p}) = \sum_{i=1}^{12} F_i(q^2) v_{i\mu}(\mathbf{p}', \mathbf{p}), \tag{5.62}$$

where $v_{i\mu}$ are the twelve vectors given above. All other combinations of Dirac matrices can be reduced to one of the twelve. To give a specific example of how this is done we can easily establish the identity, where A and B are four-vectors,

$$(\gamma A)(\gamma B) = (AB) + i\sigma_{\mu\nu}A_\mu B_\nu,$$

so that terms involving three γ's such as $(\gamma p_+)(\gamma p_+)(\gamma p_-) = p_+^2(\gamma p_-)$ reduce to terms already contained in $\mathcal{J}_\mu$. There are two additional identities that are also very valuable for this reduction process:

$$(p + p')_\nu = ((\gamma p') - im_a)\gamma_\nu + \gamma_\nu((\gamma p) - im_b) + i(m_a + m_b)\gamma_\nu + i\sigma_{\nu\mu}(p' - p)_\mu \tag{5.63}$$

and

$$\epsilon_{\mu\nu\lambda\sigma}\gamma_5\gamma_\nu = \frac{i}{3}(\sigma_{\lambda\sigma}\gamma_\mu + \sigma_{\sigma\mu}\gamma_\lambda + \sigma_{\mu\lambda}\gamma_\sigma). \tag{5.64}$$

The latter identity shows that terms of the form

$$\gamma_\mu\sigma_{\lambda\sigma}p_{+\sigma}p_{-\lambda}$$

are contained in $\mathcal{J}_\mu$. We now notice that $\mathcal{J}_\mu$ is taken in the matrix element between free Dirac spinors, so that we can use the Dirac equation to reduce many of the twelve vectors to each other simply by redefining the F_i, which are, anyway, arbitrary. We do this in a series of steps.

1. Clearly 6, 7, 8, 9 reduce to 1 and 2 by a direct application of the Dirac equation.
2. Consider the 11 and 12:

   $$\bar{u}_a((\gamma(p_a + p_b))(\gamma(p_a - p_b)))u_b = \bar{u}_a(4m_a m_b + 2(p_a p_b))u_b,$$

   and hence 11 and 12 reduce to 1 and 2.
3. By a similar argument, 4 and 5 also reduce to 1 and 2 as does 10 (using the Dirac equation and the identity).

Thus

$$\langle \mathbf{p}'|_a J_\mu{}^\gamma(0)|\mathbf{p}\rangle_b = \bar{u}(\mathbf{p}')_a[F_1(q^2)p_{+\mu} + F_2(q^2)p_{-\mu} + F_3(q^2)\gamma_\mu]u(\mathbf{p})_b. \tag{5.65}$$

Hence the requirement of Lorentz invariance, along with the fact that the initial and final particles are "on their mass shells," [11] allows us to reduce $\mathcal{J}_\mu$ to three terms, one more than for the spin-zero case.

We may now apply the various symmetry conditions to reduce this matrix

---

11. In the language of field theory this means that the initial and final particles in the matrix element are "real" as opposed to being "virtual."

still further in various cases of interest. From the Hermiticity of J_μ we have, as before

$$\begin{aligned} \langle a|\mathbf{J}^\gamma(0)|b\rangle &= \langle b|\mathbf{J}^\gamma(0)|a\rangle^*, \\ \langle a|J_0{}^\gamma(0)|b\rangle &= \langle b|J_0{}^\gamma(0)|a\rangle^*. \end{aligned} \tag{5.66}$$

Thus

$$\begin{aligned} &\bar{u}(\mathbf{p}')_a[F_1(q^2)_{b\to a}\mathbf{p}_+ + F_2(q^2)_{b\to a}\mathbf{p}_- + F_3(q^2)_{b\to a}\boldsymbol{\gamma}]u(\mathbf{p})_b \\ &\quad = \bar{u}(\mathbf{p}')_a[F_1(q^2)^*{}_{a\to b}\mathbf{p}_+ - F_2(q^2)^*{}_{a\to b}\mathbf{p}_- - F_3(q^2)^*{}_{a\to b}\boldsymbol{\gamma}]u(\mathbf{p})_b. \end{aligned} \tag{5.67}$$

In performing the operations leading to the last equation we have concealed the following steps. If $\mathbf{O}$ is any 4×4 combination of Dirac matrices and momenta, in $\mathcal{J}$ we have from Equation (5.66),

$$\begin{aligned} \bar{u}(\mathbf{p}')_a\mathbf{O}_{b\to a}u(\mathbf{p})_b &= (\bar{u}(\mathbf{p})_b\mathbf{O}_{a\to b}u(\mathbf{p}')_a)^* = (u^{*T}(\mathbf{p})_b\gamma_4\mathbf{O}_{a\to b}u(\mathbf{p}')_a)^* \\ &= u^{T*}(\mathbf{p}')_a\mathbf{O}_{a\to b}{}^\dagger\gamma_4 u(\mathbf{p})_b = \bar{u}(\mathbf{p}')_a(\gamma_4\mathbf{O}_{a\to b}{}^\dagger\gamma_4)u(\mathbf{p})_b. \end{aligned} \tag{5.68}$$

Thus the minus sign in front of F_3 reflects the anticommutivity of $\boldsymbol{\gamma}$ and γ_4. Hence, if a and b are the same particle, we have the conditions

$$\begin{aligned} F_1 &= F_1{}^*, \\ F_2 &= -F_2{}^*, \\ F_3 &= -F_3{}^*. \end{aligned} \tag{5.69}$$

Current conservation and the Dirac equation yield a condition connecting the three F's:

$$F_1(q^2)(m_b{}^2 - m_a{}^2) + F_2(q^2)q^2 + F_3(q^2)i(m_a - m_b) = 0. \tag{5.70}$$

Thus if a and b are the same particle, we can conclude from this equation that $F_2(q^2) = 0$ for all $q^2 \neq 0$ and hence, by continuity, for all q^2.

We may now discuss the effect of the other symmetries on the matrix elements of $J_\mu{}^\gamma$. The charge conjugation condition is the same as that given above for the spin-zero case:

$$\begin{aligned} CJ_\mu{}^\gamma(0)C^{-1} &= -J_\mu{}^\gamma(0), \\ \langle a|J_\mu{}^\gamma(0)|b\rangle &= -\langle\bar{a}|J_\mu{}^\gamma(0)|\bar{b}\rangle. \end{aligned} \tag{5.71}$$

We may readily define a parity transformation on the spin-$\frac{1}{2}$ fields, ψ, that guarantees that the vector $J_\mu{}^\gamma$ transform like a proper Lorentz vector. Let[12]

$$P\psi(\mathbf{r}, t)P^{-1} = \eta_p\gamma_4\psi(-\mathbf{r}, t), \qquad \eta_p{}^*\eta_p = 1. \tag{5.72}$$

Then under this transformation[13]

$$P\bar{\psi}(\mathbf{r}, t)\boldsymbol{\gamma}\psi(\mathbf{r}, t)P^{-1} = -\bar{\psi}(-\mathbf{r}, t)\boldsymbol{\gamma}\psi(-\mathbf{r}, t) \tag{5.73}$$

---

12. The extra factor of γ_4 in this definition is essential, since $\psi(\mathbf{r}, t)$ and $\psi(-\mathbf{r}, t)$ do not obey the same equation.

and

$$P\bar{\psi}(\mathbf{r}, t)\gamma_4\psi(\mathbf{r}, t)P^{-1} = \bar{\psi}(-\mathbf{r}, t)\gamma_4\psi(-\mathbf{r}, t). \tag{5.74}$$

Thus B, the baryon number, is a true scalar while $J_\mu{}^\gamma$, including both the nucleons and pions is a true vector. As in the scalar case we can ask for[14] $P|\mathbf{p}\rangle_s$.

Assuming that $P|0\rangle = |0\rangle$,

$$P|\mathbf{p}\rangle_s = \lim_{t\to\infty} \eta_p{}^* \int d\mathbf{r}\, (\bar{\psi}(-\mathbf{r}, t)\gamma_4)\gamma_4 \frac{e^{i(px)}}{(2\pi)^{3/2}} u_s(\mathbf{p})|0\rangle = \eta_p{}^*|-\mathbf{p}\rangle_s. \tag{5.75}$$

Thus the state $|\mathbf{p}\rangle_s$ does not have a definite parity unless $\mathbf{p} = 0$, in which case parity eigenvalue is $\eta_p{}^*$, which we can always choose to be real and equal to ± 1. The parity condition is, for any operator $O(0)$,

$$\begin{aligned}\langle \mathbf{p}'|_{s'a'}O(0)|\mathbf{p}\rangle_{sb} &= \langle \mathbf{p}'|_{s'a}P^{-1}PO(0)P^{-1}P|\mathbf{p}\rangle_{sb} \\ &= \langle -\mathbf{p}'|_{s'a}PO(0)P^{-1}|-\mathbf{p}\rangle_{sb}P_aP_b,\end{aligned} \tag{5.76}$$

where $P_aP_b = \pm 1$ is the relative parity of a and b. We have used above the fact that

$$\gamma_4 u(\mathbf{p})_s = u(-\mathbf{p})_s, \tag{5.77}$$

which can easily be seen, explicitly, in the representation in which

$$\gamma_4 = \begin{pmatrix} 1 & 0 & 0 & 0 \\ 0 & 1 & 0 & 0 \\ 0 & 0 & -1 & 0 \\ 0 & 0 & 0 & -1 \end{pmatrix} \tag{5.78}$$

and

$$u(\mathbf{p})_s = \begin{pmatrix} \mu_s \\ \dfrac{\boldsymbol{\sigma}\cdot\mathbf{p}}{E(\mathbf{p}) + m}\mu_s \end{pmatrix} N(\mathbf{p}), \tag{5.79}$$

with the normalization $N(\mathbf{p}) = \left(\dfrac{E(\mathbf{p}) + m}{2E(\mathbf{p})}\right)^{1/2}$ chosen so that $u^\dagger u = 1$.

Here μ_s is the usual two-component Pauli spinor $\begin{pmatrix}1\\0\end{pmatrix}$ or $\begin{pmatrix}0\\1\end{pmatrix}$. As in the spin-zero case, we complete the argument by letting $\mathbf{r} \longrightarrow -\mathbf{r}$ in the integral.

For the current this condition yields

$$\langle \mathbf{p}'|_{s'a}\mathbf{J}^\gamma(0)|\mathbf{p}\rangle_{sb} = -P_aP_b\langle -\mathbf{p}'|_{s'a}\mathbf{J}^\gamma(0)|-\mathbf{p}\rangle_{sb}. \tag{5.80}$$

This condition is satisfied for particles with even relative parities, $P_aP_b = +1$, as we can verify from the general expression for $\mathcal{J}$.

For the Dirac field, $\psi(\mathbf{r}, t)$, time reversal, T, is defined so that

---

13. We always choose the phase η to be identical for all members of an isotopic multiplet. Thus for neutron and proton, $\eta_P = \eta_N$.
14. The careful reader will check the spinor indices to see what is being multiplied by what.

$$T\psi(\mathbf{r}, t)T^{-1} = \eta_T t\psi(\mathbf{r}, -t) \tag{5.81}$$

and

$$T\bar{\psi}(\mathbf{r}, t)T^{-1} = \eta_T{}^*\bar{\psi}(\mathbf{r}, -t)t^\dagger \tag{5.82}$$

where t is a 4×4 matrix whose specific form depends on the representation of the γ's used,[15] but which is always fixed to have the properties

$$\begin{aligned} t^T &= -t \\ t^\dagger &= t^{-1} \end{aligned} \tag{5.83}$$

and

$$t^{-1}\gamma_\mu{}^T t = \gamma_\mu. \tag{5.84}$$

If O is any function of the field operators ψ, then T also complex conjugates any numerical 4×4 matrices contained in O. Thus, for the baryon current,

$$\begin{aligned} T[i\bar{\psi}(\mathbf{r}, t)\gamma_\mu\psi(\mathbf{r}, t)]T^{-1} &= -i\bar{\psi}(\mathbf{r}, -t)t^\dagger\gamma_\mu{}^* t\psi(\mathbf{r}, -t) \\ &= -i\bar{\psi}(\mathbf{r}, -t)\gamma_\mu\psi(\mathbf{r}, -t). \end{aligned} \tag{5.85}$$

Since γ is Hermitian, $\gamma_\mu{}^* = \gamma_\mu{}^T$.

Hence this operation, combined with the T operation on the pion and photon fields, guarantees that the total electromagnetic current transforms under T as

$$TJ_\mu{}^\gamma(\mathbf{r}, t)T^{-1} = -J_\mu{}^\gamma(\mathbf{r}, -t). \tag{5.86}$$

Since we know how the baryons transform under T, up to an arbitrary set of phases, we may now use the presumed T invariance of the strong interactions to fix a set of phases for the pions and nucleons.[16] We demand that

$$\begin{aligned} H_{\pi N} &= ig_{\pi N}\bar{\psi}\gamma_5\boldsymbol{\tau}\cdot\boldsymbol{\phi}\psi \\ &= ig_{\pi N}[\bar{P}\gamma_5\phi_3 P - \bar{N}\gamma_5\phi_3 N + \bar{P}\gamma_5(\phi_1 - i\phi_2)N + \bar{N}\gamma_5(\phi_1 + i\phi_2)P] \end{aligned} \tag{5.87}$$

be T-invariant, which means that

$$TH(\mathbf{r}, t)T^{-1} = H(\mathbf{r}, -t), \tag{5.88}$$

which is sufficient to guarantee that $\int d^4x\, \mathcal{L}(x)$ is invariant.[17] To satisfy Hermiticity we need the i in front of $H_{\pi N}$ if we want to choose $g_{\pi N}$ to be real.

Now

$$\begin{aligned} T(ig_{\pi N}\bar{P}\gamma_5\phi_3 \mathrm{P})T^{-1} &= -ig_{\pi N}\bar{P}(t^\dagger\gamma_5{}^*\phi_3 t)P\eta_{T\phi_3} \\ &= -ig_{\pi N}\bar{P}\gamma_5\phi_3 P\eta_{T\phi_3} \end{aligned} \tag{5.89}$$

15. In the representation in which the γ's are Hermitian and γ_2, γ_4 are real and γ_1, γ_3 are pure imaginary, $t = \gamma_1\gamma_3$.
16. Recall that the phases of π^+ and π^- have already been fixed to maintain the T invariance of the electromagnetic coupling.
17. With the machinery assembled the reader will have no trouble in seeing that $\mathcal{L}_0$, the free Lagrangian, is T-invariant.

Thus we must choose $\eta_{T\phi_2} = -1$ to keep this term invariant. Hence the electromagnetic coupling, plus the coupling of π^0 to the nucleons, fixes

$$\begin{aligned} \eta_{T\phi_1} &= -1, \\ \eta_{T\phi_2} &= +1, \\ \eta_{T\phi_3} &= -1. \end{aligned} \tag{5.90}$$

It is easy to verify if relative phases of N and P are such that

$$\eta^*_{TP}\eta_{TN} = +1, \tag{5.91}$$

this choice guarantees that the rest of $H_{\pi N}$ is invariant. Therefore we have found a consistent set of phases, and hence a T that leaves $H_0 + H_{em} + H_{\pi N}$ invariant.

Again assuming that $T|0\rangle = |0\rangle$, we may ask for $T|\mathbf{p}E(\mathbf{p})\rangle_s$. Thus

$$T|\mathbf{p}E(\mathbf{p})\rangle_s = \eta_T^* \lim_{t\to\infty} \int d\mathbf{r}\, \bar{\psi}(\mathbf{r}, -t)\gamma_4 \frac{e^{-i(px)}}{(2\pi)^{3/2}} t^\dagger u^*(\mathbf{p})_s|0\rangle. \tag{5.92}$$

To proceed, let us work in the representation with all γ's Hermitian, γ_2, γ_4 real and γ_1, γ_3 imaginary, in which $t = \gamma_1\gamma_3$. We easily show that

$$t^\dagger u^*(\mathbf{p})_s = -u(-\mathbf{p})_{-s}. \tag{5.93}$$

Performing the limiting process, we find that

$$T|\mathbf{p}E(\mathbf{p})\rangle_s = -\eta_T^*|-\mathbf{p}\rangle_{-s}. \tag{5.94}$$

Hence for any operator O and spin-$\frac{1}{2}$ particles a and b,

$$\langle\mathbf{p}'|_{s'a}O|\mathbf{p}\rangle_{sb} = \eta_{Ta}\eta_{Tb}^*\langle-\mathbf{p}'|_{-s'a}TOT^{-1}|-\mathbf{p}\rangle_{-sb}^* \tag{5.95}$$

or

$$\langle\mathbf{p}'|_{sa}\mathbf{J}^\gamma(0)|\mathbf{p}\rangle_{sb} = -\eta_{Ta}\eta_{Tb}^*\langle-\mathbf{p}|_{-sa}\mathbf{J}^\gamma(0)|-\mathbf{p}\rangle_{-sb}^*. \tag{5.96}$$

For the same particles this leads to a condition on $\mathcal{g}(p', p)$:

$$\bar{u}(\mathbf{p}')_{s'}\mathcal{g}(\mathbf{p}', \mathbf{p})u(\mathbf{p})_s = -\bar{u}(\mathbf{p}')_{s'}(t\mathcal{g}(-\mathbf{p}', -\mathbf{p})t^\dagger)^*u(\mathbf{p})_s. \tag{5.97}$$

Applying this condition we learn that

$$\begin{aligned} F_1(q^2) &= F_1^*(q^2), \\ F_2(q^2) &= F_2^*(q^2), \\ F_3(q^2) &= -F_3^*(q^2), \end{aligned} \tag{5.98}$$

which contradicts the Hermiticity condition unless $F_2(q^2) = 0$.[18] Thus, either

---

18. The CPT condition, with a and b two spin-$\frac{1}{2}$ particles, is clearly

$$\langle\mathbf{a}|_{s'}O|\mathbf{b}\rangle_s = P_aP_b\eta_{Tb}^*\eta_{Ta}\langle+\bar{\mathbf{a}}|_{-s'}(CPT)O(CPT)^{-1}|+\bar{\mathbf{b}}\rangle_{-s}^*,$$

which means, for the matrix element of the current taken between the same particles,

$$\begin{aligned} \langle\mathbf{p}'|_{s'}\mathbf{J}(0)|\mathbf{p}\rangle_s &= -\langle+\bar{\mathbf{p}}'|_{-s'}\mathbf{J}(0)|+\bar{\mathbf{p}}\rangle_{-s}^*, \\ \langle\mathbf{p}'|_{s'}J_0(0)|\mathbf{p}\rangle_s &= -\langle+\bar{\mathbf{p}}'|_{-s'}J_0(0)|+\bar{\mathbf{p}}\rangle_{-s}^*. \end{aligned}$$

from current conservation or Hermiticity and T invariance, coupled with Lorentz invariance, we see that the most general structure that the electromagnetic current operator (taken between the same spin-$\frac{1}{2}$ particles) can have is

$$\begin{aligned}\langle \mathbf{p}'|_{s'}J_\mu{}^\gamma(0)|\mathbf{p}\rangle_s &= \bar{u}(\mathbf{p}')_{s'}[F_1(q^2)p_{+\mu} + iF_3{}'(q^2)\gamma_\mu]u(\mathbf{p})_s \\ &= i\bar{u}(\mathbf{p}')_{s'}[[2mF_1(q^2) + F_3{}'(q^2)]\gamma_\mu - \sigma_{\mu\nu}(p' - p)_\nu F_1(q^2)]u(\mathbf{p})_s. \end{aligned} \quad (5.99)$$

We have explicitly extracted the factor of i from the F_3; $F_3{}' = iF_3$ to make all of the form factors manifestly real, and we have transformed the equation to write the matrix element in terms of the commonly used pair of covariants γ_μ and $\sigma_{\mu\nu}q_\nu$.

Thus after this long detour we have ended up by showing that, with both electrons on the mass shell, the most general form the photon-electron vertex, can have (again redefining the form-factors to conform to common usage) is

$$\Gamma_\mu(e', e) = ie\bar{u}(\mathbf{e}')[F_1(q^2)\gamma_\mu + \sigma_{\mu\nu}q_\nu F_2(q^2)]u(\mathbf{e}). \quad (5.100)$$

But we also showed that if for the electron we take

$$J_\mu{}^\gamma(0) = i\bar{\psi}_e(0)\gamma_\mu\psi_e(0) \quad (5.101)$$

and treat the vertex to lowest order in e, then, comparing it with Equation (5.100),

$$\Gamma_\mu{}^0(\mathbf{e}', \mathbf{e}) = ie\bar{u}(\mathbf{e}')\gamma_\mu u(\mathbf{e}). \quad (5.102)$$

Thus, to this order,

$$\begin{aligned} F_1{}^0(q^2) &= 1, \\ F_2{}^0(q^2) &= 0. \end{aligned} \quad (5.103)$$

This means that for the electron the departure of these form factors from $F_i{}^0$ is to be interpreted as representing higher-order electromagnetic corrections. Indeed, we can give a physical interpretation to these corrections by studying the nonrelativistic limit of the exact matrix elements. First, consider the limit $q^2 \longrightarrow 0$, $\mathbf{e}' \longrightarrow 0$, $\mathbf{e} \longrightarrow 0$. Thus

$$\begin{aligned}\Gamma_\mu(0, 0) &= e(2\pi)^3\langle \mathbf{0}|_{s'}J_\mu{}^\gamma(0)|\mathbf{0}\rangle_s \\ &= ie\bar{u}_{s'}(\mathbf{0})F_1(0)\gamma_\mu u_s(\mathbf{0}). \end{aligned} \quad (5.104)$$

But $\bar{u}(\mathbf{0})\gamma u(\mathbf{0}) = 0$. Thus, exactly,[19]

$$F_1(0) = \langle \mathbf{0}|J_0{}^\gamma(0)|\mathbf{0}\rangle e(2\pi)^3, \quad (5.105)$$

with

$$J_0{}^\gamma = \psi_e{}^\dagger(0)\psi_e(0). \quad (5.106)$$

However, the total observed charge, e_{obs}, is, by definition,

---

19. This matrix element is independent of the spins.
20. In this expression there is a trivial infinity since it involves $\delta^3(0)$. We have set $\delta^3(0) = 1$. This procedure can be avoided, or rigorized, by using a finite space volume for the quantization.

$$e_{\text{obs}} = e\langle \mathbf{0}|_s \int d\mathbf{r}\, J_0{}^{\gamma}(\mathbf{r})|\mathbf{0}\rangle_s$$
$$= (2\pi)^3 e\langle \mathbf{0}|_s J_0{}^{\gamma}(0)|\mathbf{0}\rangle_s = eF_1(0). \tag{5.107}$$

Thus[20]

$$F_1(0) = \frac{e_{\text{obs}}}{e}. \tag{5.108}$$

From this equation we see that the higher-order radiative corrections in the $q^2 = 0$ limit serve to renormalize the total charge. In the literature, $F_1(q^2)$ is often called the "charge" form factor, since $F_1(0)$ is related to the charge. However, this nomenclature is somewhat imprecise since the matrix element, $i\bar{u}(\mathbf{e}')\gamma_\mu u(\mathbf{e})$, for $q^2 \neq 0$, has the characteristics of a magnetic moment as well. To see this, we expand this matrix element in powers of v/c. We also multiply it by $\mathbf{A}$, an external electromagnetic potential in the radiation gauge: $\mathbf{A}\cdot(\mathbf{p}' - \mathbf{p}) = 0$. Now,

$$ie\mathbf{A}\cdot\bar{u}(\mathbf{p}')_{s'}\gamma u(\mathbf{p})_s \simeq eu_{s'}\left[\frac{\mathbf{A}\cdot\mathbf{p}}{m} + i\frac{\boldsymbol{\sigma}\cdot\mathbf{A}\times\mathbf{q}}{2m}\right]u_s, \tag{5.109}$$

where u_s is a Pauli spinor. The first term clearly represents the coupling of $\mathbf{A}$ to the current, $e(\mathbf{p}/m)$ while the second represents the coupling of the electron to an external magnetic field. Automatically, the electron in the Dirac theory, has its "normal" or Dirac magnetic moment $ge/2m$ with $g = 2$. Thus $F_1(q^2)$ is more appropriately called the Dirac form factor. The form factor $F_2(q^2)$ we may call the Pauli form factor, since to lowest order in v/c

$$\bar{u}(\mathbf{p}')_{s'}\sigma_{i\nu}q_\nu u(\mathbf{p})_s A_i \simeq u_{s'}\boldsymbol{\sigma}\cdot(\mathbf{q}\times\mathbf{A})u_s, \tag{5.110}$$

so that $F_2(0)$ is a measure of the contribution to the magnetic moment of the electron due to higher-order electromagnetic corrections to the vertex such as

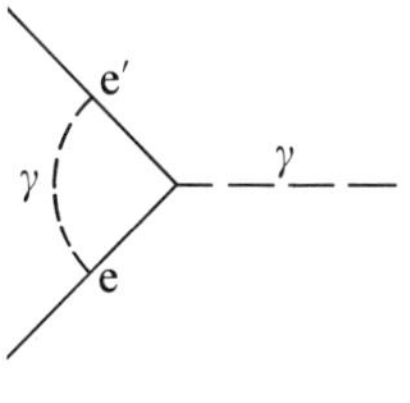

In units of the Bohr magneton, $\mu_0 = e/2m$, we may write the total magnetic moment coupling strength of the electron, μ, in the form[21]

$$\mu = \left[1 + \frac{g-2}{2}\right]2\mu_0$$

---

21. If $g = 2$, no radiative corrections, the coupling to a constant magnetic field $\mathbf{H}$ would then be $g\mu_0(\boldsymbol{\sigma}\cdot\mathbf{H}/2)$.

where the 1 comes from the γ_μ term and $(g - 2)/2$ represents the anomaly in the moment. The quantity $(g - 2)/2$ for the electron is one of the best known in all of physics, both theoretically and experimentally. The theoretical value is obtained by computing higher-order corrections using the relativistic perturbation theory (which the reader can find described in many books[22]). The result is (to order α^2)

$$\left(\frac{g-2}{2}\right)_{\text{th}} = \frac{\alpha}{2\pi} - 0.328\,\frac{\alpha^2}{\pi^2}.$$

Taking α^{-1} from experiment,[23]

$$\alpha^{-1} = 137.0388 \pm 0.0012.$$

This formula agrees very well with the experimental number,[24]

$$\left(\frac{g-2}{2}\right)_{\text{exp}} = 0.001159622 \pm 0.000000027.$$

This agreement, on the one hand, justifies using the lowest-order vertex for the electron in electron-proton scattering, since the corrections are $O(\alpha) \lesssim 1\%$, and, on the other, not including intrinsic Pauli terms in the electron current operator, since the theory gives essentially perfect agreement with experiment without them.

Before this long detour began, we were considering the graph

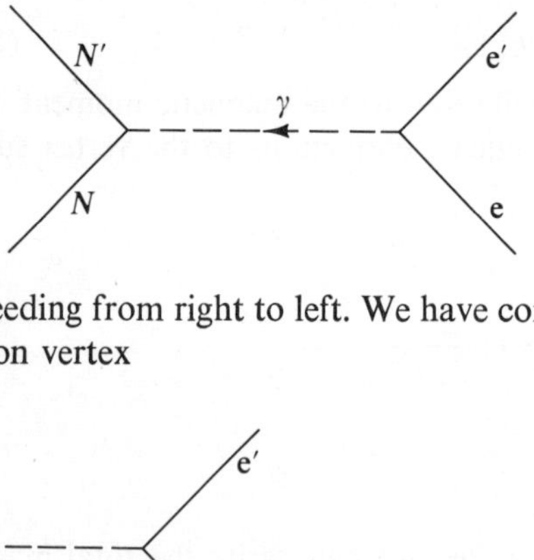

proceeding from right to left. We have completed the discussion of the electron-photon vertex

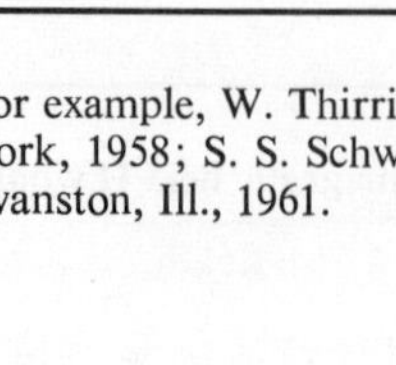

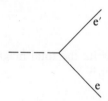

22. For example, W. Thirring, *Principles of Quantum Electrodynamics*, Academic Press, New York, 1958; S. S. Schweber, *An Introduction to Relativistic Field Theory*, Row-Peterson, Evanston, Ill., 1961.

The next term is the internal photon line $\overset{\gamma}{--}$. This line represents the photon propagator in momentum space, whose general form is

$$D_{\mu\nu}(q) = \frac{D(q^2)}{q^2}\left(\delta_{\mu\nu} - \frac{q_\mu q_\nu}{q^2}\right). \tag{5.111}$$

If we work to lowest order in e and thus drop such corrections as

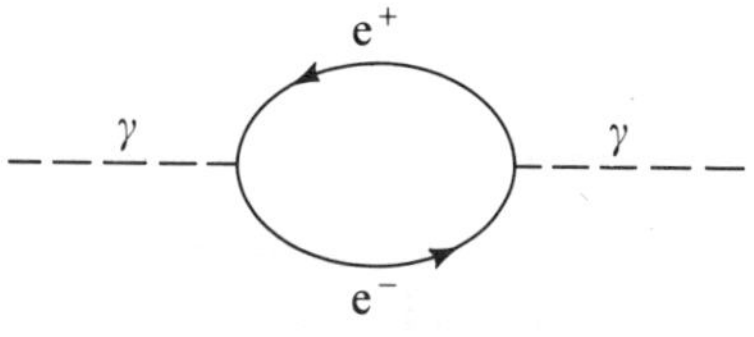

then it is well known that $D(q^2) = 1$. The term $q_\mu q_\nu$ vanishes when applied, to the right, to $\langle \mathbf{e}'|J_\mu{}^\gamma(0)|\mathbf{e}\rangle$ by current conservation. Thus, in effect, the propagator for this problem is simply written as

$$D_{\mu\nu}{}^0(q^2) = \frac{1}{q^2}\,\delta_{\mu\nu}. \tag{5.112}$$

We come next to the nucleon-photon vertex

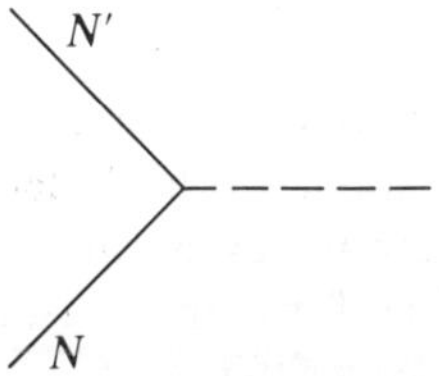

which depends essentially on[25] $\langle \mathbf{p}'|_{s'}J_\mu{}^\gamma(0)|\mathbf{p}\rangle_s$. With all that has been said it will come as no surprise that, given the same symmetry requirements used on the electron vertex and making the same general arguments as above, we may write the nucleon-photon vertex in the form

$$\Gamma_\mu(p', p) = ie\bar{u}(\mathbf{p}')_{s'}[F_1(q^2)\gamma_\mu + \sigma_{\mu\nu}q_\nu F_2(q^2)]u(\mathbf{p})_s, \tag{5.113}$$

where F_1 and F_2 are the Dirac and Pauli form factors of the nucleon. It is with this expression that the interesting physical discussion of electron scattering begins.

---

23. E. S. Dayhoff, S. Triebwasser, and W. E. Lamb, Jr., *Phys. Rev.* **89**:106 (1953).
24. For example, D. T. Wilkinson and H. R. Crane, *Phys. Rev.* **130**:852 (1963).
25. Here $J_\mu{}^\gamma(0)$ is the full electromagnetic current with pions and nucleons contained.

6

Electron-Nucleon Scattering II

Isotopic Splitting of Form Factors. The burden of the discussion in the last chapter was to show that all of the physical information contained in elastic electron-nucleon scattering is summarized in two real form factors, $F_1(q^2)$ and $F_2(q^2)$, which are the coefficients of γ_μ and $\sigma_{\mu\nu}q_\nu$ in the general construction of the photon-nucleon vertex when the nucleons are real and the photon is virtual—that is, with $q^2 \geq 0$ in our metric. The rest of the matrix element consists of kinematical factors or is determined from the well-understood photon-electron dynamics. As we shall learn shortly, there is no reliable theory that predicts the quantitative details of the nucleonic form factors precisely. This is because these form factors are determined by the strong couplings between nucleons and pions (and perhaps other particles), and these couplings present a mathematical problem that has, so far, resisted detailed solution.

We shall begin by indicating how F_1 and F_2 are to be extracted from experiment. We wish to conduct the analysis so that it applies to both neutron and proton; this we can easily do since we are working to lowest order in e so that the neutron-proton states can be thought of as forming an isotopic doublet. To this order the electromagnetic current can be written as

$$J_\mu^\gamma = J_\mu^S + J_{\mu_3}^V, \tag{6.1}$$

where[1]

$$J_\mu^S = i\,\frac{\bar{N}\gamma_\mu N}{2} \tag{6.2}$$

is a current that, as we have seen in Chapter 3, transforms like a scalar under the SU_2 group, and

$$J_{\mu_3}^V = i\bar{N}\gamma_\mu \frac{\tau_3}{2} N - \left(\boldsymbol{\phi} \times \frac{\partial}{\partial x_\mu} \boldsymbol{\phi}\right)_3 \tag{6.3}$$

transforms like the third component of an isotopic vector.

In terms of this decomposition of the current we may introduce a decomposition of the form factors. Instead of F_1^N, F_1^P, F_2^N, F_2^P we introduce four new

1. N is the nucleon field $\begin{pmatrix} \psi_P \\ \psi_N \end{pmatrix}$.

form factors F_1^S, F_1^V, F_2^S, F_2^V, where these form factors are defined in terms of the isotopic scalar and vector currents. The F_i^N, F_i^P are the quantities that are directly observed, but the F_i^S, F_i^V are more easily discussed theoretically. We now derive the relations among them. Call the proton state $|\frac{1}{2}\,\frac{1}{2}\rangle$, where the indices mean $T = \frac{1}{2}$, $T_3 = \frac{1}{2}$, and call the neutron state $|\frac{1}{2}\,-\frac{1}{2}\rangle$; that is, $T = \frac{1}{2}$, $T_3 = -\frac{1}{2}$. Let

$$\langle\tfrac{1}{2}\,\tfrac{1}{2}|J_\mu{}^S + J_{\mu_3}{}^V|\tfrac{1}{2}\,\tfrac{1}{2}\rangle = i\bar{u}(\mathbf{p}')\left[\left(\frac{F_1{}^S + F_1{}^V}{2}\right)\gamma_\mu + \frac{(F_2{}^S + F_2{}^V)}{2}\,\sigma_{\mu\nu}q_\nu\right]u(\mathbf{p}). \tag{6.4}$$

With this definition it follows that

$$F_{1,2}{}^P = \frac{F_{1,2}{}^S + F_{1,2}{}^V}{2}. \tag{6.5}$$

Now, from the transformation properties of $J_\mu{}^S$ and $J_{\mu_3}{}^V$, we may, (using T^+ and T^-, the isotopic raising and lowering operators[2]) readily prove that

$$\begin{aligned}\langle\tfrac{1}{2}\,\tfrac{1}{2}|J_\mu{}^S|\tfrac{1}{2}\,\tfrac{1}{2}\rangle &= \langle\tfrac{1}{2}\,-\tfrac{1}{2}|J_\mu{}^S|\tfrac{1}{2}\,-\tfrac{1}{2}\rangle,\\ \langle\tfrac{1}{2}\,\tfrac{1}{2}|J_{\mu_3}{}^V|\tfrac{1}{2}\,\tfrac{1}{2}\rangle &= -\langle\tfrac{1}{2}\,-\tfrac{1}{2}|J_{\mu_3}{}^V|\tfrac{1}{2}\,-\tfrac{1}{2}\rangle,\end{aligned} \tag{6.6}$$

so that

$$F_{1,2}{}^N = \frac{F_{1,2}{}^S - F_{1,2}{}^V}{2}. \tag{6.7}$$

We can now evaluate the F's at $q^2 = 0$, since the F_1's give the charge and the F_2's the anomalous magnetic moments. Thus

$$\begin{aligned}F_1{}^P(0) &= 1 = \frac{F_1{}^S(0) + F_1{}^V(0)}{2},\\ F_1{}^N(0) &= 0 = \frac{F_1{}^S(0) - F_1{}^V(0)}{2},\end{aligned} \tag{6.8}$$

so that

$$F_1{}^S(0) = F_1{}^V(0) = 1. \tag{6.9}$$

The anomalous magnetic moments, in units of the nuclear magneton $e/2m_N$, are determined from experiment to be

$$\begin{aligned}F_2{}^P(0) &= \frac{F_2{}^S(0) + F_2{}^V(0)}{2} = 1.79,\\ F_2{}^N(0) &= \frac{F_2{}^S(0) - F_2{}^V(0)}{2} = -1.91.\end{aligned} \tag{6.10}$$

Since the neutron and proton have magnetic moment anomalies that are nearly equal and opposite, we may suppose that the anomaly is due, largely, to the

2. We can also show that neither $J_\mu{}^S$ nor $J_{\mu 3}{}^V$ has matrix elements that connect the states $\langle\frac{1}{2}\,\frac{1}{2}|$ and $\langle\frac{1}{2}\,-\frac{1}{2}|$, which is consistent with charge conservation.

isotopic vector current or, roughly speaking, to the pionic current that appears in this part, although quantitative calculations using various sorts of perturbation theory have, so far, failed to reproduce the experimental numbers exactly. In this respect the neutron is especially interesting. Since the neutron is neutral it does not enter the electric current explicitly. The term $i\bar{N}\frac{(1+\tau_3)}{2}\gamma_\mu N$, in the nucleon current, projects out the proton. All of the neutron's electromagnetic properties are "induced" through the strong interactions via graphs like

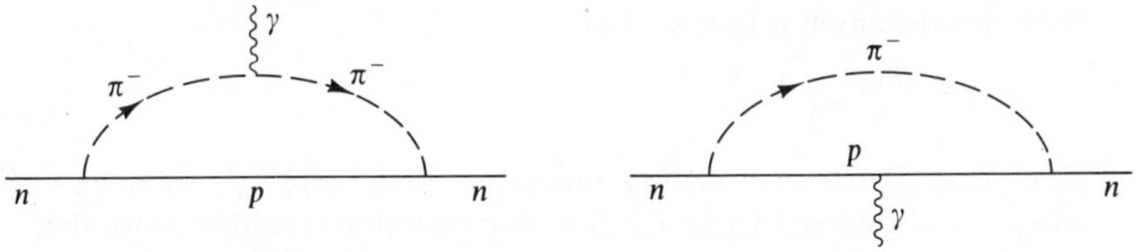

If the strong interactions were turned off (we neglect here the weak couplings) the neutron would have no electromagnetic couplings at all. This is true of all the neutral particles. For example, the π^0 is known to decay into two photons: $\pi^0 \longrightarrow \gamma + \gamma$. This decay also comes about because of the strong couplings via graphs like

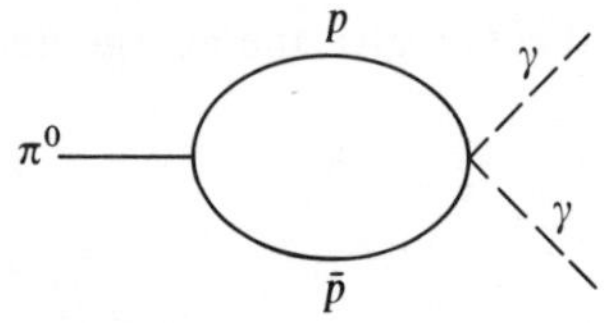

Though we have not yet discussed the weak interactions, it will be of interest to give here a very brief discussion of the electromagnetic properties of the neutrino, the prime example of a neutral particle.

As is well known, the neutrino is a zero-mass particle whose free Dirac field obeys the equation (where ν is the neutrino's four-momentum with $\nu^2 = 0$)

$$i(\gamma\nu)\psi_\nu = 0. \tag{6.11}$$

Numerous experiments have shown that this particle is "left-handed," which means that it travels with its spin antiparallel to its momentum: $\overset{s}{\Leftarrow}\!\underset{\nu}{\rightarrow}$. This is a Lorentz-invariant statement for a particle that moves with the speed of light. It is not possible for an observer to find a coordinate system with respect to which this "helicity" would appear reversed. To say that the neutrino is left-handed is to say that

$$\boldsymbol{\sigma}\cdot\hat{\nu}\psi_\nu = -\psi_\nu. \tag{6.12}$$

This is guaranteed if we impose on the neutrino field the supplementary condition (see Chapter 8 for details) that

$$\gamma_5\psi_\nu = \psi_\nu, \tag{6.13}$$

which is readily achieved by taking the usual four-component Dirac function for a particle of zero mass, ψ, and defining

$$\psi_\nu = \frac{(1 + \gamma_5)}{2}\,\psi. \tag{6.14}$$

The neutrino does not have strong interactions, but it may have induced electromagnetic couplings via graphs like[3]

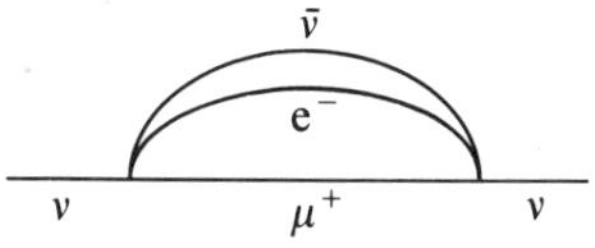

As usual,[4]

$$\langle \nu'|J_\mu{}^\gamma(0)|\nu\rangle = i\bar{u}(\nu')[\gamma_\mu F_1(q^2) + \sigma_{\mu\nu}q_\nu F_2(q^2)]u(\nu). \tag{6.15}$$

However, we have also for the neutrino the γ_5 invariance (Eq. (6.13)). From this it follows that here

$$F_2(q^2) = 0, \tag{6.16}$$

so that the two-component neutrino has no magnetic moment. Since $F_1(0) = 0$ (zero charge),

$$F_1(q^2) = q^2F'(0) + \cdots. \tag{6.17}$$

For reasons that we will make plausible in a moment, the "mean-square radius" associated with the distribution $F_1(q^2)$ is defined to be

$$\langle \mathbf{r}^2 \rangle \cdot\!\equiv\!\cdot -6F'(0). \tag{6.18}$$

The quantity $\langle \mathbf{r}^2 \rangle$ gives a measure of the size of an object as this size would be determined in electron scattering. Theoretical estimates have placed this size for the neutrino at about

$$\langle \mathbf{r}^2 \rangle_\nu \simeq 10^{-33}\ \text{cm}^2.$$

Such a radius might be studied in very accurate measurements of the process $\nu + p \longrightarrow \nu + p$.

The mean square radius. We have seen that $F_1(0)$ has a simple physical interpretation and that, for neutral particles, $F_1'(0)$ is the first nonvanishing derivative

---

3. It is also well known that there are two species of neutrino—ν_e coupled to the electron and ν_μ coupled to the muon; however, the distinction between ν_e and ν_μ does not concern us here.
4. It appears as if time reversal does not hold rigorously for the weak interactions (see Chapter 15), so in limiting the form factors we have used current conservation.

of F_1 that can enter the matrix elements of $J_\mu{}^\gamma$. We would like to motivate the connection between $F_1'(0)$ and the mean square radius.

Consider the elastic scattering in the barycentric frame so that $E(\mathbf{p}) = E(\mathbf{p}')$ or $q_4 = 0$. Thus $q^2 = \mathbf{q}^2$, and we may define $\rho_{1,2}(\mathbf{r})$ by the equation

$$F_{1,2}(q^2) = \int d\mathbf{r}\, e^{i\mathbf{q}\cdot\mathbf{r}}\rho_{1,2}(\mathbf{r}). \tag{6.19}$$

Since Hermiticity guarantees that F_i is real, $\rho_i(\mathbf{r})$ must be spherically symmetric and real. Using the relation

$$\int d\Omega\, \mathbf{A}\cdot\hat{\mathbf{r}}\, \mathbf{B}\cdot\hat{\mathbf{r}} = \frac{4\pi}{3}\mathbf{A}\cdot\mathbf{B}, \tag{6.20}$$

where $\mathbf{A}$ and $\mathbf{B}$ are any two three-vectors, we have

$$F_{1,2}(q^2) = \int d\mathbf{r}\, \rho_{1,2}(|\mathbf{r}|) - \frac{q^2}{6}\int d\mathbf{r}\, \rho_{1,2}(|\mathbf{r}|)\mathbf{r}^2 + \cdots. \tag{6.21}$$

If $\int \rho_{1,2}(|\mathbf{r}|)\, d\mathbf{r} \neq 0$, we define

$$\langle \mathbf{r}^2\rangle \cdot\!\equiv\!\cdot -6\frac{F'(0)}{F(0)}. \tag{6.22}$$

Otherwise, we define

$$\langle \mathbf{r}^2\rangle \cdot\!\equiv\!\cdot -6F'(0). \tag{6.23}$$

There is no a priori reason why $\langle \mathbf{r}^2\rangle > 0$. For example, $F(q^2)$ might look like

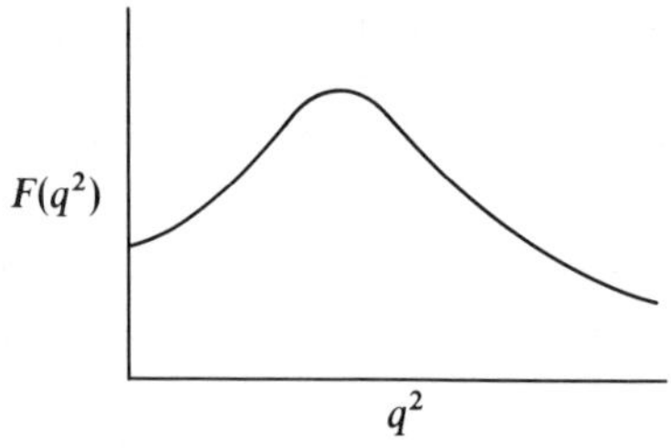

so that $F'(0) > 0$. It is a purely experimental question to decide the sign and magnitude of $F'(0)$ for the different elementary particles.

The first of the nucleons to have its electric charge radius measured was the neutron. Without going into details, the essential technique is to scatter slow (thermal) neutrons against atomic electrons. If the effect of these electrons is summarized by an external Coulomb potential A_0 and if the spin-dependence of the neutrons is averaged over, the effective matrix element of the current, that enters to order v/c, including both the charge and anomalous magnetic moment, can be written

$$\langle \mathbf{N}'|J_0{}^\gamma(0)|\mathbf{N}\rangle \simeq \frac{-q^2}{6}\left\{\langle \mathbf{r}^2\rangle_1 + \frac{1.91}{2M^2}\frac{3}{2}\right\}. \tag{6.24}$$

The second term, called the Foldy term, comes from the moment. Numerically this term contributes $.1261 \times 10^{-26}$ cm² to the matrix element, and the entire matrix element has a *measured* value of $.126 \pm 0.006 \times 10^{-26}$ cm², or

$$\langle \mathbf{r}^2 \rangle_1^N = (0 \pm .006) \times 10^{-26}\ \mathrm{cm}^2.$$

In brief, the neutron has *no* observable charge radius.

To determine the proton's charge radius, we must make a detailed analysis of electron-proton scattering. Using the matrix element indicated above and applying standard techniques of cross-section computation, one may derive the differential cross section for electron scattering (the Rosenbluth formula) in the system in which the proton is initially at rest. It is

$$\frac{d\sigma}{d\Omega} = \frac{e^2}{4(4\pi)^2 E_0^2} \frac{\cos^2(\theta/2)}{\sin^4(\theta/2)} \frac{1}{1 + \dfrac{2E_0}{M}\sin^2\left(\dfrac{\theta}{2}\right)}$$
$$\times \left\{ (F_1(q^2))^2 + \frac{q^2}{4M^2}(2(F_1(q^2) + 2MF_2(q^2))^2 \tan^2\left(\frac{\theta}{2}\right) + (2MF_2(q^2))^2) \right\}. \quad (6.25)$$

Here M is the nucleon mass, E_0 is the initial electron energy, and θ is the electron scattering angle

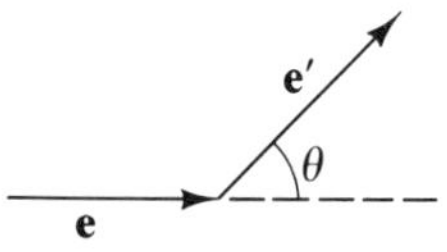

Expressed in terms of laboratory quantities, the invariant four-momentum transfer q^2 is given by

$$q^2 = \frac{4\sin^2(\theta/2)E_0^2}{1 + \dfrac{2E_0}{M}\sin^2(\theta/2)}. \quad (6.26)$$

The technique of finding the F's is simple, at least to describe; one fixes q^2, which is a function of θ and E_0; then, varying E_0, one measures $\frac{d\sigma}{d\Omega}$ at two, or more, energies for fixed q^2. One thus solves a pair of simultaneous algebraic equations for the two F's at fixed q^2 and repeats the process by varying q^2 to obtain the form factors. In this way it was discovered that

$$\langle \mathbf{r}^2 \rangle_1^P \simeq 0.66 \times 10^{-26}\ \mathrm{cm}^2.$$

At first sight this result appears to be entirely plausible, since graphs like

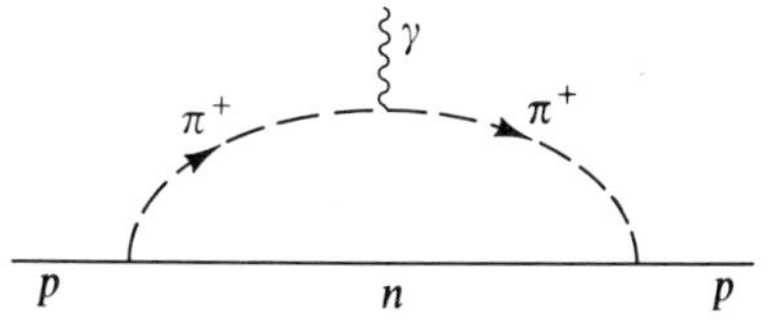

would be expected to give a proton size of about the pion compton wavelength $1/m_\pi \simeq \sqrt{2} \times 10^{-13}$ cm. However, the same graphs also contribute to *neutron*-electron scattering

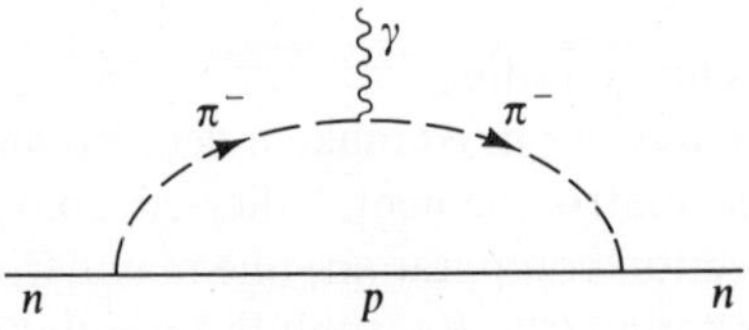

and yet the neutron has no observable charge radius. Some light is thrown on this matter by rewriting the form factor radii in isotopic spin variables. Thus, in an obvious notation,

$$\langle \mathbf{r}^2 \rangle_{1,2}{}^P = \frac{\langle \mathbf{r}^2 \rangle_{1,2}{}^S + \langle \mathbf{r}^2 \rangle_{1,2}{}^V}{2},$$
$$\langle \mathbf{r}^2 \rangle_{1,2}{}^N = \frac{\langle \mathbf{r}^2 \rangle_{1,2}{}^S - \langle \mathbf{r}^2 \rangle_{1,2}{}^V}{2}. \tag{6.27}$$

Therefore we may re-express the experimental results on $\langle \mathbf{r}^2 \rangle$ by the statement that

$$\langle \mathbf{r}^2 \rangle_1{}^S \simeq \langle \mathbf{r}^2 \rangle_1{}^V \lesssim 10^{-26} \text{ cm}^2. \tag{6.28}$$

Intermediate states. To get any insight from this observation it is essential to classify the Feynman graphs into those that contribute to the isotopic scalar form factors and those that contribute to the isotopic vector. The most elegant way of doing this is to consider, instead of the process $e^- + p \longrightarrow e^- + p$, the process $p + \bar{p} \longrightarrow e^+ + e^-$—that is, proton-antiproton annihilation into lepton pairs. This is a useful thing to do since the initial state is extremely symmetric and we can use very simple selection rules to classify the contributing diagrams. Without worrying for the moment about how this process is related to electron scattering (a point that will emerge below), we study the one-photon contribution to the annihilation; namely, the diagram

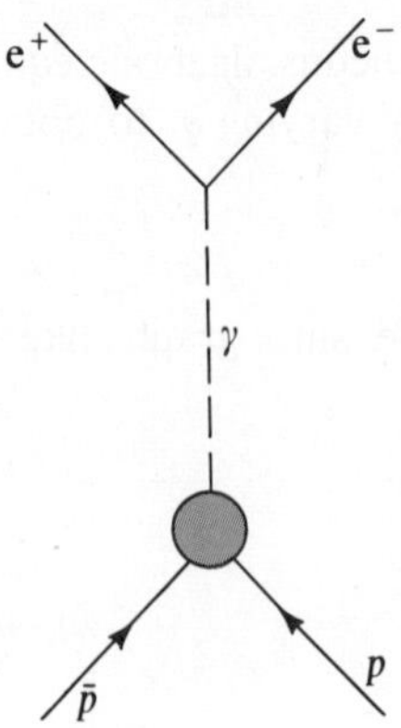

The dark bubble represents the vertex $\bar{p} + p \longrightarrow \gamma$, which corresponds to the matrix element[5] $\langle 0|J_\mu{}^\gamma(0)|\mathbf{p}\bar{\mathbf{p}}\rangle_{\text{in}}$. We may subject this matrix element to the same general analysis that we have made for $\langle \mathbf{p}'|J_\mu{}^\gamma(0)|\mathbf{p}\rangle$. Using Lorentz invariance and current conservation we arrive at the expression

$$\langle 0|J_\mu{}^\gamma(0)|\mathbf{p}\bar{\mathbf{p}}\rangle_{\text{in}} = i\bar{v}(\bar{\mathbf{p}})[\gamma_\mu \bar{F}_1(q^2) + \sigma_{\mu\nu}q_\nu \bar{F}_2(q^2)]u(\mathbf{p}). \tag{6.29}$$

The natural coordinate system in which to work here is the barycentric system in which $\bar{\mathbf{p}} = -\mathbf{p}$. In this system $q = (0, 2E(\mathbf{p}))$ so that in our metric

$$q^2 = -4(E(\mathbf{p}))^2 \leq -4M^2, \tag{6.30}$$

since for physical nucleons (antinucleons) the minimum energy per nucleon is M. The arguments using Hermiticity or time reversal invariance, that proved, above, that $F_1(q^2)$ and $F_2(q^2)$ are real, *fail* here. Hermiticity merely gives the condition

$$\langle 0|J_\mu{}^\gamma(0)|\mathbf{p}\bar{\mathbf{p}}\rangle_{\text{in}} = \langle \bar{\mathbf{p}}\mathbf{p}|_{\text{in}}J_\mu{}^\gamma(0)|0\rangle^*, \tag{6.31}$$

which says nothing about the reality of $\bar{F}_{1,2}$ since the process $\langle \bar{\mathbf{p}}\mathbf{p}|_{\text{in}}J_\mu{}^\gamma|0)|0\rangle$ is different from $\langle 0|J_\mu{}^\gamma(0)|\bar{\mathbf{p}}\mathbf{p}\rangle_{\text{in}}$. The failure of the time reversal argument is somewhat more subtle. Here we shall merely describe the essentials; the details can be found in books on scattering theory. As we have noted in Chapter 2, a one-particle state can be constructed from the Heisenberg operators acting on the vacuum by taking $t \longrightarrow \pm\infty$ in the limiting process, a fact that we described by the equation

$$|\mathbf{p}\rangle_{\text{in}} = |\mathbf{p}\rangle_{\text{out}}. \tag{6.32}$$

However, for a two-particle state there is in general a distinction between the state created at $t = -\infty$, before any scattering processes have taken place, and the state at $t = +\infty$, after the particles have scattered and separated. The incoming state, which is characterized by incoming wave boundary conditions, we have denoted by $|\mathbf{p}'\mathbf{p}\rangle_{\text{in}}$, and the outgoing state with outgoing wave boundary conditions we have denoted by $|\mathbf{p}'\mathbf{p}\rangle_{\text{out}}$. In addition to changing the spins and momenta of the particles in the states the time reversal operator changes an *in* state to an *out* state.[6] Only when the particles are free can this distinction be ignored. Thus, in general, when there are two or more particles in a state the time reversal operation does not give a reality condition, but rather a connection between incoming and outgoing processes. Hence time reversal does not lead to a reality condition on $\bar{F}_{1,2}(q^2)$, which may be, in general, complex functions of q^2.

We may ask how $F_{1,2}(q^2)$ and $\bar{F}_{1,2}(q^2)$ are connected. Under a very general set of assumptions, to be discussed in more detail below, it can be argued that

---

5. If we treat the photon dynamics to lowest order, the photon can be eliminated from the matrix element by replacing it with a photon creation operator acting on the vacuum, which then can be commuted away against the radiation field coupled to $J_\mu(0)$.
6. See, for example, J. J. Sakurai, *Invariance Principles and Elementary Particles*, Princeton University Press, 1964.

$F(q^2)$ and $\bar{F}(q^2)$ are essentially the same function, or, precisely speaking, there exists a complex function $F(z)$, where z is a complex variable, with the property that $F(z) = F(q^2)$ for $q^2 \geq 0$ and $F(z) = \bar{F}(q^2)$ for $q^2 \leq 0$. Moreover, the function $F(z)$ satisfies a *dispersion relation* that relates its real and imaginary parts. In its simplest form[7] this dispersion relation states that

$$\text{Re}\, F(q^2) = \frac{1}{\pi} P \int_{-\infty}^{\infty} \frac{\text{Im}\, F(q'^2)\, dq'^2}{q'^2 - q^2}. \tag{6.33}$$

But for $q^2 \geq 0$

$$\text{Re}\, F(q^2) = F(q^2). \tag{6.34}$$

Thus the form factor in electron-proton scattering is determined by an integral over its imaginary part, which enters into the process $p + \bar{p} \longrightarrow e^+ + e^-$. (In practice p, $\bar{p}$ annihilate, by the strong interactions, into mesons, so that the leptonic process is difficult to observe. Indeed, it may best be studied in very high-energy colliding beam experiments, which will feature the inverse reaction $e^+ + e^- \longrightarrow p + \bar{p}$ when beams of sufficient energy become available.) As we shall see in detail, this imaginary part is fixed by a chain of intermediate states of the form $p + \bar{p} \longrightarrow \pi^+ + \pi^- \longrightarrow \gamma$, $p + \bar{p} \longrightarrow \pi^+ + \pi^- + \pi^0 \longrightarrow \gamma$, and so on. Thus we can fix the set of intermediate states that contribute to the different form factors by studying the selection rules that apply to products of matrix elements of the form

$$\langle 0 | J_\mu{}^\gamma(0) | n \rangle \langle n | p\bar{p} \rangle_{\text{in}},$$

where $|n\rangle$ is any of these intermediate states, for example $\pi^+\pi^-$.

There is reason to hope that the lightest intermediate states are the most important. This is hard (impossible) to justify rigorously. Loosely speaking, the term $1/(q'^2 - q^2)$ in the dispersion integral acts like a kind of energy denominator. The quantity q'^2 represents the invariant mass of the intermediate state. The minimum mass that such a state can have is the sum of the physical masses of the particles that compose it, such as $4m_\pi{}^2$ for the two-pion state. If q^2 is not too large or if Im $F(q^2)$ does not have some special behavior near certain masses, then one is tempted to argue that the relative contributions of the different intermediate states should go, roughly, as the ratio of the squares of their masses. At least we may begin the analysis with this perspective and see how far we can go with it. Therefore we are led to consider the expression $\langle 0 | J_\mu{}^\gamma(0) | n \rangle$, where the masses of $|n\rangle$ are as small as possible. From the properties of the vacuum, $|0\rangle$, and $J_\mu{}^\gamma$, we may readily establish four properties of $|n\rangle$.

1. The electric charge of $|n\rangle$ is zero.
2. $|n\rangle$ is an eigenstate of C, charge conjugation, with $C = -1$.

---

7. No subtractions. The reader who finds these formulas mystifying is advised that a more detailed discussion follows in the next chapter.

3. The ordinary angular momentum of $|n\rangle$ is 1 and its parity is negative.
4. The total isotopic spin of $|n\rangle$ is either 0 or 1.

The proof of these properties follows.

1. The proof follows at once from the commutation relation between the electric charge Q and the current $J_\mu{}^\gamma(0)$,[8]

$$[Q, J_\mu{}^\gamma(0)]_- = 0, \tag{6.35}$$

which, formally speaking, is a consequence of the phase invariance of $J_\mu{}^\gamma(0)$. Since $Q|0\rangle = 0$ we have the condition ($|n\rangle$ can always be taken as an eigenstate of Q, since $[Q, H]_- = 0$), with q_n being the charge of $|n\rangle$:

$$q_n\langle 0|J_\mu{}^\gamma(0)|n\rangle = 0. \tag{6.36}$$

2. Since $[H, C]_- = 0$ we can always choose $|n\rangle$ to be an eigenstate of C and Q providing that $q_n = 0\,([Q, C]_+ = 0)$, which we have just established by Equation (6.36). Since $CJ_\mu{}^\gamma(0)C^{-1} = -J_\mu{}^\gamma(0)$ and $C|0\rangle = |0\rangle$, we have

$$\langle 0|J_\mu{}^\gamma(0)|n\rangle = \mp\langle 0|J_\mu{}^\gamma(0)|n\rangle, \tag{6.37}$$

depending on whether $C|n\rangle = \pm|n\rangle$. It thus follows that

$$C|n\rangle = -|n\rangle. \tag{6.38}$$

3. By the usual selection rules for quantum mechanical vectors, $J_\mu{}^\gamma(0)$ can only transform the vacuum, which has angular momentum 0, into a state with angular momentum 0 or 1. If $|n\rangle$ has angular momentum zero, then, in general,

$$\langle 0|J_\mu{}^\gamma(0)|n\rangle = p_{n\mu}F(p_n{}^2), \tag{6.39}$$

where $p_{n\mu}$ is the energy-momentum vector of $|n\rangle$. From current conservation it follows that

$$m_n{}^2F(p_n{}^2) = 0, \tag{6.40}$$

which rules out spin zero for $|n\rangle$, except perhaps for the singular case in which $m_n = 0$. However, the only known mass-zero particles are the photon and neutrino, which have spin 1 and $\frac{1}{2}$ respectively, and which have, anyway, no strong interactions to couple them to $p\bar{p}$.

4. Ignoring higher-order electromagnetic corrections, we may write

$$J_\mu{}^\gamma(0) = J_\mu{}^S(0) + J_\mu{}^V(0)_3, \tag{6.41}$$

the isotopic decomposition into scalar and vector currents. Let $|n\rangle$ be an eigenstate of T^2 and T_3. We can always choose $|n\rangle$ this way since $[H_{ST}, T^2]_- = [H_{ST}, T_3]_- = 0$. From the relation

8. The same argument shows that $|n\rangle$ must have zero baryon number.

$$Q = \frac{B}{2} + T_3, \tag{6.42}$$

which holds to the order in e to which we work, we can conclude that

$$T_3|n\rangle = 0. \tag{6.43}$$

The statement that $J_\mu{}^S(0)$ is an isoscalar implies that

$$[T_i, J_\mu{}^S(0)]_- = [T^2, J_\mu{}^S(0)]_- = 0. \tag{6.44}$$

Thus, labeling $|n\rangle$ in isotopic notation as $|T0\rangle$ and using the relation

$$0 = \langle 0|[T^2, J_\mu{}^S(0)]_-|T0\rangle = T(T+1)\langle 0|J_\mu{}^S(0)|T0\rangle, \tag{6.45}$$

we see that $J_\mu{}^S(0)$, as expected, only connects the vacuum to states of total isotopic spin zero. On the other hand, the statement that $J_{\mu 3}{}^V$ is the third component of an isotopic vector means, as we have shown earlier in the book, that

$$[T_i, J_\mu{}^V(0)_3]_- = i\epsilon_{i3k}J_\mu{}^V(0)_k, \tag{6.46}$$

where $J_\mu{}^V(0)_k$ is a general component of the isotopic spin current. Thus[9]

$$\begin{aligned} T^2J_\mu{}^V(0)_3|0\rangle &= T_i[T_i, J_\mu{}^V(0)_3]_-|0\rangle = \epsilon_{ik3}\epsilon_{ikl}J_\mu{}^V(0)_l|0\rangle \\ &= (\delta_{3l}\,\delta_{kk} - \delta_{3k}\,\delta_{kl})J_\mu{}^V(0)_l|0\rangle = 2J_\mu{}^V(0)_3|0\rangle. \end{aligned} \tag{6.47}$$

Thus $J_\mu{}^V(0)_3|0\rangle$ is an eigenstate of T^2 with eigenvalue $1(1+1) = 2$, so that the isotopic vector electromagnetic current connects the vacuum to states with $T = 1$, $T_3 = 0$. (The same formal argument can be used for ordinary spin to establish 3 above.)

We now know a great deal about the states $|n\rangle$ and we may begin to list candidates for these states from among the known particles. The obvious candidates, to begin with, are the pions, since they are the lightest known particles with strong interactions. The state $|n\rangle = |\pi^0\pi^0\rangle$ is ruled out because, by Bose statistics, two π^0's cannot have total angular momentum 1.[10] Thus the state of lowest mass with the right quantum numbers to be a candidate is the state $|\pi^+\pi^-\rangle$. We may now raise the question of whether this state contributes to F^S, or F^V, or both. Here again we can proceed by a very elegant method.

We introduce an operator G defined so that

$$G = Ce^{i\pi T_2}, \tag{6.48}$$

where C is the charge conjugation operator and $e^{i\pi T_2}$ generates a rotation of π around the 2-axis in isotopic spin space.

---

9. We use the fact that $\mathbf{T}|0\rangle = 0$. It is important to remember that even though $J_\mu{}^V(0)_3$ is an isovector it does *not* commute with T^2. Only $T_3 = \int d\mathbf{r}J_0{}^V(\mathbf{r})_3$ (and also T_1 and T_2) commute with T^2.
10. The two π^0's are identical bosons, which means that their wave function must be even under exchange of particles, whereas, if they have relative orbital angular momentum l, the wave function becomes multiplied by the factor $(-1)^l$ under exchange. A one-pion state is ruled out, for $|n\rangle$, since it has zero spin.

For the three pion fields ϕ_1, ϕ_2, ϕ_3, we have[11]

$$\begin{aligned} e^{i\pi T_2}\phi_1 e^{-i\pi T_2} &= -\phi_1, \\ e^{i\pi T_2}\phi_2 e^{-i\pi T_2} &= +\phi_2, \\ e^{i\pi T_2}\phi_3 e^{-i\pi T_2} &= -\phi_3. \end{aligned} \tag{6.49}$$

Combining this result with what we know about the C conjugation properties of the ϕ's, we have at once

$$G\pi^{\pm}_{0}G^{-1} = -\pi^{\pm}_{0}. \tag{6.50}$$

Hence, for a state of n pions with any charge,

$$G|n\pi\rangle = (-1)^n|n\pi\rangle. \tag{6.51}$$

Moreover, the currents $J_\mu{}^S(0)$ and $J_\mu{}^V(0)_3$ have very simple transformation properties under G:

$$GJ_\mu{}^S(0)G^{-1} = -J_\mu{}^S(0) \tag{6.52}$$

and

$$GJ_\mu{}^V(0)_3G^{-1} = J_\mu{}^V(0)_3. \tag{6.53}$$

Since $G|0\rangle = |0\rangle$ it follows, from the equation

$$\langle 0|J_\mu{}^S(0) + J_\mu{}^V(0)_3|n\pi\rangle = (-1)^n\langle 0| - J_\mu{}^S(0) + J_\mu{}^V(0)_3|n\pi\rangle, \tag{6.54}$$

that states with odd numbers of pions contribute to the isoscalar form factor, and states with even numbers of pions contribute to the isovector form factor.

When this was first realized, in the 1950's, it created a puzzle. We have seen that $\langle \mathbf{r}^2\rangle_1{}^S \simeq \langle \mathbf{r}^2\rangle_1{}^V$, which implies, if we think in terms of pionic graphs alone, that the two-pion and three-pion graphs should give rise to comparable charge radii and comparable contributions to $F_1(q^2)$ in general.[12] On the other hand, the structure of these graphs is entirely different and perturbation theory computations gave no clue as to why their contributions were similar. However, it was realized in the late 1950's that the strong interactions among the pions themselves might provide the clue.[13] It was well known at this time that pions and nucleons could form resonant, metastable states. In particular, the reaction

$$\pi + N \longrightarrow N^*_{3/2} \longrightarrow \pi + N,$$

where $N^*_{3/2}$ is a metastable particle with spin $\frac{3}{2}$, positive parity, isospin $\frac{3}{2}$, mass $\simeq$ 1238 Mev, and width $\simeq$ 125 Mev, had been thoroughly studied. Thus, in analogy, it was conjectured by several physicists that the pions themselves could also

---

11. Draw a three-dimensional coordinate system and carry out the rotation.
12. In fact, $F_1{}^N(q^2) \simeq 0$ for all momentum transfers so far measured, indicating that $F_1{}^S(q^2) \simeq F_1{}^V(q^2)$.
13. For a complete survey of these early papers, see the collection *Nuclear and Nucleon Structure*, edited by Robert Hofstadter, Benjamin, 1963.

form resonant states. For these states to be relevant for electron-nucleon scattering they would, as we have seen, have to have the quantum numbers $J = 1$, $P = -1, C = -1, T = 1$ or 0, and $T_3 = 0$. In fact, as mentioned in Chapter 4, in 1961, two such states, the $T = 1$, ρ meson and the $T = 0$, ω^0 meson were identified and shown to have all of the correct quantum numbers to contribute to the electromagnetic form factors. More recently a third such vector "particle," or resonance, the ϕ^0 meson, with

$$T = 0, \qquad \Gamma_{\phi^0} = 4.0 \pm 1.0,$$

has been found, and it too plays an important role in the structure of the isoscalar form factors. Since these objects, the ρ, ω^0, ϕ^0, are very short-lived ($\tau \sim 10^{-23}$ sec), it is not really correct, in a rigorous sense, to treat them as if they were stable particles. But as a first orientation we shall consider a simplified model in which ρ, ω^0, and ϕ^0 are treated as stable vector mesons.[14] In this model, which neglects all other states and the nonresonant pion-pion contributions, the entire electron scattering of nucleons (or, equivalently, the nucleon-antinucleon annihilation into leptons) is given in terms of two types of graphs (for the proton):

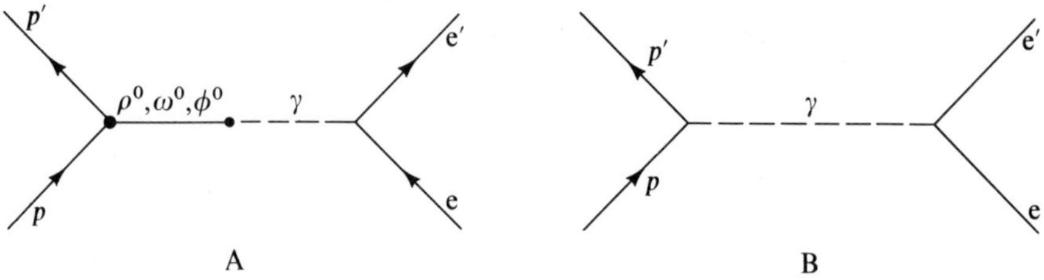

The graph on the right represents the contribution from the Dirac moment that would be present even if the proton had no strong interactions to give it a structure. This graph gives rise to the matrix element

$$M_b = e^2\bar{u}(\mathbf{e}')\gamma_\mu u(\mathbf{e}) \frac{1}{q^2} \bar{u}(\mathbf{p}')\gamma_\mu u(\mathbf{p}), \tag{6.55}$$

so that for the proton it yields the form factors

$$\begin{aligned} F_1^P(q^2) &= 1, \\ F_2^P(q^2) &= 0, \end{aligned} \tag{6.56}$$

and for the neutron (in which there is no scattering in the absence of structure),

$$F_1^N(q^2) = F_2^N(q^2) = 0, \tag{6.57}$$

the familiar results.

---

14. As we shall see later, there are other vector meson resonances, but these have nonzero "strangeness" and hence do not contribute to electron scattering.
15. For the isotopic vector meson this means the emission of $V_{\mu 3}$.

The graph on the left is much more difficult to discuss, and in the literature there is some disagreement about how it is to be treated. If we call the vector meson field V_α, where V_α may be an isovector, $\mathbf{V}_\alpha$, or an isoscalar, V_α, the field V_α must satisfy the transversality condition

$$\partial_\alpha V_\alpha = 0, \tag{6.58}$$

which assures us that it is spin 1 and not a mixture of spin 0 and spin 1. If we treat V_α as a fundamental field (as opposed to a resonant multipion system) we may couple V_α to the nucleons in the Langrangian. We write out, explicitly, the isotopic vector case; the scalar case has the same form but with no isotopic matrices. Thus

$$\mathfrak{L}_{VN} = ig_1{}^V\bar{\psi}\gamma_\mu\boldsymbol{\tau}\cdot\psi\mathbf{V}_\mu + \frac{g_2{}^V}{4M_V}\bar{\psi}\sigma_{\mu\nu}\boldsymbol{\tau}\cdot\psi(\partial_\mu\mathbf{V}_\nu - \partial_\nu\mathbf{V}_\mu). \tag{6.59}$$

The most general vertex, consistent with Lorentz and time reversal invariance, for the emission of a virtual neutral vector meson[15] by real nucleons, is, for

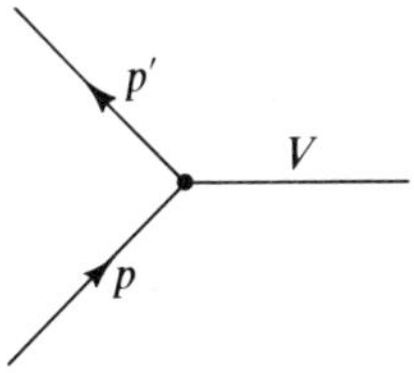

the isoscalar meson,

$$V_\mu(p', p) = i\bar{u}(\mathbf{p}')[g_1{}^S(q^2)\gamma_\mu + g_2{}^S(q^2)\sigma_{\mu\nu}q_\nu]u(\mathbf{p}), \tag{6.60}$$

where the g's are arbitrary real functions of $q^2 = (p' - p)^2$, which must be determined by the strong interaction dynamics; the u's are nucleon spinors. For the isovector,

$$V_\mu(p', p)_3 = i\bar{u}(\mathbf{p}')[\pm g_1{}^V(q^2)\gamma_\mu \pm g_2{}^V(q^2)\sigma_{\mu\nu}q_\nu]u(\mathbf{p}), \tag{6.61}$$

where $\pm$ refers to proton and neutron respectively. These vertices are automatically "conserved" in the sense that, independent of the choice of the g's,

$$q_\mu V_\mu(p', p) = 0. \tag{6.62}$$

We may therefore drop the term $q_\mu q_\nu/M_V{}^2$ in the vector meson propagator that attaches to this vertex and write the propagator simply as[16]

$$\frac{\delta_{\mu\nu}}{q^2 + M_V{}^2}. \tag{6.63}$$

16. Sometimes in the literature there is an attempt to take into account the instability of the vector meson by including a complex term in the energy denominator of the propagator; a term of the form $iM_V\Gamma_V$, where Γ_V is the width of the vector meson. Such a procedure must be followed with prudence to avoid contradictions with time-reversal invariance.

We next consider the coupling of the neutral vector meson to the photon.

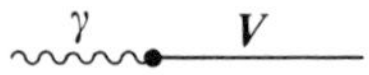

This coupling has been much discussed in the literature. We must be careful to introduce it in such a way that it does not give the neutron an electric charge; that is, $F_1^N(0) \neq 0$, via the graph

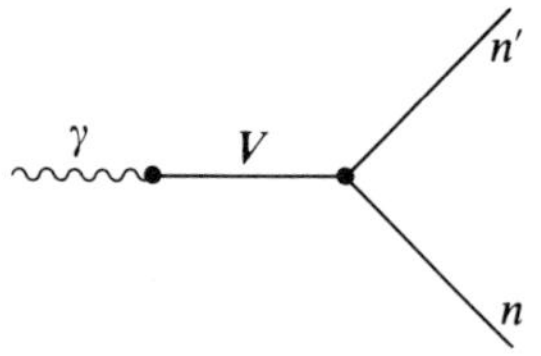

nor the photon a self-mass via the chain of graphs

etc.

For this reason we reject the most obvious Lorentz-invariant coupling between the photon and the neutral vector meson,[17]

$$\mathcal{L}_{V\gamma} = e\lambda A_\mu V_\mu. \tag{6.64}$$

If we sum up the string of photon-V graphs[18] given in the figure just above, this coupling leads to a propagator with a denominator of the form

$$\frac{1}{q^2(q^2 + M_V{}^2) - C^2},$$

where C is a real constant proportional to $e\lambda$. This denominator has two zeros.

---

17. It must be clearly understood that *all* such couplings are "phenomenological." The neutral vector meson, since it is neutral, cannot enter directly into the electromagnetic current. However, two charged pions, say, $\pi^+\pi^-$, can obviously couple to the photon. When we write $\mathcal{L}_{V\gamma}$, we have in mind that the origin of the vector meson is as a pion-pion resonance. The constant λ is an empirical constant that reflects these strong couplings and is presumably large. It must be taken from experiment. The form of $\mathcal{L}_{V\gamma}$ is dictated by the general requirements of Lorentz and gauge invariance.

    In the literature one sometimes finds the statement that the interaction of Eq. (6.64) *is* gauge-invariant. The argument given goes as follows. Let

$$L = \int d^4x \mathcal{L}_{V\gamma}(x),$$

and in the equation

$$\mathcal{L}_{V\gamma} = e\lambda A_\mu V_\mu$$

make the gauge transformation

$$A_\mu \longrightarrow A_\mu + \partial_\mu \Lambda(x).$$

One corresponds to the V mass as it is modified by the V's coupling to the photon. The other zero should correspond to the photon's mass and should, evidently, occur at $q^2 = 0$. However, in this model, it does not occur at $q^2 = 0$ and, in fact, the mass comes out pure imaginary. Hence this coupling form must be rejected. On the other hand, we can couple V to γ through the manifestly gauge-invariant coupling

$$\mathcal{L}_{V\gamma} = -\tfrac{1}{2}e\lambda(\partial_\mu A_\nu - \partial_\nu A_\mu)(\partial_\mu V_\nu - \partial_\nu V_\mu), \tag{6.65}$$

which leads to the photon-V vertex

$$e\lambda q^2 A_\mu V_\mu. \tag{6.66}$$

The extra factor of q^2 gurantees that this coupling will not contribute to the neutron charge or photon mass.[19] Putting all of the factors together, including, for illustration, a similar coupling to the isoscalar meson (we leave out the ϕ^0 since it does not change the conclusion significantly), we learn that, on this simple picture, we would predict[20]

$$\begin{aligned} F_1^S(q^2) &= 1 + \frac{q^2 g_1^{\omega^0}(q^2)\lambda_{\omega^0}}{q^2 + m_{\omega^0}^2}, \\ F_2^S(q^2) &= \frac{q^2 g_2^{\omega^0}(q^2)\lambda_{\omega^0}}{q^2 + m_{\omega^0}^2}, \\ F_1^V(q^2) &= 1 + \frac{q^2 g_1^{\rho}(q^2)\lambda_{\rho}}{q^2 + m_{\rho}^2}, \\ F_2^V(q^2) &= \frac{q^2 g_2^{\rho}(q^2)\lambda_{\rho}}{q^2 + m_{\rho}^2}. \end{aligned} \tag{6.67}$$

Several remarks about this model can be made. In the first place, if

$$\lambda_\rho g_1^{\rho}(0) \simeq \lambda_{\omega^0} g_1^{\omega^0}(0)\,, \tag{6.68}$$

Since $\partial_\mu V_\mu = 0$, it is argued, one may integrate $\partial_\mu \Lambda(x) V_\mu(x)$ by parts in L to show that L is indeed invariant under the gauge transformation. The fallacy in this argument is that $\partial_\mu V_\mu(x) = 0$ is a constraint that emerges from the equations of motion obtained by varying the full L. If we suppose that the fields in L obey constraint equations before applying the variations, then this leads to the replacement of L by a new Lagrangian whose equations of motion are different. An obvious example of this is the Dirac Lagrangian

$$\mathcal{L} = \bar{\psi}(\gamma_\mu \partial_\mu + m)\psi.$$

If the fields are assumed to obey the Dirac equation, as a constraint, then $\mathcal{L} = 0$, whose Euler-Lagrange equations are clearly different from the Dirac equation. Hence we cannot use the condition $\partial_\mu V_\mu(x) = 0$ in the L above and it is *not* gauge-invariant.

18. A reader familiar with graphs will recognize that this sum is a geometric series and is readily evaluated.
19. For these, $q^2 = 0$.
20. Since F_1 and F_2, as we have defined them, have different dimensions ($[F_1] = [F_2]/M$), we also have $[g_1] = [g_2]/M$.

then, since $m_\rho \simeq m_\omega$,

$$F_1'^S(0) \simeq F_1'^V(0). \tag{6.69}$$

This gives at least a rudimentary understanding of how it can be that the 2π and 3π states lead to charge radii of about the same magnitude—they are both dominated by vector meson resonances at about the same mass. However, experimentally[21]

$$-\langle \mathbf{r}^2 \rangle_2^N \simeq \langle \mathbf{r}^2 \rangle_2^P \simeq \langle \mathbf{r}^2 \rangle_1^P \simeq .64 f^2. \tag{6.70}$$

In other words, although the neutron's charge radius is essentially zero, its magnetic (Pauli) radius is about equal to the proton's and of opposite sign. In isotopic spin language this means that

$$\frac{\langle \mathbf{r}^2 \rangle_2^S}{2} \simeq \langle \mathbf{r}^2 \rangle_1^P, \tag{6.71}$$

but

$$\langle \mathbf{r}^2 \rangle_2^V \simeq 0. \tag{6.72}$$

If we interpret these results in terms of vector meson couplings they mean that (dividing by M_P for dimensional reasons) though

$$\lambda_\rho \frac{g_{2\rho}(0)}{M_P} \gtrsim 1, \tag{6.73}$$

we must have

$$g_2^{\omega 0}(0) \simeq 0. \tag{6.74}$$

In other words, the ω meson must have essentially zero coupling to the nucleons of Pauli form. Based on the analogy with electromagnetism,[22] it is tempting to couple the massive vector mesons to conserved currents, the baryon current and the isospin current. If we introduce the $\boldsymbol{\rho}$ as a fundamental field into the Lagrangian, then, since it is an isovector, we might couple it to $\mathbf{J}_\mu{}^V$, the isospin current, with a coupling of the form

$$\mathcal{L}_\rho = g_\rho \boldsymbol{\rho}_\mu \cdot \mathbf{J}_\mu{}^V. \tag{6.75}$$

This coupling involves the scalar product of two isovectors and is guaranteed to be invariant under SU_2 and hence to leave unaltered the conservation of $\mathbf{J}_\mu{}^V$. In this theory $\mathbf{J}_\mu{}^V$ will also contain terms involving the ρ meson, terms generated by the infinitesimal space-dependent transformation acting on the free ρ Lagrangian

$$\boldsymbol{\rho}_\mu \longrightarrow \boldsymbol{\rho}_\mu + \boldsymbol{\rho}_\mu \times \boldsymbol{\Lambda}(x). \tag{6.76}$$

---

21. 1 fermi $\cdot\!\equiv\!\cdot$ $1f \cdot\!\equiv\!\cdot$ 10^{-13} cm $\simeq 1/\sqrt{2} m_\pi$.
22. See, for example, J. J. Sakurai, *Ann. Phys.*, **11**:1 (1960).

However, the direct coupling of ρ's to nucleons will be of the form $\rho_\mu \cdot \bar{\psi}\tau\gamma_\mu\psi$ without the Pauli term. Pauli terms can arise in the exact ρ nucleon vertex through higher-order induced effects but these are extremely difficult to compute. It is clear that one is embedded in the details of the strong interaction dynamics, and without an accurate theory of the strong interactions we cannot do much better than to make approximate arguments.

As the reader has no doubt observed, there is a striking disagreement between the simple-minded vector meson model and the electron-scattering results in the magnetic moments themselves. As we have written them in Eq. (6.67),

$$F_2^P(0) = F_2^N(0) = 0, \tag{6.77}$$

whereas in fact these terms are not zero and are, indeed, the anomalous magnetic moments. We might take the position that the ordinary non-resonant pion graphs such as

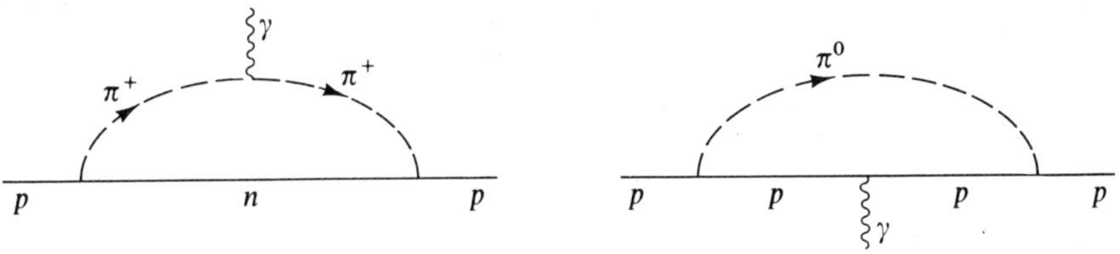

which are neglected in the model, give the $q^2 = 0$ phenomena, the moments, and the vector mesons dominate the F's for $q^2 > 0$. Some support can be given for this point of view by an actual computation of these graphs. They do not give perfect agreement with the moments, but they do give at least the correct orders of magnitude. Hence, on this picture, we might simply add on the moments by hand to the F's above and write, for example,

$$F_2^P(q^2) = \mu_P + \tfrac{1}{2}(F_2^S(q^2) + F_2^V(q^2)), \tag{6.78}$$

where the F_2^S and F_2^V are the form factors given by the vector meson model. This addition clearly does not change the radii.

However, this picture is also too simple, for it is now known, from recent experiments in which electron scatterings for q^2 at least as large as[23] $30f^{-2}$ have been measured, that the F's fall off faster with q^2 than would be predicted by Eqs. (6.78) and (6.67). Indeed, the model as it stands predicts that $F_i(q^2) \longrightarrow$ const. as $q^2 \longrightarrow \infty$, but experiments show that

$$F_i(q^2) \longrightarrow \frac{1}{q^4} \quad \text{as} \quad q^2 \longrightarrow 30f^{-2}.$$

We might again attempt to reconcile this with the vector meson model by assum-

23. See, for example, J. R. Dunning et al., *Phys. Rev.* **141**:1286 (1966). To go from f units to energies, note that $1f^{-1}$ corresponds to 197 Mev/C.

ing that, for large q^2, $g_V(q^2)$, the form factors in the nucleon-vector meson vertex are not constants but fall off as $1/q^2$. We shall return to this question later.

Sach's form factors. Here we shall give a somewhat different representation of the experimental results. It has now become customary to replace F_1 and F_2 by linear combinations[24] G_E and G_M, defined as follows (for both proton and neutron)[25]:

$$\begin{aligned} G_E(q^2) &= F_1(q^2) - \frac{q^2}{4M^2}\frac{F_2(q^2)}{\mu}, \\ G_M(q^2) &= \mu F_1(q^2) + F_2(q^2), \end{aligned} \tag{6.79}$$

where μ is the nucleon magneton, or inversely,

$$\begin{aligned} F_1(q^2) &= \frac{G_E(q^2) + \dfrac{q^2}{4M^2}G_M(q^2)/\mu}{1 + \dfrac{q^2}{4M^2}}, \\ F_2(q^2) &= \frac{G_M(q^2) - G_E(q^2)/\mu}{1 + \dfrac{q^2}{4M^2}}, \end{aligned} \tag{6.80}$$

where M is the nucleon mass. Clearly, from the definitions of F_1 and F_2, we have the following conditions on the G's:

$$\begin{aligned} G_E(0)^P &= 1, \qquad & G_M(0)^P &= 2.793\ \mu = 1 + \mu_P, \\ G_E(0)^N &= 0 \qquad & G_M(0)^N &= -1.913\ \mu = \mu_N. \end{aligned} \tag{6.81}$$

Thus, intuitively, G_E and G_M represent the "true" charge and magnetic distributions with the Dirac and Pauli factors combined so as to isolate the magnetic and electric parts respectively. Since these sets of form factors are linear combinations of each other, it is impossible to argue that one set is more "fundamental" than the other. However the G's do lead to a simpler form of the Rosenbluth formula, which aids in their experimental analysis. In terms of the G's, in the laboratory[26]

$$\frac{d\sigma}{d\Omega} = \frac{e^2}{4(4\pi)^2E_0{}^2}\frac{\cos^2\left(\dfrac{\theta}{2}\right)}{\sin^4\left(\dfrac{\theta}{2}\right)}\frac{1}{1+\dfrac{2E_0}{M}\sin^2\left(\dfrac{\theta}{2}\right)} \left[\frac{G_E{}^2(q^2) + \dfrac{q^2}{4M^2}G_M{}^2(q^2)/\mu^2}{1+\dfrac{q^2}{4M^2}} + \frac{q^2}{2M^2}\frac{G_M{}^2(q^2)}{\mu^2}\tan^2\left(\frac{\theta}{2}\right)\right]. \tag{6.82}$$

---

24. The G's are known as the Sachs' form factors, after R. G. Sachs, *Phys. Rev.* **136**:B281 (1962).
25. Clearly, from the reality of the F's for $q^2 \geq 0$, it follows that the G's are also real for $q^2 \geq 0$.

Thus the cross section in terms of the G's involves no cross-terms between G_E and G_M (unlike the F_1, F_2 expression, in which there are cross-terms). This lack of cross-terms simplifies the experimental analysis.

However, if we look at the F's expressed in terms of the G's, it is quite clear that both F_1 and F_2 have poles at $q^2 = -4M^2$ unless

$$G_E(-4M^2) = G_M(-4M^2)/\mu. \tag{6.83}$$

These poles occur at the threshold for the physically observable reaction $p + \bar{p} \longrightarrow e^+ + e^-$. A pole in the matrix element of a physically observable process is quite unattractive. It does not imply, necessarily, that the cross section diverges, since at threshold the phase space also vanishes, but it does mean that the threshold behavior must be quite complicated; for this reason the subsidiary condition above is usually adopted.

It has been possible[27] to find a simple empirical fit to all of the known data summarized in the mnemonic set of formulas (we give all the dependences in fermis, which means that we set $M^2 \simeq 90f^{-2}$)

$$G_E(q^2)^P = \frac{G_M(q^2)^P}{1+\mu_P} = \frac{G_M(q^2)^N}{\mu_N} = -\frac{360}{q^2}\,G_E(q^2)^N$$

$$= \left(\frac{1}{1+\dfrac{q^2}{18.1}}\right)^2. \tag{6.84}$$

The various mean square radii can be computed, in this fit, by the relation that holds for these parameters:

$$-6G'_E(0)^P = .66f^2. \tag{6.85}$$

It must be emphasized that this fit is by no means unique; indeed, it does not satisfy the subsidiary threshold condition given above. After we have discussed the dispersion theory of these form factors we return to the question of other fits to the data and how they might be interpreted.

The crossed channel. As a last theoretical observation in this chapter we discuss the question of why the Rosenbluth formula given, say, in terms of the G's, has the form

$$\frac{d\sigma}{d\Omega} = \frac{d\sigma}{d\Omega_{\text{point}}}\left[A(q^2) + B(q^2)\tan^2\left(\frac{\theta}{2}\right)\right], \tag{6.86}$$

where $d\sigma/d\Omega_{\text{point}}$ is the relativistic Rutherford cross section in the lab frame and is given by

26. As usual, we have set m_e/E_0 and m_e/M equal to zero.
27. J. R. Dunning et al., *Phys. Rev.* **141**:1286 (1966).

$$\frac{d\sigma}{d\Omega_{\text{point}}} = \frac{e^2}{4(4\pi)^2 E_0{}^2} \frac{\cos^2\left(\frac{\theta}{2}\right)}{\sin^4\left(\frac{\theta}{2}\right)} \frac{1}{1 + \frac{2E_0}{M}\sin^2\left(\frac{\theta}{2}\right)} \tag{6.87}$$

for a massless electron scattering from a point proton with no anomalous magnetic moment. $A(q^2)$ and $B(q^2)$ are given in terms of the G's. In essence, the fact that the angular distribution of $\frac{d\sigma}{d\Omega}\Big/\frac{d\sigma}{d\Omega_{\text{point}}}$ has this special form can be traced back to the one-photon exchange approximation used in deriving the Rosenbluth formula. Indeed a number of specific experiments have been done at the largest q^2 available to demonstrate explicitly that the ratio $(d\sigma/d\Omega)/(d\sigma/d\Omega_{\text{point}})$ remains a linear function of $\tan^2(\theta/2)$, which it does throughout the presently studied experimental range of energies. Of course, the dependence on $\tan^2(\theta/2)$ emerges directly from the computation of the Rosenbluth formula. However, we can again appeal to the "crossed" reaction $p + \bar{p} \longrightarrow e^+ + e^-$ to get a simple insight into the result.

We shall also use this discussion as an excuse to define a set of kinematical variables (the Mandelstam variables) that are used extensively in relating processes in different channels. Consider the graph

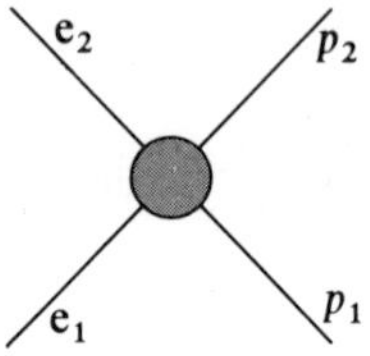

This graph, in which the e's are electron lines and the p's are nucleon lines, can stand for several processes; for example, $e^+ + p \longrightarrow e^+ + p$, $p + \bar{p} \longrightarrow e^+ + e^-$, $e^- + \bar{p} \longrightarrow e^- + \bar{p}$, depending on which lines are taken as incoming or outgoing and which stand for particles or antiparticles. The entire graph is assumed to be characterized by a single complex function with well-defined analyticity properties, a function that can be evaluated in different regions of the arguments to give the several processes described by the graph. There are three Lorentz scalars (the Mandelstam variables) that are used in the literature for this discussion:

$$\begin{aligned} s &= -(e_1 + p_1)^2, \\ t &= -(e_1 - e_2)^2, \\ u &= -(p_1 - e_2)^2. \end{aligned} \tag{6.88}$$

They are not independent, and for the electron scattering problem we have (in our metric) the relation

$$s + t + u = 2m_e{}^2 + 2M^2. \tag{6.89}$$

We define the s, t, or u "channels" to be the scattering channel for which s, t or u is the total center of mass energy in that channel. Thus electron-proton scattering $e^- + p \longrightarrow e^- + p$ occurs in the s-channel, while $p + \bar{p} \longrightarrow e^+ + e^-$ occurs in the t-channel. If we have an expression in one channel in terms of s, t and u we can re-interpret it in other channels by studying what the variables mean when re-read in the appropriate channel.

As an illustration we can consider this translation between the process $p + \bar{p} \longrightarrow e^+ + e^-$ in the center of mass and $p + e^- \longrightarrow p + e^-$ in, say, the lab. We define the angle θ by the diagram

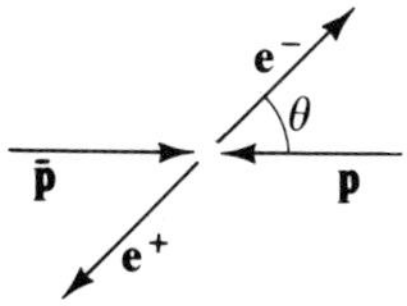

In other words, θ is the angle at which the electron-positron momenta point relative to the line of the nucleon-antinucleon momenta. It is easy to see that $\cos\theta$ can be defined in terms of s, t, and u. If we set $m_e = 0$, then

$$\cos\theta = \frac{s - u}{\sqrt{(t - 4M^2)t}}. \tag{6.90}$$

We may now ask what is the most general possible angular distribution for $p + \bar{p}$ annihilation in the $p + \bar{p}$ center of mass, if the initial state is restricted, by the one photon exchange condition, to have spin one. This is in essence a problem in the addition of angular momenta.[28] The initial state has spin one and it is transformed into a final e^+, e^- state by the exchange of a virtual photon, which can have either spin zero or spin one. Thus the final state will have angular momentum 0, 1, or 2. Therefore the angular distribution, where E is the proton (or antiproton) c.o.m. energy, can be written in the form

$$\left.\frac{d\sigma}{d\Omega}\right|_{p\bar{p}} = \bar{A}_0(E) + \bar{A}_1(E)\cos\theta + \bar{A}_2(E)\cos^2\theta. \tag{6.91}$$

The $\bar{A}_i$'s are arbitrary complex functions of E.

However, since parity is conserved in the reaction, the differential cross section must be invariant under the transformation

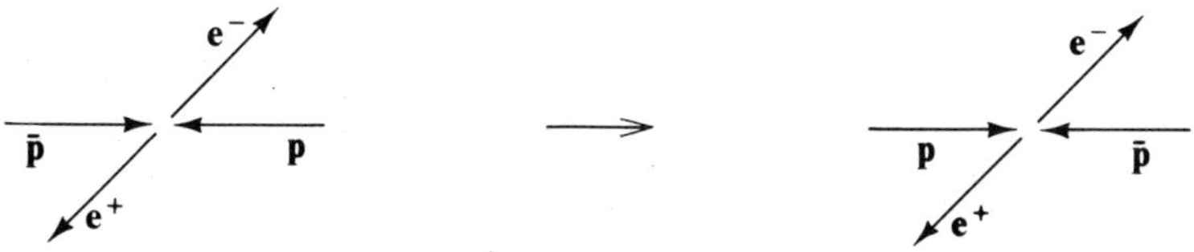

28. For a more rigorous discussion of this problem, see M. Gourdin and A. Martin, *Analysis of Reaction Cross Sections in Partial Waves of a Crossed Channel*, CERN/4804/Th.261, 1962.

That is, the differential cross section must be invariant under $\theta \longrightarrow \pi - \theta$, which means that $\bar{A}_1(E) = 0$, so that[29]

$$\left.\frac{d\sigma}{d\Omega}\right|_{p\bar{p}} = \bar{A}_0(E) + \bar{A}_2(E)\cos^2\theta. \tag{6.92}$$

We now wish to re-express $\cos\theta$ in terms of the physical variables appropriate to electron-proton scattering in, say, the lab system. Thus

$$\cos\theta = \frac{2M(E + E')}{\sqrt{(q^2 + 4M^2)q^2}}. \tag{6.93}$$

The variable t, which is the total energy for $p\bar{p}$ annihilation, is the negative of the invariant 4-momentum transfer for electron scattering. To finish the discussion, we must re-express $E + E'$ in terms of the laboratory scattering angle for electron scattering. From the kinematics in the lab system, with $m_e = 0$, we easily show that

$$q^2 = 4EE' \sin^2\left(\frac{\theta}{2}\right) \tag{6.94}$$

and

$$\frac{E'}{E} = \frac{1}{1 + \dfrac{2E}{M}\sin^2\left(\dfrac{\theta}{2}\right)}. \tag{6.95}$$

Thus

$$(E - E') = \frac{q^2}{2M} \tag{6.96}$$

and

$$\operatorname{ctn}^2\left(\frac{\theta}{2}\right) = \frac{4EE' - q^2}{q^2}, \tag{6.97}$$

or

$$(E + E')^2 = \frac{q^2(q^2 + 4M^2)}{4M^2} + q^2\operatorname{ctn}^2\left(\frac{\theta}{2}\right). \tag{6.98}$$

It follows that $\left.\dfrac{d\sigma}{d\Omega}\right|_{\text{lab}}$, for electron scattering, is of the form

---

29. The explicit computation of this cross section, in the $p\bar{p}$ c.o.m. (see A. Zichici et al., *Nuovo Cimento*, **24**:170 (1962)) gives, in terms of $\bar{F}_1$ and $\bar{F}_2$,

$$\frac{d\sigma}{d\Omega_{p\bar{p}}} = \frac{\pi}{8}\frac{\alpha^2}{E(E^2 - M^2)^{1/2}}\left[|\bar{F}_1(E) + 2M\bar{F}_2(E)|^2(1 + \cos^2\theta) + \left|\frac{M}{E}\bar{F}_1(E) + 2E\bar{F}_2(E)\right|^2 \sin^2\theta\right],$$

which is clearly of the general form given above.

$$\left.\frac{d\sigma}{d\Omega}\right|_{\text{lab}} \propto A(q^2) + B(q^2)\,\text{ctn}^2\left(\frac{\theta}{2}\right), \tag{6.99}$$

where the proportionality involves a kinematical factor that depends on the transformation from the $p\bar{p}$ center of mass system to the laboratory system in *ep* scattering. The fact that a formula of this form holds up to the highest q^2 yet measured means that the one-photon approximation is still adequate at these momentum transfers.

We now turn to another and perhaps more sophisticated method of analyzing electron-proton scattering (and many other processes involving currents): the dispersion relations.

7
Dispersion Relations and Electron Scattering

The simpleminded vector meson model, with constant-vector meson-nucleon form factors, has led us to form factors of the equivalent forms

$$F(q^2) = B + \frac{q^2 A}{q^2 + m_V^2} = B' + \frac{A'}{q^2 + m_V^2}, \tag{7.1}$$

where the equivalence is given by the relations

$$\frac{A'}{m_V^2} + B' = B, \qquad A + B = B'. \tag{7.2}$$

The behavior of $F(q^2)$ as $q^2 \longrightarrow \infty$ depends on the relations among the constants. Of course, in reality, the physics of the situation at $q^2 \longrightarrow \infty$ may be completely different from that inferred from the experiments done so far at finite, and relatively small, q^2, and may have nothing to do with the behavior at infinity suggested by the model. However, if we proceed naively we may try to use the model and perturbation theory, more generally, to guess at the analytic structure of the F's. Letting $q^2 = z$, a complex variable, we can summarize the analytic properties of the F's in the model as follows.

1. $F(z)$ is analytic in the upper half-plane.
2. $F(z)/z \longrightarrow 0$ in the upper half-plane. (Of course, with suitable A and B, $F(z)$ might itself go to zero in the upper half-plane but we take care of all contingencies by dividing $F(z)$ by z.)
3. At $z = -m_V^2$, $F(z)$ has a pole.

Although we might imagine that 1 and 2 could persist as general properties of $F(z)$, the pole condition, given by 3, is clearly a very special feature of the vector meson model. Among other things, it completely ignores the nonresonant contribution of the pion graphs, which will alter the analytic character of the F's. Indeed, we may get an insight into the more general analytic character of the F's by the following heuristic considerations.[1] Consider, for example, the

1. A reader who would like to see a discussion of this question from the point of view of perturbation theory will find a clear one in S. D. Drell and F. Zachariasen, *Electromagnetic Structure of Nucleons*, Oxford University Press, 1961.
2. As we shall see below, $A(m^2)$ need not be positive. For example, if there are several vector mesons at various masses, then $A(m^2)$ evaluated at these masses is related to the coupling constants of these vector mesons to the nucleons and these coupling constants can have either sign.

pionic contribution to $F^V(q^2)$, which comes from $J = 1, 2\pi$ graphs. The minimum effective invariant mass such a 2π state can have is $m^2 = 4m_\pi^2$ and the maximum is $+\infty$, corresponding to two pions at infinite energy. In analogy to the case in which the two π's can actually be taken to be a stable vector meson, we might expect that we could write the general 2π contribution to $F^V(q^2)$ as a weighted sum over the emissions of all $J = 1, 2\pi$ states with effective masses satisfying $4m_\pi^2 \le m^2 \le \infty$. Thus we would expect a form like

$$F^V(q^2) = \sum_{m^2} \frac{A(m^2)}{q^2 + m^2},$$

where $A(m^2)$ is related to the matrix element[2] for the emission by, say, the $p\bar{p}$ system of 2π's in a $J = 1$ state with an effective mass m^2. In particular, if $A(m^2) = g\delta(m^2 - m_V^2)$, we return to the "pole" model for $F^V(q^2)$. Thus we would expect that $F^V(z)$ has, in general, a line of singularities extending from $q^2 = -4m_\pi^2$ to $q^2 = -\infty$. This conclusion is also borne out in the perturbation theory. A similar argument holds for $F^S(q^2)$, which picks up contributions from the 3π graphs. We may argue that it has a line of singularities, a branch cut, from $q^2 = -9m_\pi^2$ to $q^2 = -\infty$.[3] Thus we may amend statement 3 above to read

3′. $F(z)$ has a branch cut on the negative real axis.

We may now examine what general structure a function satisfying 1, 2, 3′ has—a consideration that will lead us to the dispersion relations. Let us begin by considering the case in which $F(z) \longrightarrow 0$ in the upper half-plane (UHP). Later we can easily extend the argument to cover the case in which $F(z) \longrightarrow C \neq 0$, so that $F(z)/z \longrightarrow 0$ and, more generally, to the case in which $F(z)/z^n \longrightarrow 0$.

We proceed by considering a contour Γ in the upper half-plane as shown below.

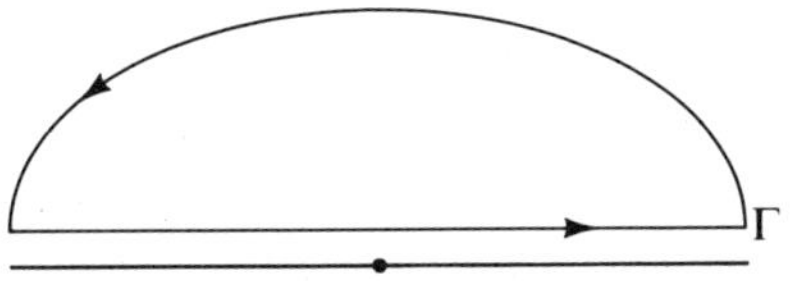

The fact that $F(z)$ is singular along the negative real axis means that we must stay away from this axis by some infinitesimal but positive amount. By Cauchy's theorem, under the assumptions listed above,

$$F(z) = \frac{1}{2\pi i} \oint_\Gamma \frac{F(z')}{z' - z}\, dz'. \tag{7.3}$$

3. The pions are the lightest strongly interacting particles. Since there are no strongly interacting vector mesons with masses less than $m = 3m_\pi$, the lower limit on the branch cut is determined by the π meson masses.
4. For $q^2 < 0$, $F(q^2)$ does not necessarily exist and for $q^2 > 0$ the integral is defined by a limiting process in which the contour is allowed to approach the real axis from above. The procedure we follow here is hardly very rigorous but it enables us to indicate the origin of known and widely used results.

Thus for $z = q^2 + i\epsilon$, with $\epsilon \longrightarrow 0$,[4]

$$F(q^2) = \lim_{\epsilon \to 0} \frac{1}{2\pi i} \int_{-\infty}^{\infty} \frac{F(q'^2)\, dq'^2}{q'^2 - q^2 - i\epsilon}. \tag{7.4}$$

We may now apply the identity (to be understood as taken under an integration)

$$\lim_{\epsilon \to 0} \frac{1}{x - i\epsilon} = P\frac{1}{x} + i\pi\, \delta(x), \tag{7.5}$$

where $P(1/x)$ is a shorthand for the principal part integral. Thus

$$F(q^2) = \frac{F(q^2)}{2} + \frac{1}{2\pi i} P \int_{-\infty}^{\infty} \frac{F(q'^2)}{q'^2 - q^2}\, dq'^2 \tag{7.6}$$

or, remembering that from Hermiticity and current conservation (or time reversal), it follows that $\text{Im}\, F(q^2) = 0$ for $q^2 \geq 0$, and

$$\begin{aligned} F(q^2) &= \frac{1}{\pi} P \int_{-\infty}^{\infty} \frac{\text{Im}\, F(q'^2)\, dq'^2}{q'^2 - q^2} = \frac{1}{\pi} P \int_{-\infty}^{0} \frac{\text{Im}\, F(q'^2)}{q'^2 - q^2}\, dq'^2 \\ &= \frac{1}{\pi} \int_{-\infty}^{\infty} \frac{\text{Im}\, F(q'^2)\, dq'^2}{q'^2 - q^2 - i\epsilon}. \end{aligned} \tag{7.7}$$

For electron scattering, $q^2 \geq 0$, the dispersion relations are always used in this form. However, it is possible to define $\text{Im}\, F(q^2)$ unambiguously on the real line so that the integral can be taken on the real line instead of resorting to a limiting process with the contour. Indeed, consider,

$$\begin{aligned} &\lim_{\epsilon \to 0} [F(q^2 + i\epsilon) - F(q^2 - i\epsilon)] \\ &\qquad = \frac{1}{2\pi i} \lim_{\epsilon \to 0} \int_{-\infty}^{\infty} F(q'^2) \left[\frac{1}{q'^2 - q^2 - i\epsilon} - \frac{1}{q'^2 - q^2 + i\epsilon}\right] \\ &\qquad = 2i\, \text{Im}\, F(q^2). \end{aligned} \tag{7.8}$$

Equation 7.8 can be taken as the definition of the imaginary part in general. Thus we see that the existence of an imaginary part for $F(z)$ is related to the development of the branch cut in $F(z)$ along the negative real axis. The existence of this branch cut is a consequence of the fact that the particles are in interaction, since we have seen that for free particles the form factors in a matrix element such as $\langle 0|J_\mu|\bar{\mathbf{p}}\mathbf{p}\rangle$ are, as can be seen from time-reversal invariance, real for *all* values of q^2. At this point the reader may be somewhat puzzled about the analytic structure of the "pole" model in which the form factors have the form

$$F(q^2) = a + \frac{b}{q^2 + m_V{}^2}.$$

The constants a and b must be real for $q^2 \geq 0$, so it would appear that $F(q^2)$ is everywhere real and has an actual pole on the real axis, a pole which could, in principle, be in the physical region for $p\bar{p}$ annihilation if the vector mesons were heavy enough. In actuality, this form for F can only hold far away from the pole. Indeed, if the vector meson is regarded as a metastable multipion state, a resonance, then, for $q^2 < 0$, $F(q^2)$ is a complex function[5] involving the pion-pion interaction. Even in the "pole" approximation we must, for q^2 in the neighborhood of the vector meson mass, take into account the fact that the resonance has a width. This adds an imaginary part to the denominator and removes the pole from the real axis. Only far away from the resonance, $q^2 \geq 0$, can this function be approximated by a pole with a real residue.

Before turning to the question of how Im $F(q^2)$ is determined in terms of matrix elements, we extend the discussion above to the case in which it is $F(z)/z$ that goes to zero at infinity. Using the same contour as above we write

$$\frac{F(z)}{z} = \frac{1}{2\pi i}\oint_\Gamma \frac{F(z')}{z'(z' - z)}\, dz'. \tag{7.9}$$

Thus

$$F(q^2) = \frac{q^2}{2\pi i}\int_{-\infty}^{\infty} \frac{F(q'^2)}{(q'^2 - q^2 - i\epsilon)(q'^2 + i\epsilon)}\, dq'^2. \tag{7.10}$$

By manipulating the denominator and using the identity above, we find that

$$F(q^2) = F(0) + \frac{q^2}{\pi} P \int_{-\infty}^{\infty} \frac{\text{Im}\, F(q'^2)}{(q'^2 - q^2)q'^2}\, dq'^2. \tag{7.11}$$

This "subtracted" dispersion relation implies that $F(q^2) \longrightarrow$ const as $q^2 \longrightarrow \infty$. The experiments argue against this, and if the present trend continues for large q^2 then the "unsubtracted" dispersion relations would be favored.

The dispersion relations alone merely express the analytical structure of $F(q^2)$ but do not determine the function explicitly. We must supplement them in the electron scattering case with an independent evaluation of Im $F(q^2)$. In doing this we proceed in several steps.

We begin by making a number of mathematical definitions.

1. We define a function $\epsilon(x)$ such that

$$\epsilon(x) = \begin{cases} 1, & x \geq 0, \\ -1, & x < 0, \end{cases} \tag{7.12}$$

and $\theta(x)$ such that

$$\theta(x) = \begin{cases} 1, & x \geq 0, \\ 0, & x < 0. \end{cases} \tag{7.13}$$

Thus

$$\theta(x) = \frac{1 + \epsilon(x)}{2}. \tag{7.14}$$

If $A(t)$ and $B(t)$ are any[6] operator functions of t, we define P, the "time-ordered" product of A and B, by the equation

---

5. See, for example, W. R. Frazer and J. R. Fulco, *Phys. Rev. Letters* **2**:365 (1959).
6. They need not commute or anticommute.

$$P(A(t)B(t')) = \begin{cases} A(t)B(t'), & t \geq t', \\ B(t')A(t), & t < t'. \end{cases} \tag{7.15}$$

Thus

$$P(A(t)B(t')) = \frac{[A(t), B(t')]_+}{2} + \frac{\epsilon(t-t')}{2}[A(t), B(t')]_-. \tag{7.16}$$

We use these definitions and the general construction of the one-particle states to derive a compact formula that relates a matrix element like $\langle \mathbf{p}'|O(x)|\mathbf{p}\rangle$ ($O(x)$ is a general operator function of the four-position x) to a matrix element involving $|0\rangle$, the vacuum state, on either side. We give the spin-$\frac{1}{2}$ case as an illustrative example, but the same method applies to states of arbitrary spin. Thus, if ψ is the interacting Heisenberg field,

$$\langle \mathbf{p}'|_{s'} O(x)|\mathbf{p}\rangle_s = \lim_{t\to\pm\infty} \int d\mathbf{r}' \, \langle \mathbf{p}'|_{s'} O(x)\bar{\psi}(\mathbf{r}', t)\gamma_4 f_{\mathbf{p}}(x')_s|0\rangle, \tag{7.17}$$

with

$$f_{\mathbf{p}}(x)_s = \frac{e^{i(px)}u(\mathbf{p})_s}{(2\pi)^{3/2}}. \tag{7.18}$$

Clearly, if we call m the physical nucleon mass,[7]

$$(\partial_\mu \gamma_\mu + m) f_{\mathbf{p}}(x)_s = 0. \tag{7.19}$$

This limiting process can be expressed in a manifestly covariant form by introducing the four-dimensional spacelike surface element vector $d\sigma_\mu$ (Eq. (1.16)). (In the formulae derived in previous chapters we have made use of the special spacelike surfaces at constant time.) Thus we may write

$$\langle \mathbf{p}'|_{s'} O(x)|\mathbf{p}\rangle_s = \lim_{\sigma\to-\infty} \int d\sigma_\mu' \, \langle \mathbf{p}'|_{s'} O(x)\bar{\psi}(x')\gamma_\mu f_{\mathbf{p}}(x')_s|0\rangle, \tag{7.20}$$

where x' is a point located on σ'.

If $A(x')$ is any operator function of x' and d^4x' is the four-dimensional volume element, we may, by integrating over all four-space and using the four-dimensional Gauss theorem, as well as the Dirac equation applied to $f_{\mathbf{p}}(x')_s$, derive the identity

$$\int d^4x' \, (-\partial_\mu' A(x')\gamma_\mu + A(x')m) f_{\mathbf{p}}(x')_s$$
$$= \int_{\sigma'\to-\infty} A(x')\gamma_\mu f_{\mathbf{p}}(x')_s \, d\sigma_\mu' - \int_{\sigma'\to+\infty} A(x')\gamma_\mu f_{\mathbf{p}}(x')_s \, d\sigma_\mu'. \tag{7.21}$$

If we denote by

$$A(x')\overleftarrow{\partial_\mu}' = \partial_\mu' A(x'), \tag{7.22}$$

7. We have used m for various masses in the text. When there is some chance of confusion we use different subscripts to distinguish the masses. In general, we try not to clutter the notation with festoons of subscripts.

we may use Equation (7.21) and the work above to show, where P is the time-ordering symbol, that

$$\int d^4x'\, P(\langle \mathbf{p}'|_{s'} O(x)\bar{\psi}(x')|0\rangle(-\overleftarrow{\partial}_\mu{}'\gamma_\mu + m) f_{\mathbf{p}}(x')_s$$
$$= -\int_{\sigma'\to+\infty} d\sigma_\mu{}'\, \langle \mathbf{p}'|_{s'}\bar{\psi}(x')O(x)|0\rangle\gamma_\mu f_{\mathbf{p}}(x')_s + \langle \mathbf{p}'|_{s'}O(x)|\mathbf{p}\rangle_s. \quad (7.23)$$

We first deal with the first term in the right side of Eq. (7.23). The momentum in $f_{\mathbf{p}}(x')_s$ and the limiting process pick out of $\bar{\psi}(x)$ the Fourier component corresponding to one physical nucleon. When acting to the left on $\langle \mathbf{p}'|_{s'}$, $\bar{\psi}$ becomes its Hermitian conjugate, which annihilates physical nucleons. Thus this term is proportional to

$$\delta^3(\mathbf{p} - \mathbf{p}')\langle 0|O(x)|0\rangle. \quad (7.24)$$

In most cases of interest to us the $O(x)$ are such that

$$\langle 0|O(x)|0\rangle = \langle 0|O(0)|0\rangle = 0. \quad (7.25)$$

For example, if $O(0) = J_\mu{}^\gamma(0)$, the electromagnetic current, then

$$\langle 0|J_\mu{}^\gamma(0)|0\rangle = 0 \quad (7.26)$$

for any number of reasons; for example, using charge conjugation,

$$\langle 0|J_\mu{}^\gamma(0)|0\rangle = -\langle 0|J_\mu{}^\gamma(0)|0\rangle. \quad (7.27)$$

Thus if we restrict our choice of O's to have this property, letting $x = 0$ we have the identity

$$\langle \mathbf{p}'|_{s'}O(0)|\mathbf{p}\rangle_s = \int d^4x\, P[\langle \mathbf{p}'|_{s'}O(0)\bar{\psi}(x)|0\rangle](-\overleftarrow{\partial}_\mu\gamma_\mu + m) f_{\mathbf{p}}(x)_s. \quad (7.28)$$

We may carry out the differentiation to the left explicitly and, noting that

$$\frac{d}{dt}\epsilon(t) = 2\delta(t), \quad (7.29)$$

and using Eq. (7.16)

$$\partial_\mu P[O(0)\bar{\psi}(x)] = P[O(0)\partial_\mu\bar{\psi}(x)] - \delta(t)\delta_{\mu 4}[O(0), \bar{\psi}(x)]_-. \quad (7.30)$$

We define the "nucleon source" $j_N(x)$ by the equation

$$(\partial_\mu\gamma_\mu + m)\psi(x) = j_N(x). \quad (7.31)$$

In a classical field theory, $j_N(x)$ would play the same role as the charge density in the Poisson equation. For a free field, $j_N(x) = 0$, and

$$\bar{\psi}(x)(-\overleftarrow{\partial}_\mu\gamma_\mu + m) = \bar{j}_N(x). \quad (7.32)$$

Thus

$$\langle \mathbf{p}'|_{s'}O(0)|\mathbf{p}\rangle_s = \int d^4x\, \langle \mathbf{p}'|_{s'}P[O(0)\bar{j}_N(x)]|0\rangle f_{\mathbf{p}}(x)_s$$
$$- \int d^4x \langle \mathbf{p}'|_{s'}[O(0), \bar{\psi}(x)]_-|0\rangle f_{\mathbf{p}}(x)_s\, \delta(t)\gamma_4. \quad (7.33)$$

This is as far as we can go without knowing the equal-time commutator between $O(0)$ and $\bar{\psi}$. In many cases this commutator is either zero or a relatively simple function of the fields. We will now consider as a fundamental example

$$O_\mu(0) = J_\mu{}^\gamma(0), \tag{7.34}$$

the electromagnetic current. Using the equal-time commutation relations among the fields to evaluate expressions such as

$$[\bar{\psi}(0)\gamma_\mu\psi(0), \bar{\psi}(\mathbf{r}, 0)]_-,$$

we arrive at the relation[8]

$$[J_\mu{}^\gamma(0), \bar{\psi}(\mathbf{r}, 0)]_- = i\bar{\psi}(\mathbf{r}, 0)\gamma_4\gamma_\mu\delta^3(\mathbf{r}). \tag{7.35}$$

Thus

$$\int d^4x \langle \mathbf{p}'|_{s'}[J_\mu{}^\gamma(0), \bar{\psi}(\mathbf{r}, 0)]_-|0\rangle f_{\mathbf{p}}(x)_s\,\delta(t) = \frac{i}{(2\pi)^{3/2}}\langle \mathbf{p}'|_{s'}\psi^\dagger(0)|0\rangle\gamma_\mu u(\mathbf{p})_s. \tag{7.36}$$

We may now consider

$$\langle 0|\psi(0)|\mathbf{p}\rangle_s = e^{-i(px)}\langle 0|\psi(x)|\mathbf{p}\rangle_s. \tag{7.37}$$

Since $p^2 = -m^2$, the physical nucleon mass,[9] we have

$$(\Box^2 - m^2)\langle 0|\psi(x)|\mathbf{p}\rangle_s = 0. \tag{7.38}$$

Thus $\langle 0|\psi(0)|\mathbf{p}\rangle_s$, which is, essentially, the Fourier transform of $\langle 0|\psi(x)|\mathbf{p}\rangle_s$, satisfies

$$(p^2 + m^2)\langle 0|\psi(0)|\mathbf{p}\rangle_s = 0. \tag{7.39}$$

From the transformation properties of ψ and $|\mathbf{p}\rangle_s$ under the Lorentz group, it follows that

$$\langle 0|\psi(0)|\mathbf{p}\rangle_s = Nu(\mathbf{p})_s, \tag{7.40}$$

where N is an undetermined Lorentz scalar; a constant. Thus

$$(i\gamma p + m)\langle 0|\psi(0)|\mathbf{p}\rangle_s = 0. \tag{7.41}$$

Therefore we may write

$$\int d^4x \langle \mathbf{p}'|_{s'}[J_\mu{}^\gamma(0), \bar{\psi}(\mathbf{r}, 0)]_- f_{\mathbf{p}}(x)_s\,\delta(t) = N' i\bar{u}(\mathbf{p}')_{s'}\gamma_\mu u(\mathbf{p})_s, \tag{7.42}$$

where N' is another constant. Hence we have the important identity

$$\langle \mathbf{p}'|_{s'}J_\mu{}^\gamma(0)|\mathbf{p}\rangle_s + iN'\bar{u}(\mathbf{p}')_{s'}\gamma_\mu u(\mathbf{p})_s = \int d^4x \langle \mathbf{p}'|_{s'}P[J_\mu{}^\gamma(0)\bar{j}_N(x)]|0\rangle f_{\mathbf{p}}(x)_s. \tag{7.43}$$

But we know that the general covariant structure of $\langle \mathbf{p}'|_{s'}J_\mu{}^\gamma(0)|\mathbf{p}\rangle_s$ is

---

8. $J_\mu{}^\gamma$ must be the "minimal" current, with no Pauli terms; otherwise Equation (7.35) isn't true.
9. We assume that $|\mathbf{p}\rangle_s$ represents one *physical* nucleon.

$$\langle \mathbf{p}'|_{s'} J_\mu{}^\gamma(0)|\mathbf{p}\rangle_s = i\bar{u}(\mathbf{p}')_{s'}[\gamma_\mu F_1(q^2) + \sigma_{\mu\nu}q_\nu F_2(q^2)]u(\mathbf{p})_s. \tag{7.44}$$

Thus we may redefine $F_1(q^2)$ so that

$$F'_1(q^2) = F_1(q^2) + N', \tag{7.45}$$

and identify $F'_1(0)$ with the observed electric charge, which enters the theory as a parameter that must be determined from experiment in any case. With this understanding we may write

$$\langle \mathbf{p}'|_{s'} J_\mu{}^\gamma(0)|\mathbf{p}\rangle_s = \int d^4x\, \langle \mathbf{p}'|_{s'} P[J_\mu{}^\gamma(0)\bar{j}_N(x)]|0\rangle f_{\mathbf{p}}(x)_s. \tag{7.46}$$

This expression may be reduced further as follows.

First, we give an alternate definition of the P symbol in terms of the function $\theta(t)$,

$$P[A(0)B(t)] = \theta(t)B(t)A(0) + \theta(-t)A(0)B(t), \tag{7.47}$$

so that, writing out $f_{\mathbf{p}}(x)_s$ explicitly,

$$\langle \mathbf{p}'|_{s'} J_\mu{}^\gamma(0)|\mathbf{p}\rangle_s = \int d^4x \frac{e^{i(px)}}{(2\pi)^{3/2}} \langle \mathbf{p}'|_{s'}[\theta(t)\bar{j}_N(x)J_\mu{}^\gamma(0) + \theta(-t)J_\mu{}^\gamma(0)\bar{j}_N(x)]|0\rangle u(\mathbf{p})_s. \tag{7.48}$$

We may next make use of the assumption that the eigenstates of P_μ the energy-momentum operator form a complete set. This assumption is fundamental to the theory of elementary particles and we make it throughout the discussion that follows. It means that any matrix element of the form $\langle a|AB|b\rangle$, where $\langle a|$ and $|b\rangle$ are arbitrary states and A and B are *any* operators can be written as (the sum is over all the eigenstates of P_μ; $|n\rangle$)

$$\langle a|AB|b\rangle = \sum_n \langle a|A|n\rangle\langle n|B|b\rangle. \tag{7.49}$$

Thus

$$\begin{aligned}\langle \mathbf{p}'|_{s'} J_\mu{}^\gamma(0)|\mathbf{p}\rangle_s &= \int d^4x \frac{e^{i(px)}}{(2\pi)^{3/2}} \sum_n \{\theta(t)\langle \mathbf{p}'|_{s'}\bar{j}_N(x)|n\rangle\langle n|J_\mu{}^\gamma(0)|0\rangle \\ &\quad + \theta(-t)\langle \mathbf{p}'|_{s'}J_\mu{}^\gamma(0)|n\rangle\langle n|\bar{j}_N(x)|0\rangle\}\, u(\mathbf{p})_s \\ &= \int \frac{d^4x}{(2\pi)^{3/2}} \{\sum_n \theta(t)e^{i((p'-p+p_n)x)}\langle \mathbf{p}'|_{s'}\bar{j}_N(0)|n\rangle\langle n|J_\mu{}^\gamma(0)|0\rangle \\ &\quad + \sum_n \theta(-t)\langle \mathbf{p}'|_{s'}J_\mu{}^\gamma(0)|n\rangle\langle n|\bar{j}_N(0)|0\rangle e^{i((p-p_n)x)}\}\, u(\mathbf{p})_s\end{aligned} \tag{7.50}$$

We do the space and time integrals, in order, using the two well-known identities

$$\int e^{i\mathbf{p}\cdot\mathbf{r}}\, d\mathbf{r} = (2\pi)^3\, \delta^3(\mathbf{p}) \tag{7.51}$$

and

$$\int_0^\infty dt\, e^{i\alpha t} = \lim_{\epsilon\to 0} \frac{i}{\alpha + i\epsilon}, \tag{7.52}$$

and remembering that $dx_4 = i\,dt$. Thus ($\lim_{\epsilon\to 0}$ understood)

$$\langle \mathbf{p}'|_{s'} J_\mu{}^\gamma(0)|\mathbf{p}\rangle_s = (2\pi)^{3/2} \sum_n \left\{ \frac{\delta^3(\mathbf{p}_n + \mathbf{p}' - \mathbf{p})\langle \mathbf{p}'|_{s'} \bar{j}_N(0)|n\rangle\langle n|J_\mu{}^\gamma(0)|0\rangle}{E(\mathbf{p}') + E(\mathbf{p}_n) - E(\mathbf{p}) - i\epsilon} \right.$$
$$\left. - \frac{\langle \mathbf{p}'|_{s'} J_\mu{}^\gamma(0)|n\rangle\langle n|\bar{j}_N(0)|0\rangle}{E(\mathbf{p}) - E(\mathbf{p}_n) + i\epsilon} \delta^3(\mathbf{p} - \mathbf{p}_n) \right\} u(\mathbf{p})_s \quad (7.53)$$

This formula expresses the matrix element for electron scattering in terms of a perturbation theory-like sum over intermediate states $|n\rangle$. It gives us a rigorous justification of the method that we used earlier in which we made use of the expression $\langle n|J_\mu{}^\gamma(0)|0\rangle$ and the symmetries of $J_\mu{}^\gamma(0)$ to determine the states $\langle n|$, which contribute to the scattering process. We shall shortly use this formula to compute the imaginary parts of the form factors and we now prove that the terms in the sum involving $\langle \mathbf{p}'|_{s'} J_\mu{}^\gamma(0)|n\rangle$ do not contribute to them. In fact, as we have shown, the imaginary parts of the F's are determined by the singularities in $\langle \mathbf{p}'|_{s'} J_\mu{}^\gamma(0)|\mathbf{p}\rangle_s$ for $q^2 < 0$. Hence the sum over states will only contribute to Im F if it becomes singular for some value of q^2. We will now show that

$$\sum_n \frac{\langle \mathbf{p}'|_{s'} J_\mu{}^\gamma(0)|n\rangle\langle n|\bar{j}_N(0)|0\rangle}{E(\mathbf{p}) - E(\mathbf{p}_n) + i\epsilon} \delta^3(\mathbf{p} - \mathbf{p}_n)u(\mathbf{p})_s \quad (7.54)$$

is never singular; that is, if the denominator vanishes, so does the numerator. But the denominator vanishes only if

$$E(\mathbf{p}) = E(\mathbf{p}_n),$$
$$\mathbf{p} = \mathbf{p}_n, \quad (7.55)$$
$$p^2 = p_n{}^2 = -m^2,$$

which means, since the nucleon is the only state with the nucleon mass, that the denominator vanishes only if $|n\rangle$ in Eq. (7.54) is a one-nucleon state. However, we have just argued that $\langle \mathbf{p}|_s \bar{\psi}(x)|0\rangle$ satisfies the free-particle Dirac equation for a nucleon with its physical mass. Thus, for this state,

$$\langle \mathbf{p}|_s \bar{j}_N(0)|0\rangle = 0, \quad (7.56)$$

and the sum is nonsingular. We may now turn to

$$\sum_n \left\{ \delta^3(\mathbf{p}_n + \mathbf{p}' - \mathbf{p}) \frac{\langle \mathbf{p}'|_{s'} \bar{j}_N(0)|n\rangle\langle n|J_\mu{}^\gamma(0)|0\rangle}{E(\mathbf{p}') + E(\mathbf{p}_n) - E(\mathbf{p}) - i\epsilon} \right\}. \quad (7.57)$$

Here, the denominator vanishes only if

$$E(\mathbf{p}_n) = E(\mathbf{p}) - E(\mathbf{p}'),$$
$$\mathbf{p}_n = \mathbf{p} - \mathbf{p}'. \quad (7.58)$$

But

$$p' - p = q, \quad (7.59)$$

where q is the four-momentum transfer in electron scattering. Thus the denominator vanishes for all

$$q^2 = p_n{}^2 - E(\mathbf{p}_n)^2 < 0. \quad (7.60)$$

Hence, unless the terms in the numerator vanish for certain special values of q, the sum is singular for this range of q^2 and hence the F's will have imaginary parts in this range. But it is precisely for $q^2 < 0$ that the time-reversal reality condition no longer applies. Hence it is consistent with the time-reversal character of the theory that this sum is singular. Indeed we may now use the relation

$$\lim_{\epsilon\to 0} [F(q^2 + i\epsilon) - F(q^2 - i\epsilon)] = 2i \operatorname{Im} F(q^2) \tag{7.61}$$

for the F's to compute their imaginary parts. For (neglecting the second sum, which does not contribute to the singularity)

$$\begin{aligned}\langle \mathbf{p}'|_{s'}J_\mu{}^\gamma(0)|\mathbf{p}\rangle_{s+i\epsilon} &= i\bar{u}(\mathbf{p}')_{s'}[\gamma_\mu F_1(q^2 + i\epsilon) + \sigma_{\mu\nu}q_\nu F_2(q^2 + i\epsilon)]u(\mathbf{p})_s \\ &= \sum_n (2\pi)^{3/2}\delta^3(\mathbf{p}_n - \mathbf{q} - i\epsilon)\frac{\langle \mathbf{p}'|_{s'}\bar{j}_N(0)|n\rangle\langle n|J_\mu{}^\gamma(0)|0\rangle u(\mathbf{p})_s}{E(\mathbf{p}_n) - (E(\mathbf{p}) - E(\mathbf{p}')) - i\epsilon}\end{aligned} \tag{7.62}$$

and

$$\begin{aligned}\langle \mathbf{p}'|_{s'}J_\mu{}^\gamma(0)|\mathbf{p}\rangle_{s-i\epsilon} &= i\bar{u}(\mathbf{p}')_{s'}[\gamma_\mu F_1(q^2 - i\epsilon) + \sigma_{\mu\nu}q_\nu F_2(q^2 - i\epsilon)]u(\mathbf{p})_s \\ &= \sum_n (2\pi)^{3/2}\frac{\delta^3(\mathbf{p}_n - \mathbf{q} + i\epsilon)\langle \mathbf{p}'|_{s'}\bar{j}_N(0)|n\rangle\langle n|J_\mu{}^\gamma(0)|0\rangle u(\mathbf{p})_s}{E(\mathbf{p}_n) - (E(\mathbf{p}) - E(\mathbf{p}')) + i\epsilon}.\end{aligned} \tag{7.63}$$

Thus

$$\begin{aligned}&\lim_{\epsilon\to 0}[\langle \mathbf{p}'|_{s'}J_\mu{}^\gamma(0)|\mathbf{p}\rangle_{s+i\epsilon} - \langle \mathbf{p}'|_{s'}J_\mu{}^\gamma(0)|\mathbf{p}\rangle_{s-i\epsilon}] \\ &\quad = \lim_{\epsilon\to 0}(2\pi)^{3/2}\sum_n \delta^3(\mathbf{p}_n - \mathbf{q})\langle \mathbf{p}'|_{s'}\bar{j}_N(0))|n\rangle\langle n|J_\mu{}^\gamma(0)|0\rangle u(\mathbf{p})_s \\ &\quad\quad \times\left[\frac{1}{E(\mathbf{p}_n) - (E(\mathbf{p}) - E(\mathbf{p}')) - i\epsilon} - \frac{1}{E(\mathbf{p}_n) - (E(\mathbf{p}) - E(\mathbf{p}')) + i\epsilon}\right].\end{aligned} \tag{7.64}$$

We may evaluate the limit by using the identity

$$\lim_{\epsilon\to 0}\frac{1}{x - i\epsilon} = P\frac{1}{x} + i\pi\,\delta(x) \tag{7.65}$$

and arrive at the identity

$$\begin{aligned}&2i\bar{u}(\mathbf{p}')_{s'}[\gamma_\mu \operatorname{Im} F_1(q^2) + \sigma_{\mu\nu}q_\nu \operatorname{Im} F_2(q^2)]u(\mathbf{p})_s \\ &\quad = (2\pi)^{5/2}\sum_n \delta^4(p_n - q)\langle \mathbf{p}'|_{s'}\bar{j}_N(0)|n\rangle\langle n|J_\mu{}^\gamma(0)|0\rangle u(\mathbf{p})_s.\end{aligned} \tag{7.66}$$

We may now recover the results of the pole approximation[10] by choosing for $|n\rangle$ one of the vector meson states. We label this state $|\mathbf{v}\rangle_\lambda$, where λ refers to the spin polarization.[11] We may make a general analysis of the matrix elements $\langle \mathbf{v}|_\lambda J_\mu{}^\gamma(0)|0\rangle$ and $\langle \mathbf{p}'|_{s'}\bar{j}_N(0)|\mathbf{v}\rangle_\lambda$. Thus (no sum on λ)

$$\langle \mathbf{v}|_\lambda J_\mu{}^\gamma(0)|0\rangle = \epsilon_\lambda[\delta_{\lambda\mu}\rho_1(v^2) + v_\mu\rho_2(v^2)]. \tag{7.67}$$

---

10. For a detailed treatment of many of the other low-mass states that enter this sum and hence the dispersion integral, see P. Federbush et al., *Phys. Rev.* **112**:642 (1958).
11. The state $|\mathbf{v}\rangle_\lambda$ can be created out of the vacuum using the same techniques by which the pion and nucleon states were constructed.

Current conservation yields the condition

$$(v\epsilon)\rho_1(v^2) + v^2\rho_2(v^2) = 0. \tag{7.68}$$

We recall that all of the one-particle states that enter the sum over $|n\rangle$ are "real" in the sense that they are characterized by the actual invariant rest mass of the corresponding particle:

$$p^2 = -m^2. \tag{7.69}$$

Hence the sum, which is a function of q, will, in the approximation in which we neglect everything but the vector meson treated as a stable particle, only contribute when $q^2 = v^2 = -m_V{}^2$. But a real vector meson obeys the transversality condition

$$(v\epsilon) = 0. \tag{7.70}$$

Thus current conservation, in this case, implies that[12]

$$\langle \mathbf{v}|_\lambda J_\mu{}^\gamma(0)|0\rangle = \epsilon_\lambda \delta_{\lambda\mu}\rho_1(-m_V{}^2) \tag{7.71}$$

with no sum on λ.

We may note that this constant $\rho_1(-m_V{}^2)$ will enter in any electromagnetic process of the form

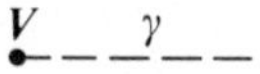

in which the vector meson decays into a single real or virtual photon. Thus $\rho_1(-m_V{}^2)$ can be determined experimentally from processes such as $V \longrightarrow e^+ + e^-$ or $V \longrightarrow u^+ + u^-$ providing that the only coupling of the vector meson to the photon is of this form. We shall return to these decays later when we discuss unitary symmetry and the quark models.

As we have seen in Chapter 6 we may write the nucleon-vector meson coupling Lagrangian in the form (for simplicity we consider an isotopic-scalar vector meson)

$$\mathcal{L}_{VN} = if_1{}^V\bar{\psi}\gamma_\mu\psi V_\mu + f_2{}^V\bar{\psi}\sigma_{\mu\nu}\psi(\partial_\mu V_\nu - \partial_\nu V_\mu) \tag{7.72}$$

This means that the contribution to $\bar{j}_N(0)$ from the vector meson will be of the form $if_1{}^V\bar{\psi}\gamma_\mu V_\mu + f_2{}^V\bar{\psi}\sigma_{\mu\nu}(\partial_\mu V_\nu - \partial_\nu V_\mu)$. Thus we can write

---

12. Note that, if the electric charge operator Q is given by

$$Q = \int d\mathbf{r} J_0{}^\gamma(\mathbf{r}, 0),$$

then

$$0 = \langle \mathbf{v}|_\lambda Q|0\rangle = (2\pi)^3\,\delta^3(\mathbf{v})\langle \mathbf{v}|_\lambda J_0{}^\gamma(0)|0\rangle.$$

This is consistent with the transversality condition, which states that

$$\frac{\boldsymbol{\epsilon}\cdot\mathbf{v}}{v_0} = \epsilon_0$$

or that $\epsilon_0 = 0$ if the vector meson is at rest.

$$\langle \mathbf{p}'|_{s'}\bar{j}_N(0)|\mathbf{v}\rangle_{\lambda_{q^2=-m_V^2}}$$
$$= i\bar{u}(\mathbf{p}')_{s'}[g_1(-m_V^2)\gamma_\lambda + g_2(-m_V^2)\sigma_{\lambda\nu}q_\nu]u(\mathbf{p})_{s_{q^2=-m_V^2}}. \quad (7.73)$$

The g's and the f's need not be the same constants, since these particles have strong interactions that change the magnitude of these constants in the matrix element. We then conclude that in this approximation

$$\text{Im } F_i(q^2) = \delta(q^2 + m_V^2)g_i(-m_V^2)f_i(-m_V^2). \quad (7.74)$$

This expression enters a dispersion integral that, for $q^2 \geq 0$, will be of the form

$$F(q^2) = \frac{1}{\pi}\int_{-\infty}^{0} \frac{\text{Im } F(q'^2)\,dq'^2}{q'^2 - q^2}, \quad (7.75)$$

or, perhaps,

$$F(q^2) = F(0) + \frac{q^2}{\pi}\int_{-\infty}^{0} \frac{\text{Im } F(q'^2)}{(q'^2 - q^2)q'^2}\,dq'^2. \quad (7.76)$$

From the vector meson approximation alone we cannot determine which of these forms, if either, is correct. For this, we need a detailed dynamical theory, or experimental guidance. As remarked in Chapter 6, the F's appear to fall off for large q^2 as $1/q^4$. In the dispersion theory this could be arranged if there were two vector mesons with nearby masses whose coupling constants, the f's and g's, were nearly equal and opposite. In this case F might look like

$$F(q^2) \propto \left[\frac{1}{q^2 + m_V^2} - \frac{1}{q^2 + m_V'^2}\right] = \frac{m_V'^2 - m_V^2}{(q^2 + m_V^2)(q^2 + m_V'^2)} \quad (7.77)$$

for $q^2 > 0$, which is of the right form at large q^2 to agree with experiment. There are two such related isotopic scalar mesons, the ω^0 and ϕ^0 with masses of about 783 and 1020 MeV, so that the isotopic scalar form factor's decrease at large q^2 might be accounted for in this way. However, there is no known counterpart to the isotopic vector ρ meson. Hence, until such a meson turns up, this picture fails to account for the rapid falloff of the isotopic vector form factors. This does not imply a breakdown in the dispersion relations, but it means that the approximation in which they are evaluated solely in terms of stable vector meson intermediate states is too simpleminded, which is, perhaps, not surprising.

We turn in the next chapter to the currents that enter the weak interactions and return to the vector mesons later.

8

The Weak Interactions I: Mostly Leptons

When Fermi invented the first field theoretic treatment of the process $n \longrightarrow p + e^- + \bar{\nu}_e$, ordinary β-decay of the neutron, in the early 1930's, it was natural for him to pattern his treatment after the current couplings of quantum electrodynamics, which were the only elementary particle couplings then known. The closest relevant analogue to the nucleon electromagnetic current is, for β-decay, a current that has the property of transforming neutron into proton—that is, which changes the charge; for example,

$$V_\mu(x) = i\bar{\psi}_P(x)\gamma_\mu\psi_N(x). \tag{8.1}$$

Indeed, Fermi assumed that this current described the nucleonic part of the weak coupling. At that time, of course, it was believed that parity was conserved in the weak interactions, so to make a Lorentz invariant, parity conserving Lagrangian, V_μ must be multiplied in the Lagrangian by another proper Lorentz vector, say, $W_\mu(x)$. To complete the analogy it would be tempting to suppose that $W_\mu(x)$ is actually a vector meson field analogous to the photon field, A_μ. Thus β-decay would become, like electron scattering, a second-order process, described by the diagram[1]

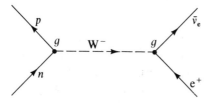

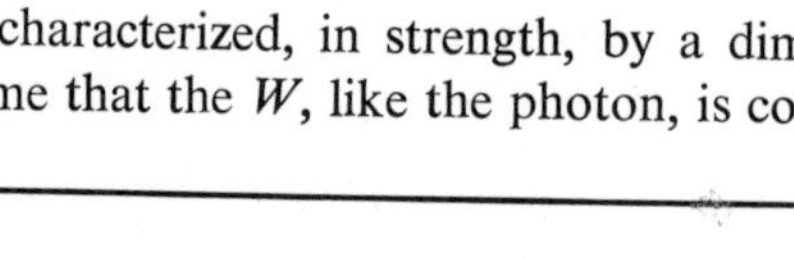

and characterized, in strength, by a dimensionless[2] coupling constant g. We assume that the W, like the photon, is coupled with the same intrinsic strength

1. To simplify the diagram we have, in fact, drawn it for the related process $n + e^+ \longrightarrow p + \bar{\nu}_e$. We distinguish the two distinct neutrinos, ν_e, associated with electrons, and ν_μ, associated with muons, by appropriate subscripts.
2. Note that, in the units $\hbar = c = 1$, each current has the dimension M^3 (M is a mass). The W field has dimension M. Thus $V_\mu W_\mu$ has dimension M^4, or Energy/volume, the dimension of a Hamiltonian density. Thus the coupling constant of the W to the currents must be dimensionless.

at each vertex—leptonic, nucleonic, and so on. This matter is discussed in detail below.

Many of the most important properties that the W must have can be inferred simply from the fact that it has been invented to mediate the known weak interactions. We list the properties.

1. The W has spin 1. (It is a vector meson.)

2. The W is electrically charged and has charge $\pm e$, the electron charge, since it mediates both the process $n \longrightarrow p + e^- + \bar{\nu}_e$ and $e^- + p \longrightarrow n + \nu_e$ (so-called K-capture), and these processes are known, experimentally, to conserve the electric charge. There has been some discussion[3] in the literature as to whether a *neutral* W meson exists. It turns out that the introduction of such a meson always leads to some difficult theoretical problems since, if it couples to pairs of neutral leptons such as e^+, e^- or ν, $\bar{\nu}$ with its normal coupling strength, g, this predicts the existence of weak processes involving the emission of such pairs that have, so far, not been seen. There do not appear to be any phenomena that demand the introduction of neutral, weakly coupled, vector mesons, so we shall not consider them further here.[4]

3. The W must be massive. This follows from the fact that the electron spectrum in β-decay, $N(E)$, the relative number of electrons emitted with energy E, is given quite well, theoretically, by the "phase space" spectrum alone. In general, the spectrum is computed by using Fermi's Golden Rule for the transition probability per unit time,

$$W = 2\pi |H_{WK}|^2 \rho_F(E). \tag{8.2}$$

The phase space, or "allowed," spectrum is obtained simply by taking the phase space factor,[5] essentially the product of $d^3\mathbf{p}/(2\pi)^3$ for all particles in the final state, and using the conservation of energy-momentum, $\delta^4(n - p - e - \nu)$ to integrate over all variables, except the electron energy. In the approximation in which the proton is nonrelativistic, if the neutrino mass is set equal to zero, apart from a constant factor, it is given by

$$E_e(E_e^2 - m_e^2)(E_{\max} - E)^2.$$

Here $E_{\max}$ is the maximum available electron energy which is, for neutron β-decay, the neutron-proton mass difference. Apart from small corrections, some of which are discussed below, this spectrum is in good agreement with experiment. This means that $|H_{WK}|^2$ must be very nearly energy independent. However, if the weak interactions are mediated by a W meson, there is always an energy

---

3. By theorists.
4. Recent measurements on the upper limit for the branching ratio for the K^0 decaying into lepton pairs (e.g., M. Bott-Bodenhanson et al., *Phys. Letters* **23**:277 (1966), imply that the neutral W, if it exists, must be coupled to leptons with a coupling constant of order, at most, of $10^{-4}\, g$, where g is the charged lepton-W coupling constant.
5. The matrix element $|H_{WK}|$ is, in this approximation, assumed not to vary with energy.

dependence introduced in $|H_{WK}|^2$ simply from the W propagator. The propagator, $W_{\mu\nu}$, can be written

$$W_{\mu\nu} = \frac{\delta_{\mu\nu} + \dfrac{q_\mu q_\nu}{m_W{}^2}}{q^2 + m_W{}^2}. \tag{8.3}$$

Here q^2 is the 4-momentum transferred to the leptons.[6] Hence there will be an additional energy dependence in $|H_{WK}|^2$ of the form $1/(q^2 + m_W{}^2)^2$, which will spoil the agreement with experiment unless $m_W{}^2 \gg |q^2| \sim m_e{}^2$.

4. The fact that $m_W > 0$ means that the W is unstable. It can decay into a large variety of modes, depending on its mass. Among the possible modes involving particles of light mass we list

$$W^\pm \longrightarrow e^\pm + \nu_e,$$

$$W^\pm \longrightarrow \mu^\pm + \nu_\mu,$$

$$W^\pm \longrightarrow \pi^\pm + \pi^0.$$

We defer the discussion of the last mode and outline the theory of the leptonic modes. For the W to participate in weak reactions involving leptons, it must be coupled to them. Taking into account the fact that the W is a vector particle and that both neutrinos, ν_μ and ν_e, are two-component left-handed mass zero particles,[7] we can write the W-lepton coupling in the form

$$H_l = gW_\mu L_\mu + \text{h.c.}, \tag{8.4}$$

where g is the dimensionless constant discussed above and L_μ is the "leptonic current," which we write in the form

$$L_\mu(x) = i[\bar{\psi}_e(x)\gamma_\mu(1 + \gamma_5)\psi_{\nu_e} + \bar{\psi}_\mu\gamma_\mu(1 + \gamma_5)\psi_{\nu_\mu}]. \tag{8.5}$$

In Eq. (8.4) h.c. stands for the Hermitian conjugate current, which must be added on in this case, since terms like $i\bar{\psi}_a\gamma_\mu\psi_b$ are not Hermitian if a and b are different particles. Several features of the weak interactions have been built into this current. In the first place it is clear that L_μ as written is not a proper vector but, rather, a fifty-fifty mixture of vector and axial vector. This means that no matter how the parity of the W meson is defined, and no matter how the relative parities of the leptons and neutrinos are chosen,[8] H_l will not be a scalar. In brief, H_l does not conserve parity, which of course is by now a well-known experimental result. In writing the mixture as $(1 + \gamma_5)$, we embody the fact that

---

6. For neutron β-decay, in the frame in which the neutron is at rest,

$$q^2 = -(M_N{}^2 + M_P{}^2) + 2M_N E_P < 0.$$

7. Experimentally it is known that $m_{\nu_e} < 200$ eV, and $m_{\nu_\mu} < 3.5$ MeV. All experiments are consistent with zero mass for both neutrinos and since this leads to, far and away, the simplest theoretical description, we make this choice.
8. We adopt the convention that these relative parities are even, so that the γ_μ term transforms like a vector and the $\gamma_\mu\gamma_5$ term like a pseudovector.

this nonconservation is "maximum," which is sometimes expressed by saying that the neutrinos are two-component left-handed objects. If $\psi_\nu(x)$ is a 4-component spinor corresponding to a particle with zero mass, then[9] $(1 + \gamma_5)\psi_\nu = \psi_\nu'$ is a spinor satisfying

$$\gamma_5\psi_\nu' = \psi_\nu'. \tag{8.6}$$

From the identity

$$i\gamma_4\gamma_5\boldsymbol{\gamma} = \begin{pmatrix} \boldsymbol{\sigma} & 0 \\ 0 & \boldsymbol{\sigma} \end{pmatrix} \tag{8.7}$$

and the Dirac equation for ψ_ν', or

$$i(\gamma\nu)\psi_\nu' = 0, \tag{8.8}$$

we see at once that

$$\boldsymbol{\sigma}\cdot\hat{\boldsymbol{\nu}}\psi_\nu' = -\psi_\nu', \tag{8.9}$$

which is to say that the neutrino described by ψ_ν' has negative "helicity," or is left-handed.

The current $L_\mu(x)$ also has the "muon-electron symmetry" built into it. This symmetry states that the weak interaction Lagrangian must be invariant under the transformation

$$\begin{aligned} \psi_\mu &\rightleftarrows \psi_e, \\ m_\mu &\leftrightarrows m_e, \\ \psi_{\nu_\mu} &\leftrightarrows \psi_{\nu_e}. \end{aligned} \tag{8.10}$$

The free leptonic Lagrangian,

$$\mathfrak{L}_0 = -[\bar\psi_\mu(\partial_\mu\gamma_\mu + m_\mu)\psi_\mu + \bar\psi_e(\partial_\mu\gamma_\mu + m_e)\psi_e + \bar\psi_{\nu_\mu}\gamma_\mu\partial_\mu\psi_{\nu_\mu} + \bar\psi_{\nu_e}\gamma_\mu\partial_\mu\psi_{\nu_e}] \tag{8.11}$$

clearly has this property, and so does $L_\mu W_\mu$ providing the W couples to muon and electron quantities with the same constant g. The implications of this symmetry are especially strong in relating decay processes of the form

$$a \longrightarrow \mu + \nu_\mu,$$

$$a \longrightarrow e + \nu_e.$$

It implies that the formulas for the rates of such processes should be related by the simple substitution $m_e \leftrightarrows m_\mu$. There is excellent evidence that this is true, and we return to it later when we discuss the leptonic decays of the π meson.

9. In our representation for the γ's,

$$(1 + \gamma_5) = \begin{pmatrix} 1 & 0 & -1 & 0 \\ 0 & 1 & 0 & -1 \\ -1 & 0 & 1 & 0 \\ 0 & -1 & 0 & 1 \end{pmatrix}.$$

Using the leptonic Lagrangian it is elementary to compute the rate for $W \longrightarrow l + \nu_e$. Neglecting the lepton mass it turns out to be

$$R_{W\to l+\nu} = \frac{g^2 m_W}{6\pi}. \tag{8.12}$$

We will shortly discuss how g^2 is determined experimentally from β-decay and other weak processes. In terms of the W mass, g^2 is given by the formula

$$g^2 \simeq \frac{1}{\sqrt{2}}\left(\frac{m_W}{m_P}\right)^2 \times 10^{-5}. \tag{8.13}$$

(The reason for the curious factors in this formula will be made clear below.)

Many experiments have been performed to look for the W in reactions such as

$$\nu + p \longrightarrow p + l + W,$$

$$\pi + p \longrightarrow p + W + \pi.$$

No W's have been identified in these experiments and the conclusion is, that if the W exists, it must be too massive to have been produced in these reactions, which involve incident particles with energies limited by the size of the present generation of accelerators; proton energies of, say, 30 BeV. In particular, from experiments done at CERN and Brookhaven with incident neutrinos in the BeV range, we may conclude that if the W exists,

$$m_W > 2 \text{ BeV} \tag{8.14}$$

and probably even greater. Hence from Eq. (8.12) and Eq. (8.13) we must have

$$R_{W\to l+\nu} > 5 \times 10^{18} \text{ sec}^{-1}. \tag{8.15}$$

In other words, even though the W has only weak and electromagnetic interactions, it is extremely unstable because there is so much available phase space for its allowed decay modes.

From the form of the leptonic weak Lagrangian we can see at once that it is invariant under the transformations

$$\begin{aligned} W_\mu &\longrightarrow W_\mu, \\ \psi_e &\longrightarrow e^{i\Lambda}\psi_e, \\ \psi_{\nu_e} &\longrightarrow e^{i\Lambda}\psi_{\nu_e}, \end{aligned} \tag{8.16}$$

and, *separately*,

$$\begin{aligned} W_\mu &\longrightarrow W_\mu, \\ \psi_\mu &\longrightarrow e^{i\Lambda}\psi_\mu, \\ \psi_{\nu_\mu} &\longrightarrow e^{i\Lambda}\psi_{\nu_\mu}, \end{aligned} \tag{8.17}$$

where Λ is an arbitrary real number. From the work of Chapter 2, we can therefore conclude that this Lagrangian admits two separately conserved currents,[10]

$$j_\mu{}^e = i\bar{\psi}_e\gamma_\mu\psi_e + i\bar{\psi}_{\nu_e}\gamma_\mu\psi_{\nu_e} \tag{8.18}$$

and

$$j_\mu{}^\mu = i\bar{\psi}_\mu\gamma_\mu\psi_\mu + i\bar{\psi}_{\nu_\mu}\gamma_\mu\psi_{\nu_\mu}. \tag{8.19}$$

Hence there are two associated conserved, Lorentz-scalar, "charges" that we can call the muon and electron numbers:

$$N_\mu = n_{\mu^-} + n_{\nu_\mu} - n_{\mu^+} - n_{\bar{\nu}_\mu} \tag{8.20}$$

and

$$N_e = n_{e^-} + n_{\nu_e} - n_{e^+} - n_{\bar{\nu}_e}. \tag{8.21}$$

So long as the full Lagrangian is invariant under these phase transformations, N_μ and N_e will be absolutely conserved quantities,[11] which means, for example, that a muon can never transform into an electron without the emission of at least one muon neutrino. Since N_μ and N_e involve leptonic fields, alone, we know from the work of Chapters 2 and 3 that, applying equal time commutators when needed, we have

$$N_\mu|a\rangle = N_e|a\rangle = 0, \tag{8.22}$$

where $|a\rangle$ is any state not containing leptons. This, along with the conservation of the muon and electron numbers (see Table below),

Lepton	*Lepton number*
e^+	-1
e^-	$+1$
ν_e	$+1$
$\bar{\nu}_e$	-1
μ^+	-1
μ^-	$+1$
ν_μ	$+1$
$\bar{\nu}_\mu$	-1

leads to several selection rules all of which have been verified to an extremely high degree of accuracy. In particular, relative to the "allowed" and well-studied process,[12]

$$\mu^- \longrightarrow e^- + \bar{\nu}_e + \nu_\mu,$$

10. These currents would presumably be the ones that would couple to the neutral W if there were such a particle.
11. Leptons have, so far as is known, only weak and electromagnetic interactions. It is easy to see that the electromagnetic couplings, which are diagonal in the lepton fields, conserve the lepton numbers.
12. In this reaction, leptonic number conservation implies that the electron will be emitted with a right-handed electron-antineutrino and a left-handed muon-neutrino.

the experimental limits[13] on some processes that might violate lepton number conservation are:

$$\mu^{\pm} \longrightarrow e^{\pm} + \gamma < 2 \times 10^{-8},$$
$$\mu^{\pm} \longrightarrow e^{\pm} + e^{+} + e^{-} < 1.5 \times 10^{-7},$$
$$\mu^{-} + Z \longrightarrow e^{-} + Z < 2.5 \times 10^{-7}.$$

(The latter reaction refers to the imaginable, but unobserved, conversion of a muon into an electron in the Coulomb field of a nucleus of charge Z.) In addition there are experiments (measuring the helicity of the charged leptons, from which we infer the helicity of the neutral and unobserved neutrino) that show convincingly that

$$\pi^{-} \longrightarrow \mu^{-} + \bar{\nu}_{\mu}$$

but *not*

$$\pi^{-} \longrightarrow \mu^{-} + \nu_{\mu},$$

and

$$n \longrightarrow p + e^{-} + \bar{\nu}_{e}$$

but *not*

$$n \longrightarrow p + e^{-} + \nu_{e}.$$

There are also high-energy neutrino experiments making use of muon neutrinos from the decay of energetic pions,

$$\pi^{+} \longrightarrow \nu_{\mu} + \mu^{+},$$

that show that

$$\nu_{\mu} + n \longrightarrow p + e^{-}$$

does not occur. All in all, the evidence for the separate conservation of muon and electron number appears to be extremely strong.[14]

Up to this point we have not discussed the assignment of strong interaction quantum numbers, such as parity and isotopic spin, to the W meson. Indeed, these assignments are essentially arbitrary and have no physical content. To understand the point we may return to the photon. As we have seen, when the photon field is introduced into the strong interaction Lagrangian, SU_2, or isotopic spin symmetry, no longer holds. In the presence of the photon field no particle has a well-defined isotopic spin, including the photon. However, the

---

13. These numbers are taken from G. Feinberg and L. M. Lederman in *Ann. Rev. Nucl. Sci.* **13**:431 (1963), where references to the experimental papers are given.
14. For a very complete and up-to-date review of all this evidence and of the weak interactions, in general, see T. D. Lee and C. S. Wu, *Annual Review of Nuclear Science*, Vol. 15 (1965), p. 385.

electromagnetic coupling is rather weak, so that it is a good approximation to consider the electromagnetic transitions between states that *are* eigenstates of the strong couplings and hence have a well-defined isotopic spin. Acting between such states the photon carries away either no isotopic spin or one unit of isotopic spin. Thus one can say in this approximate context, if one likes, that the photon is a mixed isoscalar, isovector particle. With the W meson the situation is more complicated. A process like $n \longrightarrow p + e^- + \bar{\nu}_e$ cannot take place at all if the neutron and proton have the same mass. However, the mass difference is probably of electromagnetic origin. Indeed, the whole isotopic spin philosophy is built on the assumption that the mass differences, in the isotopic multiplets, vanish if there is no electromagnetic coupling. Hence, in a strict sense, there would be no weak decays among the members of an isotopic multiplet without the presence of the electromagnetic coupling. However, it is an extremely useful approximation to ignore these mass differences in the *matrix elements* for the weak processes, but, of course, to take them into account in the phase space factors. This means that a process like $n \longrightarrow p + e^- + \bar{\nu}_e$ can be viewed, in this approximation, as a transition in which the 3-component of iso-spin is raised by one unit.[15] We can also express this fact by saying that, in this approximation, the W meson carries away one unit of isotopic spin.

A similar situation arises if we attempt to define the parity of the W. So far as is known the W has only weak and electromagnetic couplings. The weak couplings do not conserve parity, and hence the parity of the W can never be determined in any weak process; in particular, it can never be determined in any of the W decays. However, the electromagnetic interactions appear to be parity conserving, and hence we might imagine that these interactions could be used to fix the W parity relative to the photon parity. (We note that the decay, $\pi^0 \longrightarrow \gamma + \gamma$, can be used to determine the π^0 parity by measuring the correlations between the linear polarizations of the photons, which must be perpendicular if the π^0 is a pseudo-scalar and if parity is conserved by the strong and electromagnetic forces responsible for the decay.) This method also fails if the W photon coupling is *minimal*, which implies that the W current always contains the W *bilinearly* in terms like $W^\dagger \partial_\mu W$, so that the intrinsic W parity cancels when one determines the parity of the current. To call the W a "vector" meson as opposed to a pseudovector meson is a pure convention.

Not only do the weak interactions fail to conserve P but they also fail to conserve C, the charge conjugation, and T, the time reversal.[16] This raises the

---

15. There is no useful isotopic spin assignment that can be given to leptons since they only take part in nonisotopic spin conserving interactions.
16. The violation of T was discovered in 1964 (see J. H. Christenson, J. W. Cronin, V. L. Fitch, and R. Turley, *Phys. Rev. Letters* **13**:138 (1964)) in decays of the neutral K meson that will be discussed later in the book. The precise origin of the violation, whether it is a new weak interaction, or something else, is, at least at the time of this writing, unclear. However, there is no doubt that the exact Hamiltonian of the world does not commute with T.

question of how the antiparticles, such as the anti-W, are to be defined in a universe in which C, the particle-antiparticle transformation, is not conserved. In fact the problem is still deeper; what does the operator C *mean* in a theory in which it is not conserved? Leaving aside the weak interactions, C has had historically two quite different meanings. For the strongly interacting particles it has denoted the transformation that takes particles into antiparticles. "Antiparticles" refer here to the states, degenerate in mass to the particle, that are inevitably creatable in any Lorentz invariant field theory, by using Hermitian conjugate components in the Fourier decomposition of the fields. This particle-antiparticle transformation we may call C_{ST}. On the other hand, charge conjugation has also referred to the process of transforming electric currents and charges into their negatives. This process we can call C_γ. We can express charge conjugation invariance in strong and electromagnetic interactions by the statements that

$$C_{ST} = C_\gamma = C \tag{8.23}$$

and

$$[C, H_0 + H_{ST} + H_{em}]_- = 0. \tag{8.24}$$

In other words, in a theory in which C is conserved by $H_{ST} + H_{em}$, the operations of transforming particle into antiparticle and changing the signs of electromagnetic quantities are identical.

If only the weak interactions, and no others, were present, we would be free to try to define a weak charge conjugation operator[17] C_{WK} with the property that

$$[C_{WK}, H_{WK}]_- = 0. \tag{8.25}$$

The fact that the Hamiltonian of the world is not charge conjugation invariant can be expressed by the statement that there exists *no* definition of C_{WK} such that

$$[C_{WK}, H]_- = 0 \tag{8.26}$$

and

---

17. For a more systematic exposition of this point of view see T. D. Lee, *Proceedings of the Oxford International Conference on Elementary Particles*, September 1965, p. 225. (Published by the Rutherford High Energy Laboratory.)
18. T. D. Lee, op. cit., gives a specific example of how parity nonconservation can be regarded as a clash among different definitions of parity associated with different interactions. In the presence of electromagnetism the observed charged pion weak decays into muons and neutrinos (to be treated in detail below) are

$$\pi^+ \longrightarrow \mu^+ + \nu_\mu,$$

$$\pi^- \longrightarrow \mu^- + \bar{\nu}_\mu.$$

Since it is known, experimentally, that the neutrinos (antineutrinos) that emerge from this decay are respectively left-handed (right-handed), it is known that parity is violated since the neutrino helicity (handedness) is a pseudoscalar quantity which would vanish if parity were conserved. If, however, the electric field were turned off, we would be quite free to take as the "real" pion states

$$C_{WK} = C. \tag{8.27}$$

(A similar set of remarks can be made for P_{WK} and T_{WK}.[18])

However, it is now known experimentally, to a very high degree of accuracy,[19] that

$$C_{WK}P_{WK}T_{WK} = CPT \tag{8.28}$$

and, even though $C_{WK} \neq C$, $P_{WK} \neq P$, $T_{WK} \neq T$,

$$[CPT, H]_- = 0, \tag{8.29}$$

where H is the Hamiltonian of the world. In fact,[20] it is impossible to construct a local Lorentz invariant field theory that is not invariant under *CPT*. *CPT* invariance appears to be on almost as sound a footing experimentally, and theoretically, as Lorentz invariance itself. In view of this, it is very natural to define the antiparticle state of a given state $|a\rangle$ as

$$|\bar{a}\rangle = CPT|a\rangle. \tag{8.30}$$

If $|a\rangle$ represents a stable one-particle state, which is, therefore, an eigenstate of H with eigen-value $E(\mathbf{a}) = \sqrt{m_a^2 + \mathbf{a}^2}$, it then follows from the *CPT* invariance of H, if we let $\mathbf{a} = 0$, that

$$m_a = \langle \mathbf{0}|_a H|\mathbf{0}\rangle_a = \langle \mathbf{0}|_{\bar{a}} H|\mathbf{0}\rangle_{\bar{a}} = m_{\bar{a}}. \tag{8.31}$$

This relation has been checked experimentally wherever it is applicable.[21]

Of course most of the particles in nature are unstable. Insofar as the decays of these particles can be computed in perturbation theory by taking as unperturbed states the eigenstates of H_{ST} and treating H_{WK} as a perturbation, to be computed in lowest order, then from the fact that

$$CPT = C_{WK}P_{WK}T_{WK} \tag{8.32}$$

and

$$[H_{WK}, C_{WK}P_{WK}T_{WK}]_- = 0, \tag{8.33}$$

$$\phi_1 = \frac{\pi^+ + \pi^-}{\sqrt{2}},$$

$$\phi_2 = \frac{\pi^+ - \pi^-}{i\sqrt{2}}.$$

These states decay into a fifty-fifty mixture of left- and right-handed neutrinos and so, from the point of view of this choice of states, parity would appear to be conserved for this decay. We could never tell whether parity was or was not conserved without turning on the electric field, which then fixes the decaying states as being the eigenstates of the electric charge. The decay of these charged states violates parity.

19. We shall give a brief discussion of the nature of the evidence below, and also in Chapter 15.
20. See, for example, R. Jost, *Theoretical Physics in the Twentieth Century*, Interscience, p. 107 ff.
21. The best check is in the $\bar{K}^0$, K^0 system which we treat in detail in Chapter 15.

it follows that

$$|\langle a|H_{WK}|b\rangle|^2 = |\langle \bar{a}|H_{WK}|\bar{b}\rangle|^2. \tag{8.34}$$

In other words, the lifetimes (or rates of transition), as defined in perturbation theory, are identical for particle and antiparticle. This has also been checked experimentally to high accuracy for the pairs π^+, π^-; μ^+, μ^-, among others.[22] If we call the W_μ^- the particle field, then we will define, W_μ^+, the anti-W^- field, as

$$W_\mu^+ = CPTW_\mu^-(CPT)^{-1}. \tag{8.35}$$

This definition is meaningful even though the W is involved in C violating interactions.

As we have remarked, in the W meson theory the decay $\mu^- \longrightarrow e^- + \bar{\nu}_e + \nu_\mu$ is characterized as a second-order process whose matrix element (assuming μ, e, universality) will assume the form

$$g^2\langle \nu_\mu|L_\alpha(0)|\mu^-\rangle W_{\alpha\beta}(q^2)\langle 0|L_\beta(0)|e^-\bar{\nu}_e\rangle, \tag{8.36}$$

with

$$q^2 = (p_\mu - p_{\nu_\mu})^2. \tag{8.37}$$

Here, $W_{\alpha\beta}$ is the W propagator (Eq. (8.3)). If the W couplings are treated to lowest order in g and if we assume that $(m_W/q)^2 \gg 1$, we may write Eq. (8.36) simply as

$$\frac{g^2}{m_W{}^2}(\bar{u}(\nu_\mu)\gamma_\mu(1 + \gamma_5)u(\mu^-))(\bar{u}(e^-)\gamma_\mu(1 + \gamma_5)u(\bar{\nu}_e)). \tag{8.38}$$

This is the same matrix element that would arise from the "local" four-Fermi coupling

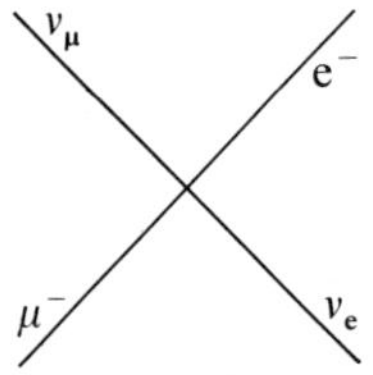

in which the leptonic currents interact directly at a single point in space time

22. From the experimental references cited in T. D. Lee and C. S. Wu, op. cit., we have the measured lifetime ratios

$$\frac{\tau(\mu^+)}{\tau(\mu^-)} = 1.000 \pm 0.001 \quad \text{and} \quad \frac{\tau(\pi^+)}{\tau(\pi^-)} = 1.00 \pm 0.08.$$

23. R. Feynman and M. Gell-Mann, *Phys. Rev.* **109**:193 (1958).
24. See G. Källen, *Elementary Particle Physics*, Addison-Wesley, Chapter 14, for a very complete discussion of the calculational techniques involved.

with an effective coupling constant, G, which is defined, for historical reasons,[23] as

$$\frac{G}{\sqrt{2}} = \frac{g^2}{m_W^2}. \tag{8.39}$$

Indeed for the experimentally known weak interaction processes we can always replace the W theory by such a direct current-current coupling if we want to. It would appear as if G were directly open to experimental determination if we make use of the observed mass and life time of the muon:

$$m_\mu = (206.788 \pm 0.003)m_e$$

and

$$\tau_\mu = (2.198 \pm 0.001) \times 10^{-6} \text{ sec.}$$

In fact, neglecting the mass of the electron, and using the matrix element given in Eq. (8.38) and the standard techniques for computing decay rates,[24] we find that the rate $1/\tau_\mu$ is given by[25]

$$\frac{1}{\tau_\mu} = \frac{G^2 m_\mu^5}{192\pi^3}. \tag{8.40}$$

Using this formula naively and substituting the experimental numbers, we find that

$$G = (1.4320 \pm 0.0011) \times 10^{-49} \text{ erg cm}^3 \simeq \frac{10^{-5}}{m_P^2}, \tag{8.41}$$

where the mass of the proton has been introduced for the convenience of giving Gm_P^2 a simple mnemonic value. We see that knowing G does not fix g^2 since m_W is still unknown. Thus, g could be quite large if m_W were large enough. If, for the fun of it, we were to take $g^2 = e^2$, the electric charge, then $m_W \simeq 30\ m_P$. If a W were to turn up at this mass it might imply a connection between weak and electromagnetic interactions. (There is another apparent connection to be discussed shortly.)

However, the determination of G is not as straightforward as it might seem. There are electromagnetic radiative corrections that modify the lifetimes and decay spectra in the decays involving electrically charged particles. For muon

---

25. The electron spectrum in units of $2|\mathbf{p}_e|/m_\mu = x$, with $m_e = 0$, is given by

$$I(x) = \frac{m_\mu^5}{16\pi^3} G^2[1 - x + \tfrac{2}{3}\rho(\tfrac{4}{3}x - 1)].$$

The quantity ρ is known as the Michel parameter and is exactly $\frac{3}{4}$ for the two-component theory of neutrinos. According to Lee and Wu, op. cit., the most accurate experimental determination of ρ gives

$$\rho_{\text{exp}} = .747 \pm .005,$$

in good agreement with the two-component theory.

decay these corrections are given, primarily, by diagrams (in the local theory) of the form

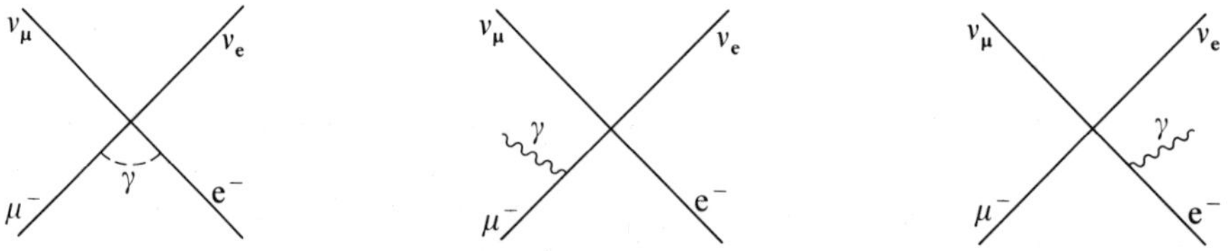

The first diagram represents the standard radiative correction due to the exchange of virtual photons and the other diagrams show radiative emission of real photons (*bremsstrahlung*), which cannot be separated from the nonradiative μ decay process in the infrared limit. A computation of these diagrams[26] yields for τ_μ the corrected formula

$$\tau_{\mu\ \mathrm{corr}} = \frac{G^2 m_\mu{}^5}{192\pi^3}\left[1 - \frac{\alpha}{4\pi}\left(\pi^2 - \frac{25}{4}\right)\right], \tag{8.41}$$

where G, as before, is the weak coupling constant. Hence the G determined using this formula will be somewhat larger than before. Indeed the G found from the experimental lifetime, using Eq. (8.41), turns out to be[27]

$$G = 1.4350 \pm 0.0011 \times 10^{-49}\ \text{erg-cm}^3.$$

If other purely leptonic processes such as

$$\nu_e + e^- \longrightarrow \nu_e + e^-$$

could be studied experimentally, then their rates and cross sections would be determined by this G.[28]

In the next chapter we study how the theory of weak interactions must be modified if the weak interactions involve strongly interacting particles such as pions and nucleons.

---

26. See, for example, T. Kinoshita and A. Sirlin, *Phys. Rev.* **113**:1652 (1959).
27. Lee and Wu, op. cit., p. 406.
28. From the experimental value of G one may show that the cross section for elastic electron-neutrino scattering for, say, 5 MeV incident neutrinos is about 8×10^{-44} cm^2.

9

The Weak Interactions II: Nucleons

In the last chapter we saw, in the limit of large W mass, that is "large" compared to the invariant momentum transfer, q^2, exchanged between the pairs of particles A, B and C, D in a weak reaction like

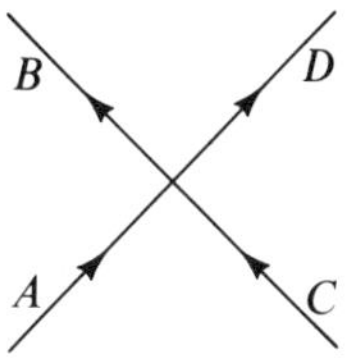

that the effective weak interaction H_{WK} could be written as a product of currents (say, J_μ and J_μ')

$$H_{WK} = \frac{G}{\sqrt{2}}\left[\frac{J_\mu J_\mu'^* + \text{h.c.}}{2}\right]. \tag{9.1}$$

The simplest H_{WK} of this form[1] is to take $J'_\mu = J_\mu$. Thus we would write:[2]

---

1. R. Feynman and M. Gell-Mann, op. cit. The history of the V-A theory; the theory in which the weak couplings are attributed to vector and axial vector currents alone is a very interesting one. As previously mentioned, Fermi first wrote β-decay in the vector current form. But his model fell out of favor in the 1950's when some β-decay experiments, notably the electron-neutrino angular correlation in He^6, appeared to require a scalar and tensor coupling. By 1957 almost total chaos ensued since some experiments appeared to require V and A while others required S and T. In 1957 R. G. Marshak and E. C. G. Sudarshan (*Proceedings of the Padua-Venice Conference on Mesons and Newly Discovered Particles*), guided by considerations of symmetry—namely the idea that the parity violation was "maximal," which could be expressed by making the interaction $\gamma_\mu(1 + \gamma_5)$—proposed that, despite the conflicting experiments, the V-A theory was correct. Indeed, when the experiments were redone, it turned out that they all agreed with the V-A theory, which is to say that the weak interactions do their best, given parity violation, to look like electromagnetism. Feynman and Gell-Mann (op. cit.) then speculated that the V current was identical with the isotopic vector part of the electromagnetic current. In this book, when we refer to the "Feynman-Gell-Mann theory," we mean the particular V-A theory in which V has this special form.
2. For reasons that will be discussed in Chapter 12, we must use J_μ^* and not $J_\mu^\dagger$ in Eq. (9.2).

$$\begin{aligned} J_\mu^* &= \quad J_\mu^\dagger; \qquad \mu = 1, 2, 3 \\ &= -J_\mu^\dagger; \qquad \mu = 4. \\ H_{WK} &= \frac{G}{\sqrt{2}} J_\mu J_\mu^* \end{aligned} \tag{9.2}$$

The only part of J_μ that we have considered in detail so far is the leptonic current. It can be written in the form, discussed in the last chapter,

$$\begin{aligned} L_\mu &= 2i\left[\bar{\psi}_e\gamma_\mu \frac{(1+\gamma_5)}{2}\psi_{\nu_e} + \bar{\psi}_\mu\gamma_\mu \frac{(1+\gamma_5)}{2}\psi_{\nu_\mu}\right] \\ &= 2i[\bar{\psi}_e\gamma_\mu a_+ \psi_{\nu_e} + \bar{\psi}_\mu\gamma_\mu a_+\psi_{\nu_\mu}]. \end{aligned} \tag{9.3}$$

The quantities

$$a_\pm = \frac{(1 \pm \gamma_5)}{2} \tag{9.4}$$

are orthogonal "projection operators" in the sense that

$$\begin{aligned} a_+^2 &= a_-^2 = 1, \\ a_+a_- &= a_-a_+ = 0. \end{aligned} \tag{9.5}$$

The current-current form for H_{WK} predicts that "diagonal" processes arising from terms in the products of currents of the form

$$(\bar{\psi}_a\gamma_\mu a_+\psi_b)(\bar{\psi}_a\gamma_\mu a_+\psi_b) \tag{9.6}$$

occur with the same coupling constant G as do the "off-diagonal" processes involving products of terms containing different particles. Moreover, the two-component theory has another special symmetry which we can summarize by the identity, a so-called Fierz transformation, in which a, b, c, d are four different spinor particles:

$$(\bar{\psi}_a\gamma_\mu a_+\psi_b)(\bar{\psi}_c\gamma_\mu a_+\psi_d) = (\bar{\psi}_c\gamma_\mu a_+\psi_b)(\bar{\psi}_a\gamma_\mu a_+\psi_d). \tag{9.7}$$

This means, for example, that

$$(\bar{\psi}_e\gamma_\mu a_+\psi_{\nu_e})(\bar{\psi}_{\nu_e}\gamma_\mu a_+\psi_e) = (\bar{\psi}_{\nu_e}\gamma_\mu a_+\psi_{\nu_e})(\bar{\psi}_e\gamma_\mu a_+\psi_e), \tag{9.8}$$

which implies that the weak process $e^+ + e^- \longrightarrow \bar{\nu}_e + \nu_e$, important for astrophysics,[3] is, in this theory, governed by the same coupling constant and Hamiltonian that govern $\nu_e + e^- \longrightarrow \nu_e + e^-$.

Ordinary neutron β-decay $n \longrightarrow p + e^- + \bar{\nu}_e$ will be described, in the theory, by a coupling of the form

$$\frac{G}{\sqrt{2}} J_\mu L_\mu^*, \tag{9.9}$$

with L_μ being the electron part of the leptonic current and J_μ a nonleptonic[4] current operator which has first-order matrix elements between neutron and

3. For a review of such applications, see M. A. Ruderman, *Reports on Progress in Physics*, XXVIII (1965): 411.
4. J_μ is sometimes called the hadronic current since it has first-order matrix elements between heavy or "hadronic" particles as opposed to L_μ, which has first-order matrix elements between light or "leptonic" particles.

proton. The novel features of the β-decay problem begin with the determination of the structure of J_μ.

We may make several remarks about the general form of J_μ.

1. All known weak interaction processes are consistent with a description in which the *change* in electric charge, ΔQ, caused by applying J_μ to some initial one-particle state, is in units of e, $\Delta Q = 1$. In other words, J_μ taken between one-particle states of the strongly interacting particles has first-order matrix elements only between states whose electric charge differs by one unit. The matrix elements $\langle P|J_\mu(0)|N\rangle$ and $\langle 0|J_\mu|\pi^-\rangle$, which describe the neutron β-decay and the weak decay of the pion (to be discussed later), are illustrations of this feature of J_μ. We can summarize this property by the commutation relation (Q is the electric charge operator):

$$[Q, J_\mu]_- = J_\mu. \tag{9.10}$$

Thus J_μ is the current operator that raises the electric charge by one unit. From the Hermiticity of Q it follows that

$$[Q, J_\mu{}^*]_- = -J_\mu{}^*, \tag{9.11}$$

so that the current that lowers charge by one unit is $J_\mu{}^*$. It clearly follows from these commutation relations that J_μ cannot be Hermitian.

2. If we restrict our discussion to pions and nucleons, then, as we have argued earlier in the book, so long as electromagnetic couplings are treated to lowest order in e, so that SU_2 remains a valid group for hadronic states, we may write, for these states,

$$Q = \frac{B}{2} + T_3. \tag{9.12}$$

So far as one knows, baryon number is absolutely conserved, even by the weak interactions. Hence from remark 1 it follows that for weak transitions involving pions and nucleons alone (or any of the resonant pion-nucleon states)

$$\Delta Q = \Delta T_3 = \pm 1. \tag{9.13}$$

This selection rule does not fix the functional form of J_μ; indeed, any field operator that transforms under SU_2 like an isovector will obey it. If T_2 is the isotopic rotation generator around the 2 axis in isotopic space, then it is useful to define

$$I_2 = e^{i\pi T_2}. \tag{9.14}$$

I_2 generates a rotation through π around the 2-axis in isospace. We now make an assumption about the behavior of J_μ under the transformation I_2. This assumption is motivated by the fact that J_μ acts like an isotopic "raising operator." In neutron β-decay, for example, treated in the good i-spin approximation, J_μ

---

5. In the good i-spin limit (baryon number is always conserved),

$$[J_\mu, Q]_- = \left[J_\mu, \frac{B}{2} + T_3\right]_- = [J_\mu, T_3]_- = J_\mu.$$

raises the 3-component of i spin by one unit. If we suppose that J_μ transforms like $T_+ = T_1 + iT_2$, the isotopic spin T_3 raising operator,[5] we can conclude that under I_2 we have,

$$e^{i\pi T_2} J_\mu e^{-i\pi T_2} = -J_\mu{}^*, \tag{9.15}$$

since the rotation $e^{i\pi T_2}$ takes an operator that transforms like T_+ into an operator that transforms like $-T_- = -T_1 + iT_2$. This operator is then the charge lowering operator which, as we have seen, is $J_\mu{}^*$. Any current that transforms like Eq. (9.15) we call "charge symmetric." We can derive an important reality condition for matrix elements of such currents, remembering that $e^{i\pi T_2}$, acting on the isospinor N, P states replaces N by P and vice versa:

$$\begin{aligned} e^{i\pi T_2}|\mathbf{p}\rangle_{sP} &= -|\mathbf{p}\rangle_{sN}, \\ e^{i\pi T_2}|\mathbf{p}\rangle_{sN} &= +|\mathbf{p}\rangle_{sP}. \end{aligned} \tag{9.16}$$

Thus

$$\langle \mathbf{p}'|_{s'P}\mathbf{J}(0)|\mathbf{p}\rangle_{sN} = \langle \mathbf{p}'|_{s'N}\mathbf{J}(0)^\dagger|\mathbf{p}\rangle_{sP} = \langle \mathbf{p}|_{sP}\mathbf{J}(0)|\mathbf{p}'\rangle^*{}_{s'N}. \tag{9.17}$$

This is the generalization to β-decay of the Hermiticity reality condition derived for electron-proton scattering on the basis of assuming that the electric current operator is Hermitian. The derivation here depends essentially on the fact that we work in the good i-spin limit, so that neutron and proton belong to an isodoublet.

3. The weak interactions do not conserve parity. Indeed, the leptonic current is a fifty-fifty mixture of vector and pseudovector components. It is now a well-known feature of all of the weak interactions, whether or not they involve leptons, that they violate parity conservation. We can take account of the known experimental results (many are discussed in what follows) if we write

$$J_\mu = V_\mu + A_\mu, \tag{9.18}$$

where V_μ is assumed to transform like a Lorentz four-vector and A_μ like a Lorentz pseudovector. In particular, we may analyze the general form of the neutron β-decay matrix element, to lowest order in the weak couplings, in much the same way that we analyzed the matrix element of the electric current for elastic electron-proton scattering. In fact, using the same arguments as given in Chapter 5 we may introduce six invariant form factors defined as follows,[6] where $q^2 = (n - p)^2 = (e + \nu_e)^2$:

$$(2\pi)^3\langle \mathbf{p}'|_{s'P}V_\mu(0)|\mathbf{p}\rangle_{sN} = i\bar{u}(\mathbf{p}')_{s'P}[\gamma_\mu f_V(q^2) + \sigma_{\mu\nu}q_\nu f_M(q^2) + iq_\mu f_S(q^2)]u(\mathbf{p})_{sN} \tag{9.19a}$$

6. We introduce the $(2\pi)^3$ here so that we can identify the matrix elements of $\int d\mathbf{r} V_0(\mathbf{r}, t)$ with $f_V(0)$, without the $(2\pi)^3$ factor.

7. There is also a condition for V_0 and A_0, but the ones above are sufficient to make the argument.

and

$$(2\pi)^3\langle \mathbf{p}'|_{s'P}A_\mu(0)|\mathbf{p}\rangle_{sN}$$
$$= i\bar{u}(\mathbf{p}')_{s'P}[\gamma_5[\gamma_\mu g_A(q^2) + iq_\mu g_P(q^2) + i(p'+p)_\mu g_E(q^2)]]u(\mathbf{p})_{sN}, \quad (9.19)^{\text{b}}$$

with

$$q_\mu = (p' - p)_\mu.$$

Below we give the motivation for the particular choice of the form factors given here. (Any other choice can be reduced to one of these g's or f's.) But first we shall use the assumed "charge symmetry" of J_μ and hence of V_μ and A_μ to discuss their reality properties.[7] From Eq. (9.17),

$$\begin{aligned}\langle \mathbf{p}'|_{s'P}\mathbf{V}(0)|\mathbf{p}\rangle_{sN} &= \langle \mathbf{p}|_{sP}\mathbf{V}(0)|\mathbf{p}'\rangle^*_{s'N},\\ \langle \mathbf{p}'|_{s'P}\mathbf{A}(0)|\mathbf{p}\rangle_{sN} &= \langle \mathbf{p}|_{sP}\mathbf{A}(0)|\mathbf{p}'\rangle^*_{s'N}.\end{aligned} \quad (9.20)$$

Thus we must have

$$i\bar{u}(\mathbf{p}')_{s'}[\gamma_i f_V(q^2) + \sigma_{i\nu}q_\nu f_M(q^2) + iq_i f_S(q^2)]u(\mathbf{p})_s$$
$$= (i\bar{u}(\mathbf{p})_s[\gamma_i f_V(q^2) - \sigma_{i\nu}q_\nu f_M(q^2) - iq_i f_S(q^2)]u(\mathbf{p}')_{s'})^*, \quad (9.21)$$

where $i = 1, 2, 3$. (We have dropped the nucleon labels on the spinors since, in the good i-spin limit, $u(\mathbf{p})_{sN} = u(\mathbf{p})_{sP} \equiv u(\mathbf{p})_s$.) We may now use the condition (O is any combination of γ matrices)

$$(u^\dagger(\mathbf{p})_s\gamma_4 O u(\mathbf{p}')_{s'})^* = u^\dagger(\mathbf{p}')_{s'}O^\dagger\gamma_4 u(\mathbf{p})_s \quad (9.22)$$

to establish[8] the relations

$$\begin{aligned}f_V(q^2) &= f_V(q^2)^*,\\ f_M(q^2) &= f_M(q^2)^*,\\ f_S(q^2) &= -f_S(q^2)^*.\end{aligned} \quad (9.23)$$

For $A_\mu(0)$ we have

$$i\bar{u}(\mathbf{p}')_{s'}\gamma_5[\gamma_i g_A(q^2) + iq_i g_P(q^2) + i(p+p')_i g_E(q^2)]u(\mathbf{p})_s$$
$$= (i\bar{u}(\mathbf{p})_s\gamma_5[\gamma_i g_A(q^2) - iq_i g_P(q^2) + i(p+p')_i g_E(q^2)]u(\mathbf{p}')_{s'})^*, \quad (9.24)$$

and from Eq. (9.24) we show that

$$\begin{aligned}g_A &= g_A^*,\\ g_P &= g_P^*,\\ g_E &= -g_E^*.\end{aligned} \quad (9.25)$$

We now discuss, at least in a preliminary way, the physical significance of the six form factors. In the limit in which $\mathbf{p} = \mathbf{p}' \longrightarrow \mathbf{0}$ we have[9]

$$(2\pi)^3\langle \mathbf{p}'|_{S'}V_\mu(0)|\mathbf{p}\rangle_S \simeq i\bar{u}(\mathbf{0})_{s'}\gamma_4 f_V(0)u(\mathbf{0})_s\delta_{\mu 4}. \quad (9.26)$$

---

8. In dealing with $\sigma_{i\nu}q_\nu$, keep in mind that $q_4^* = -q_4$.

9. In fact, from the large and small component reduction of the u's we see that

$$\bar{u}(\mathbf{0})\gamma u(\mathbf{0}) = 0.$$

Hence the factor of $\delta_{\mu 4}$.

Thus $f_V(0)$ plays the role of the renormalization factor for the vector current coupling constant. If we write, for the vector part,

$$H_V{}^\beta = \frac{G}{\sqrt{2}} V_\mu L_\mu{}^*, \tag{9.27}$$

then, in analogy to the way in which the *observed* charge was defined for electron scattering, we define G_{obs}^V by the equation

$$G_{\text{obs}}^V = Gf_V(0), \tag{9.28}$$

where G is the coupling constant defined in Eq. (9.2), which is measured in μ-decay. There is, however, an important technical point to be observed here. In neutron β-decay

$$q^2 = (e + \nu)^2 = -m_e{}^2 + 2(\boldsymbol{\nu}\cdot\mathbf{e} - \nu E(\mathrm{e})). \tag{9.29}$$

This quantity is a maximum when

$$\hat{\boldsymbol{\nu}}\cdot\hat{\mathbf{e}} = \cos\theta = 1.$$

Therefore, in β-decay,

$$q^2 < 0$$

as $m_e \neq 0$. Thus $f_V(0)$ is strictly speaking not measurable in β-decay. Even so, it is customary to define G_{obs}^V in terms of $f_V(0)$. We shall show, shortly, that there are strong arguments to indicate that the variation of $f_V(q^2)$ between the limits[10]

$$-(M_N - M_P)^2 \leq q^2 \leq 0 \tag{9.30}$$

is completely negligible. Even taking this for granted, it requires a good deal of theoretical and experimental work to extract G_{obs}^V from the β-decay data of the neutron and other nuclei.[11] Deferring a discussion of what is involved, we simply quote the experimental result

$$G_{\text{obs}}^V = (1.4149 \pm 0.0022) \times 10^{-49} \text{ erg-cm}^3, \tag{9.31}$$

to be compared with the value of G, from μ-decay:

$$G = (1.4350 \pm 0.0011) \times 10^{-49} \text{ erg-cm}^3. \tag{9.32}$$

This gives

$$f_V(0) \simeq .986. \tag{9.33}$$

The striking discovery here is, of course, that $f_V(0) \simeq 1$. A priori there is no

---

10. The minimum value of q^2 in β-decay is $-(M_N - M_P)^2$.
11. In fact, G_{obs}^V is most accurately determined from the decay rate measured for the $0^+ \longrightarrow 0^+$ transition from the ground state of O^{14} to the first excited state of N^{14}. Since this transition involves no change of spin parity (it is "allowed" Fermi), its rate, apart from small corrections, is fixed by G_{obs}^V alone, so that, here, one is not involved in sorting this constant out from the others, such as $g_A(0)$, that enter in neutron β-decay, and the β-decay of heavy nuclei.

reason to expect this. The β-decay of the neutron, for example, differs fundamentally from the decay of the muon because the neutron has strong interactions but the muon does not. These interactions would be expected to renormalize G through graphs such as

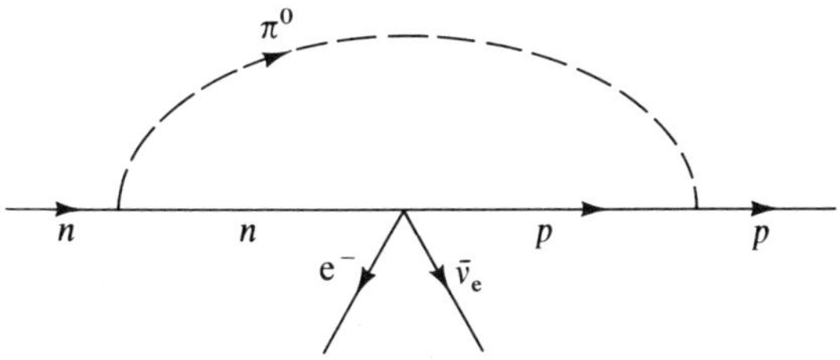

and yet the experimental result is that such a graph has somehow conspired to cancel itself out to give G essentially no renormalization. This "conspiracy" is similar to the conspiracy in electron scattering in which graphs such as

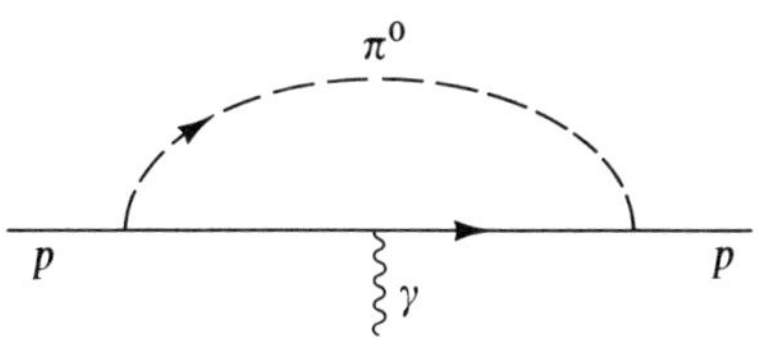

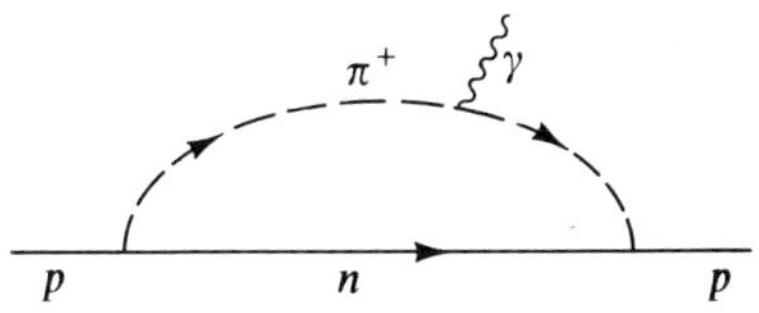

also add up in such a way as to give no renormalization for the proton charge, in agreement with the experimental fact that the charge of the strongly interacting proton is identical to the charge of the positron, which has no strong interactions. In this case, as we have seen in Chapter 4, we can explain the result by the electromagnetic current conservation. In the β-decay case, as we shall discuss in detail, it is explained also by supposing that $\partial_\mu V_\mu = 0$.

There are electromagnetic radiative corrections for the neutron β-decay and, more generally, for the β-decay of nuclei such as O^{14}. However, here the theoretical situation is complicated by the strong interactions. The computations contain divergences that must be treated by cutoff methods.[12] The general result of these computations is to introduce radiative corrections of a few percent and to decrease $f_V(0)$ from .986 to, perhaps, .980.[13] We shall come back later to the

12. See, for example, T. Kinoshita and A. Sirlin, *Phys. Rev.* **113**:1652 (1959); S. M. Berman and A. Sirlin, *Ann. Phys.* **20**:20 (1962); or L. Durand, L. F. Landowitz, and R. B. Marr, *Phys. Rev.* **130**:1188 (1963).
13. See V. L. Telegdi, *Proceedings of the International Conference on Weak Interactions*, Argonne National Laboratories (1965), p. 337.

possible significance of the departure of $f_V(0)$ from unity. We now continue the discussion of the physical significance of the form factors.

The quantity $f_M(q^2)$, the coefficient of $\sigma_{\mu\nu}q_\nu$, in the matrix element of the vector current, is called the "weak magnetic" form factor since it has the same functional form in β-decay as the anomalous magnetic moment term, $F_2(q^2)$, has in electron scattering. It is unobservable in neutron β-decay, since q_μ is too small, but it has been carefully measured in the β-decay of certain nuclei. We shall discuss the significance of these measurements in the next chapter.

The function $f_S(q^2)$ is usually called the "induced scalar" form factor. The reason why it is a "scalar" is seen simply if we couple this term to the leptonic current matrix element and use the Dirac equation and the 4-momentum conservation equation $q_\mu = (e + \nu_e)_\mu$, so that

$$\begin{aligned} iq_\mu\bar{u}(\mathbf{p}')f_S(q^2)u(\mathbf{p})i\bar{u}(\mathbf{e})\gamma_\mu(1 + \gamma_5)u(\boldsymbol{\nu}) \\ = i\bar{u}(\mathbf{p}')f_S(q^2)u(\mathbf{p})i(e + \nu)_\mu\bar{u}(\mathbf{e})\gamma_\mu(1 + \gamma_5)u(\boldsymbol{\nu}) \\ = -i\bar{u}(\mathbf{p}')m_e f_S(q^2)u(\mathbf{p})\bar{u}(\mathbf{e})(1 + \gamma_5)u(\boldsymbol{\nu}). \end{aligned} \tag{9.34}$$

Thus, even though we begin with an underlying Lagrangian that contains only vector and axial vector couplings, the matrix elements have "induced" in them (by the strong interactions which, in principle, determine the f's) a term that has all of the features of a scalar coupling of nucleons to leptons. As we shall see later, there are excellent reasons for believing that this coupling must vanish in neutron β-decay.

We next consider the three g's defined by the matrix element of A_μ. In the limit in which $\mathbf{p} = \mathbf{p}' = \mathbf{0}$, the only one of the g's to survive is the coefficient of $\gamma_5\gamma_\mu$; g_A. However,

$$\bar{u}(\mathbf{0})\gamma_5 u(\mathbf{0}) = 0. \tag{9.35}$$

Thus, in this limit, corresponding to the limit in which G^V_{obs} is defined,

$$\lim_{\mathbf{p}',\mathbf{p}\to 0} \langle\mathbf{p}'|_{s'}A_0(0)|\mathbf{p}\rangle_s = 0. \tag{9.36}$$

and from the identity $i\gamma_4\gamma_5\boldsymbol{\gamma} = \boldsymbol{\sigma}$,

$$\lim_{\mathbf{p}',\mathbf{p}\to 0} (2\pi)^3\langle\mathbf{p}'|_{s'}\mathbf{A}(0)|\mathbf{p}\rangle_s = u^\dagger(\mathbf{0})_{s'}\boldsymbol{\sigma}u(\mathbf{0})_s g_A(0). \tag{9.37}$$

Experimentally, $g_A(0)$ can be determined from the β-decay of the neutron combined with the knowledge of G from muon decay and $f_V(0)$ from O^{14}. In the simplest approximation, in an obvious notation, we may write the effective Hamiltonian for neutron β-decay in the small $\mathbf{p}$, $\mathbf{p}'$ limit as

$$\frac{G}{\sqrt{2}}\, iu^\dagger(\mathbf{0})_{s'}[f_V(0) - i\boldsymbol{\sigma}g_A(0)]u(\mathbf{0})_s, \tag{9.38}$$

with the understanding that each term is to be multiplied into the corresponding term in the leptonic current. The neutron lifetime which is, experimentally, 11.7 ± 0.3 min, depends, in the theory, on the combination

$$f_V(0)^2 + 3g_A(0)^2, \tag{9.39}$$

so that knowing $f_V(0)$ we can find $g_A(0)$. Indeed, from the data,

$$\frac{g_A(0)}{f_V(0)} = -1.18 \pm .02. \tag{9.40}$$

That these two numbers are so nearly the same in absolute value, and that both are so close to one, is remarkable.

The term $g_P(q^2)$ is known as the "induced" pseudoscalar, since, following the same line of argument as used in showing that f_S was effectively a scalar constant, we can show that $g_P(0)$ measures the effective pseudoscalar coupling induced by the strong nucleonic interactions. There is a good theoretical prediction for g_P that we shall give below. In the next section we shall show that strong theoretical arguments indicate that the last of the six form factors $g_E(q^2)$ is zero in neutron β-decay. This conclusion is also in agreement with all the experimental evidence.

4. Although the weak interactions are known to violate parity, all the experimental evidence is consistent with the assumption that the Hamiltonian responsible for β-decay is time-reversal invariant. There is a time-reversal violation in the decay of the neutral K meson, but it appears to be very small, on the scale of the usual weak interactions, and has not been observed at all in ordinary β-decay.[14] To say that the β-decay coupling is T-conserving is to say that the Hamiltonian density $H_{WK}(\mathbf{r}, t)$ transforms under T—the time-reversal operator as follows:

$$TH_{WK}(\mathbf{r}, t)T^{-1} = H_{WK}(\mathbf{r}, -t). \tag{9.41}$$

But in the current-current theory

$$H_{WK} = \frac{G}{\sqrt{2}} J_\mu J_\mu{}^*, \tag{9.42}$$

which means that the transformation properties of J_μ under T must be such as to assure Eq. (9.41). We do not know the precise functional form of V_μ and A_μ, but their transformation under T can be deduced from the transformation of L_μ, the leptonic current, whose form we do know. Take the electronic part of L_μ, for example,

$$L_\mu{}^e = i\bar{\psi}_e(\mathbf{r}, t)\gamma_\mu(1 + \gamma_5)\psi_{\nu_e}(\mathbf{r}, t), \tag{9.43}$$

and apply T. Thus[15]

---

14. See F. P. Calaprice et al., *Phys. Rev. Letters* **18**:918 (1967) for a recent experiment and a discussion of other experiments.
15. Here the arbitrariness of phase in T is related to the conservation of lepton number N_l. If T is a time-reversal operator, then so is

$$T' = e^{i\theta(N_\mu + N_e)}T$$

for arbitrary θ.

$$Ti\bar{\psi}_e(\mathbf{r}, t)\gamma_\mu(1+\gamma_5)\psi_{\nu_e}(\mathbf{r}, t)T^{-1}$$
$$= -i\eta_e{}^*\eta_{\nu_e}\bar{\psi}_e(\mathbf{r}, -t)t^\dagger(\gamma_\mu{}^*(1+\gamma_5{}^*))t\psi_{\nu_e}(\mathbf{r}, -t). \quad (9.44)$$

We fix the phases of e and ν_e so that $\eta_e{}^*\eta_\nu = 1$. Thus from the Hermiticity of the γ's and γ_5,

$$TL_\mu{}^e(\mathbf{r}, t)T^{-1} = -i\bar{\psi}_e(\mathbf{r}, -t)t^\dagger(\gamma_\mu{}^T(1+\gamma_5{}^T))t\psi_\nu(\mathbf{r}, -t). \quad (9.45)$$

But, by definition,

$$t^\dagger\gamma_\mu{}^Tt = \gamma_\mu \quad (9.46)$$

and

$$t^\dagger\gamma_\mu{}^T\gamma_5{}^Tt = t^\dagger\gamma_\mu{}^Ttt^\dagger\gamma_5{}^Tt = \gamma_\mu\gamma_5. \quad (9.47)$$

Thus, with suitable choice of the relative phases,

$$TL_\mu(\mathbf{r}, t)T^{-1} = -L_\mu(\mathbf{r}, -t). \quad (9.48)$$

Hence we demand that, to keep $(V_\mu + A_\mu)L_\mu$ invariant,

$$\begin{aligned} TV_\mu(\mathbf{r}, t)T^{-1} &= -V_\mu(\mathbf{r}, -t), \\ TA_\mu(\mathbf{r}, t)T^{-1} &= -A_\mu(\mathbf{r}, -t). \end{aligned} \quad (9.49)$$

Therefore we have the condition (see Chapter 5)

$$\langle \mathbf{p}'|_{s'}\mathbf{J}(0)|\mathbf{p}\rangle_s = -\langle -\mathbf{p}'|_{-s'}\mathbf{J}(0)|-\mathbf{p}\rangle_{-s}^* \quad (9.50)$$

or, in terms of the free spinor representation of this matrix element,

$$\bar{u}(\mathbf{p}')_{s'}\mathscr{J}(\mathbf{p}', \mathbf{p})u(\mathbf{p})_s = -\bar{u}(\mathbf{p}')_{s'}(t\mathscr{J}(-\mathbf{p}', -\mathbf{p})t^\dagger)^*u(\mathbf{p})_s, \quad (9.51)$$

or

$$i(\gamma_i f_V(q^2) + \sigma_{i\nu}q_\nu f_M(q^2) + iq_i f_S(q^2))$$
$$= i(\gamma_i f_V{}^*(q^2) + \sigma_{i\nu}q_\nu f_M{}^*(q^2) + iq_i f_S{}^*(q^2)). \quad (9.52)$$

Thus time reversal implies that

$$\begin{aligned} f_V &= f_V{}^*, \\ f_M &= f_M{}^*, \\ f_S &= f_S{}^*, \end{aligned} \quad (9.53)$$

whereas charge symmetry implied that

$$\begin{aligned} f_V &= f_V{}^*, \\ f_M &= f_M{}^*, \\ f_S &= -f_S{}^*. \end{aligned} \quad (9.54)$$

The two conditions, together, imply that there is no induced scalar in the β-decay of the neutron. If we analyze the matrix element of $A_\mu(0)$ in the same way we find that

$$i[\gamma_5[\gamma g_A(q^2) + i\mathbf{q}g_P(q^2) + i(\mathbf{p}+\mathbf{p}')g_E(q^2)]]$$
$$= i[\gamma_5[\gamma g_A(q^2)^* + i\mathbf{q}g_P{}^*(q^2) + i(\mathbf{p}+\mathbf{p}')g_E{}^*(q^2)]], \quad (9.55)$$

or that

$$\begin{aligned} g_A &= g_A^*, \\ g_P &= g_P^*, \\ g_E &= g_E^*. \end{aligned} \tag{9.56}$$

Charge symmetry required that

$$\begin{aligned} g_A &= g_A^*, \\ g_P &= g_P^*, \\ g_E &= -g_E^*, \end{aligned} \tag{9.57}$$

so that the two requirements, together, imply that

$$g_E = 0. \tag{9.58}$$

Hence, assuming time reversal and charge symmetry, the neutron β-decay depends on four form factors, which can all be chosen real.

There is an essentially equivalent way of presenting these results, due to Weinberg.[16] Putting the time reversal and parity requirements together, we deduce at once

$$\begin{aligned} (PT)\mathbf{V}(0)(PT)^{-1} &= \mathbf{V}(0), \\ (PT)\mathbf{A}(0)(PT)^{-1} &= -\mathbf{A}(0). \end{aligned} \tag{9.59}$$

We now wish to fix the C character of the currents. We cannot do this directly from the CPT invariance of the weak interactions. From the CPT invariance of H_{WK} we know only that we must have

$$(CPT)(J_\mu J_\mu^* + \text{h.c.})(CPT)^{-1} = (J_\mu J_\mu^* + \text{h.c.}). \tag{9.60}$$

But this does not uniquely fix the transformation property of $J_\mu(0)$ under CPT, since J_μ and J_μ^* enter H_{WK} together in a product. However, we can specify this transformation by studying once again how the leptonic current, whose functional form we do know, transforms. Under PT we have

$$(PT)(i(\bar{\psi}_e(0)\gamma(1+\gamma_5)\psi_{\nu_e}(0)))(PT)^{-1} = i\bar{\psi}_e(0)\gamma(1-\gamma_5)\psi_{\nu_e}. \tag{9.61}$$

Hence it only remains to determine how L_μ^e transforms under C. In the Hermitian representation of the γ's, choosing all the lepton C phases to be 1, we can write

$$\begin{aligned} C\psi_l(\mathbf{r}, t)C^{-1} &= \bar{\psi}_l(\mathbf{r}, t)c, \\ C\bar{\psi}_e(\mathbf{r}, t)C^{-1} &= -c^{\dagger}\psi_e(\mathbf{r}, t), \end{aligned} \tag{9.62}$$

with, in this representation of the γ's,

$$\begin{aligned} c &= \gamma_4\gamma_2, \\ (\gamma_4\gamma_2)\gamma_\mu^{t}(\gamma_4\gamma_2)^\dagger &= -\gamma_\mu, \end{aligned} \tag{9.63}$$

---

16. S. Weinberg, *Phys. Rev.* **112**:1375 (1958).

and prove (∴ understood) that

$$C(i\bar{\psi}_e\gamma\psi_{\nu_e})C^{-1} = -(i\bar{\psi}_e\gamma\psi_{\nu_e})^\dagger \tag{9.64}$$

and

$$C(i\bar{\psi}_e\gamma\gamma_5\psi_{\nu_e})C^{-1} = +(i\bar{\psi}_e\gamma\gamma_5\psi_{\nu_e})^\dagger. \tag{9.65}$$

Thus

$$(CPT)\mathbf{L}^e(0)(CPT)^{-1} = -(\mathbf{L}^e)^\dagger. \tag{9.66}$$

Hence the *CPT* invariance of the Feynman-Gell-Mann theory will be assured if we have

$$\begin{aligned} C\mathbf{V}(0)C^{-1} &= -\mathbf{V}(0)^\dagger, \\ C\mathbf{A}(0)C^{-1} &= \mathbf{A}(0)^\dagger, \end{aligned} \tag{9.67}$$

so that[17]

$$(CPT)\mathbf{J}(0)(CPT)^{-1} = -\mathbf{J}(0)^\dagger. \tag{9.68}$$

This result is also consistent with the commutation relation between J_μ and the electric charge Q,

$$[\mathbf{J}, Q]_- = \mathbf{J}, \tag{9.69}$$

since, with $(CPT)Q(CPT)^{-1} = -Q$, it implies that

$$[\mathbf{J}^\dagger, Q]_- = -\mathbf{J}^\dagger. \tag{9.70}$$

Weinberg classifies currents according to how they transform under

$$G = Ce^{i\pi T_2}. \tag{9.71}$$

If a current is a first-class Weinberg vector current, then by definition[18]

$$GV_\mu{}^{(1)}G^{-1} = V_\mu{}^{(1)}. \tag{9.72}$$

If it is a second-class vector, then

$$GV_\mu{}^{(2)}G^{-1} = -V_\mu{}^{(2)}. \tag{9.73}$$

If it is a first-class axial vector, then

$$GA_\mu{}^{(1)}G^{-1} = -A_\mu{}^{(1)}, \tag{9.74}$$

---

17. The *CPT* transformation of **J** looks like the transformation of **J** under I_2, assuming charge symmetry. However, we cannot use *CPT* alone to deduce the reality condition since, under C, a particle state is transformed into its antiparticle state. Under *CPT*, we have

$$\langle \mathbf{P}'|_{s'P}\mathbf{J}(0)|\mathbf{P}\rangle_{sN} = -(\langle \mathbf{P}'|_{-s'\bar{P}}\mathbf{J}(0)^\dagger|\mathbf{P}\rangle_{-s\bar{N}})^*,$$

which relates N and $\bar{N}$ lifetimes but does not give a reality condition.

18. Any current can be decomposed into first- and second-class Weinberg currents; namely, for arbitrary V_μ and A_μ let

$$V_\mu{}^{(1)} = \tfrac{1}{2}(V_\mu + GV_\mu G^{-1}),$$

$$V_\mu{}^{(2)} = \tfrac{1}{2}(V_\mu - GV_\mu G^{-1}),$$

and

and the second-class axial currents have

$$GA_\mu{}^{(2)}G^{-1} = A_\mu{}^{(2)}. \tag{9.75}$$

From the transformation property under *CPT* it follows that, with $G' = Te^{i\pi T_2}$,

$$\begin{aligned} G'\mathbf{V}^{(1)}(0)G'^{-1} &= \mathbf{V}^{(1)}(0)^\dagger, \\ G'\mathbf{A}^{(1)}(0)G'^{-1} &= \mathbf{A}^{(1)}(0)^\dagger, \end{aligned} \tag{9.76}$$

and

$$\begin{aligned} G'\mathbf{V}^{(2)}(0)G'^{-1} &= -\mathbf{V}^{(2)}(0)^\dagger, \\ G'\mathbf{A}^{(2)}(0)G'^{-1} &= -\mathbf{A}^{(2)}(0)^\dagger. \end{aligned} \tag{9.77}$$

Hence, if we assume all currents are first-class Weinberg, we can conclude that for $0 \leq q^2 \leq \infty$

$$\begin{aligned} f_V &= f_V{}^*, \\ f_M &= f_M{}^*, \\ f_S &= 0, \end{aligned} \tag{9.78}$$

and

$$\begin{aligned} g_A &= g_A{}^*, \\ g_P &= g_P{}^*, \\ g_E &= 0, \end{aligned} \tag{9.79}$$

the same conditions we derived from T invariance and charge symmetry. It is a welcome, simplifying character of the weak interactions that these conditions are experimentally valid,[19] at least at small q^2, where they are tested.

In the next chapter we turn to the detailed study of V_μ.

$$\begin{aligned} A_\mu{}^{(1)} &= \tfrac{1}{2}(A_\mu - GA_\mu G^{-1}), \\ A_\mu{}^{(2)} &= \tfrac{1}{2}(A_\mu + GA_\mu G^{-1}). \end{aligned}$$

There is only one minor subtlety here. If $e^{i\pi T_2}$ is applied to a quantity that transforms like an isospinor, then, because of the half-angle character of such spinor transformations, $(e^{i\pi T_2})^2 = -1$ effectively. However, for any operator O,

$$G^2OG^{-2} = O.$$

19. The neutron β-decay experiments show that

$$\frac{g_A(0)}{f_V(0)} = \frac{|g_A(0)|}{|f_V(0)|}e^{i\phi},$$

with $\phi = 180° \pm 8°$; M. T. Burgy et al., *Phys. Rev.* **110**:1214 (1958).

10
Weak Currents: The Vector

One of the most significant results presented in the last chapter was the experimental observation that

$$G^V_{\text{obs}} = Gf_V(0) \simeq G, \tag{10.1}$$

meaning that the strong interactions do not succeed in renormalizing, in a significant way, the weak coupling constant of the vector current. As we have remarked, this situation is analogous to electromagnetism, where the strong interactions fail to renormalize the electric charge,[1] a circumstance that is explained by the conservation of electric current in the presence of the strong interactions. Hence it is tempting to postulate something like

$$\partial_\mu J_\mu = 0 \tag{10.2}$$

for the weak current. We begin this chapter by showing that this is *not* possible as an *exact* dynamical equation valid to all orders in all of the interactions. For the moment, we confine the argument to V_μ and then later extend it to A_μ.

We give two separate arguments for the nonconservation of V_μ.

1. Suppose $\partial_\mu V_\mu(x) = 0$ exactly. Then

$$Q_V = \int d\mathbf{r}\, V_0(\mathbf{r}, t) \tag{10.3}$$

would commute with the Hamiltonian of the world. Let $|\mathbf{N}\rangle$ and $|\mathbf{P}\rangle$ be the physical neutron and proton states with their actual masses. (We treat the weak interaction to first order but include the electromagnetic and strong couplings to all orders.) Thus[2]

$$0 = \langle \mathbf{n}|[H, Q_V]_-|\mathbf{p}\rangle = (M_N - M_P)\langle \mathbf{n}|Q_V|\mathbf{p}\rangle. \tag{10.4}$$

But, letting $\mathbf{n} = \mathbf{p} = \mathbf{0}$, we see that

$$\langle \mathbf{0}|Q_V|\mathbf{0}\rangle = f_V(0) \neq 0. \tag{10.5}$$

Thus we must have $M_N = M_P$, which we know is not true exactly since the electromagnetic couplings (presumably) split the nucleon masses. This gives us

1. The electromagnetic interactions themselves do renormalize the charge despite the conserved current. But this is a special feature of the infinite range, zero-mass photon field.
2. To keep the distinction between nucleon and lepton masses as clear as possible, we shall use capital letters for nucleon masses.

a hint that the weak current conservation may be expected to hold only in the same approximation in which the neutron and proton are degenerate in mass—that is, the good *i*-spin limit.

2. The second argument is, perhaps, more experimental in character. Consider

$$(2\pi)^3\langle \mathbf{p}'|_N V_\mu(0)|\mathbf{p}\rangle_P = i\bar{u}(\mathbf{p}')[\gamma_\mu f_V(q^2) + \sigma_{\mu\nu}q_\nu f_M(q^2) + iq_\mu f_S(q^2)]u(\mathbf{p}). \tag{10.6}$$

Since we are not working in the good *i*-spin limit, for the moment, we cannot exclude the induced scalar $f_S(q^2)$. If we apply q_μ to this equation and set the result equal to zero, we find[3]

$$m_e(M_N - M_P)\frac{f_V(q^2)}{q^2} = f_S(q^2)m_e. \tag{10.7}$$

However, in β-decay, $0 < |q^2| \leq (M_N - M_P)^2$. Therefore

$$|m_e f_S(q^2)| \gtrsim \frac{m_e}{M_N - M_P} f_V(0) \sim .5. \tag{10.8}$$

An induced scalar of such magnitude is ruled out by the experiments on the β-decay of the neutron. But even apart from this, Eq. (10.8) means that $f_S(0)$ is infinite unless $f_V(0) = 0$, which would imply an extremely bizarre behavior for f_S and f_V in the neighborhood of $q^2 = 0$. Here again, the argument leading to Eq. (10.8) breaks down in the good *i*-spin limit, where $M_N = M_P$.

We may conclude from this discussion that, if we are to postulate current conservation of the vector current, it can only be done approximately, and in an approximation that coincides with the good *i*-spin limit—that is, with the electromagnetic field turned off. However, there *is* one current that we know is conserved in this limit and that is the isotopic current itself. The proposal of Feynman and Gell-Mann[4] is to identify the current V_μ of β-decay with the isotopic current. This means, specifically, that V_μ, the vector current that raises electric charge by one unit in β-decay, is to be identified with

$$V_\mu = J^V_{\mu 1} + iJ^V_{\mu 2}. \tag{10.9}$$

The isotopic "raising" current, while

$$V_\mu{}^\dagger = J^V_{\mu 1} - iJ^V_{\mu 2}, \tag{10.10}$$

is the isotopic lowering current. This hypothesis, entirely confirmed, as we shall see, by experiment, has been one of the most fruitful ever made in the weak interactions.

To begin with, we note that this V_μ has all of the desirable properties discussed in the last chapter.

---

3. We multiply here by m_e since the effective induced scalar constant (see last chapter) is $m_e f_S(q^2)$.
4. R. Feynman and M. Gell-Mann, *Phys. Rev.* **109**:193 (1958). See also S. Gershstein and J. Zeldovich, *JETP* **29**:698 (1955). [Translation: *Soviet Physics, JETP* **2**:576 (1956).]

1. From the relation (good *i*-spin limit)[5] for the electric charge Q,

$$Q = \frac{B}{2} + T_3, \tag{10.11}$$

and the commutation relation,

$$[J_\mu^V(0)_i, T_3]_- = i\epsilon_{i3j} J_\mu^V(0)_j, \tag{10.12}$$

we have explicitly

$$[Q, V_\mu]_- = V_\mu, \tag{10.13}$$

so that $\Delta Q = \Delta T_3 = 1$ in transitions that are first order in V_μ.

2. This V_μ is manifestly charge-symmetric.

3. Since the pion-nucleon interactions have been constructed to be time-reversal invariant, $\mathbf{J}_\mu^V(0)$, transforms as a T-symmetric current should. Thus we know that

$$(2\pi)^3 \langle \mathbf{p}' | V_\mu(0) | \mathbf{p} \rangle = i\bar{u}(\mathbf{p}')[\gamma_\mu f_V(q^2) + \sigma_{\mu\nu} q_\nu f_M(q^2)] u(\mathbf{p}) \tag{10.14}$$

with

$$f_S = 0 \tag{10.15}$$

and

$$\begin{aligned} f_V &= f_V^*, \\ f_M &= f_M^*. \end{aligned} \tag{10.16}$$

The vanishing of the induced scalar is consistent with current conservation, and is, in fact implied by it, in the good *i*-spin limit.

We can now, however, use the Feynman-Gell-Mann hypothesis to make a large number of specific new predictions. We may begin by computing $f_V(0)$. We shall again proceed in two quite different ways, each of which illustrates a technique with wide application.

1. Using the same arguments as in Chapter 5 we see that

$$f_V(0) = \langle \mathbf{0}\, \tfrac{1}{2}\, \tfrac{1}{2} | \int d\mathbf{r}\, V_0(\mathbf{r}, 0) | \mathbf{0} \tfrac{1}{2} - \tfrac{1}{2} \rangle, \tag{10.17}$$

where we have explicitly exhibited the isotopic indices T and T_3 for neutron and proton. However, since V_μ is the + component of the isotopic current,

$$\int d\mathbf{r}\, V_0(\mathbf{r}, 0) = T_1 + iT_2 = T_+, \tag{10.18}$$

where T_1 and T_2 are the isotopic charges, or the generators of SU_2. But, in general,

---

5. $[\mathbf{J}_\mu^V, B]_- = 0$.
6. See J. Bernstein, M. Gell-Mann, and L. Michel, *Il Nuovo Cimento* **16**:560 (1960).
7. The student of field theory will wonder whether we are referring to the renormalized or unrenormalized propagator. For the moment the distinction is irrelevant and we clarify it shortly. $|0\rangle$ is the vacuum.

$$T_+|TT_3\rangle = \sqrt{(T - T_3)(T + T_3 + 1)}\,|TT_3 + 1\rangle \tag{10.19}$$

or

$$f_V(0) = \langle \tfrac{1}{2}\tfrac{1}{2}|T_+|\tfrac{1}{2} - \tfrac{1}{2}\rangle = 1. \tag{10.20}$$

This argument contains an implicit assumption that has been concealed in the statement of Eq. (10.19). As we saw in Chapter 4, a charge derived from a conserved current is only conserved in time if the fields involved do not have interactions of infinite range; otherwise there is a renormalization effect due to these couplings. Since we are working in the approximation in which the photon field is turned off, we can ignore such effects and Eq. (10.20) is correct.

2. In making this second argument[6] we introduce a number of new concepts. In particular we shall need a general definition of the propagator of a Heisenberg field. To give it, we make a small, but important, modification of the definition of time ordering. We introduce the subscript + as follows:

$$(\phi(x)\phi(y))_+ = \theta(x_0 - y_0)\phi(x)\phi(y) + \theta(y_0 - x_0)\phi(y)\phi(x), \tag{10.21}$$

providing that $\phi(x)$ and $\phi(y)$ commute at equal times like free Boson fields, and we call

$$(\psi(x)\bar{\psi}(y))_+ = \theta(x_0 - y_0)\psi(x)\bar{\psi}(y) - \theta(y_0 - x_0)\bar{\psi}(y)\psi(x), \tag{10.22}$$

if $\psi(x)$ and $\bar{\psi}(y)$ anticommute, at equal times, like free Fermion fields. Thus the + symbol acts like P ordering for Bosons, but for Fermions a factor of (-1) is introduced whenever ψ and $\bar{\psi}$ are permuted. The exact nucleon (or Boson) Feynman propagator is defined to be[7]

$$\mathbf{S}_F(x - y) = \langle 0|(\psi(x)\bar{\psi}(y))_+|0\rangle. \tag{10.23}$$

We are now going to use symmetry arguments of the type used before in connection with matrix elements of the current to deduce the general structure of $\mathbf{S}_F(p)$, the Fourier transform of[8] $\mathbf{S}_F(x - y)$. The discussion is somewhat involved and we just outline it here.[9] The object (α and β are Dirac indices),

$$\langle 0|\psi_\alpha(x)\bar{\psi}_\beta(0)|0\rangle, \tag{10.24}$$

can, by general Lorentz covariance, be expanded in the following covariant form

$$\langle 0|\psi_\alpha(x)\bar{\psi}_\beta(0)|0\rangle = \delta_{\alpha\beta}A^S(x) + (\gamma_\mu)_{\alpha\beta}A_\mu{}^V(x) + (\gamma_5)_{\alpha\beta}A^P(x) + (\sigma_{\mu\nu})_{\alpha\beta}A^T_{\mu\nu}(x) + i(\gamma_5\gamma_\mu)_{\alpha\beta}A_\mu{}^A(x), \tag{10.25}$$

where the five A's are functions of x that must transform under Lorentz transformations in the same way that their associated γ matrices do. By using the

---

8. Note that

$$\langle 0|\psi(x)\bar{\psi}(y)|0\rangle = \langle 0|\psi(0)e^{i(\mathrm{P}(x-y))}\bar{\psi}(0)|0\rangle,$$

so that S_F is really just a function of $x - y$.

9. The reader who wants the details may study S. Schweber, *Relativistic Quantum Field Theory*, Harper and Row, 1961, p. 672 ff.

translation invariance of $\langle 0|\psi_\alpha(x)\bar{\psi}_\beta(0)|0\rangle$ and taking Hermitian conjugates, we have

$$\langle 0|\psi_\alpha(x)\bar{\psi}_\beta(0)|0\rangle = (\gamma_4)_{\beta\delta}\langle 0|\psi_\delta(-x)\bar{\psi}_{\delta'}(0)|0\rangle^*(\gamma_4)_{\delta'\alpha}. \tag{10.26}$$

Neglecting the effect of the weak couplings on the fields we can invoke *PT* invariance to show that

$$\langle 0|\psi_\alpha(x)\bar{\psi}_\beta(0)|0\rangle = (\gamma_4 t)_{\alpha\delta}\langle 0|\psi_\delta(-x)\bar{\psi}_{\delta'}(0)|0\rangle^*(\gamma_4 t)^\dagger_{\delta'\beta}, \tag{10.27}$$

where $t_{\alpha\beta}$ is the time-reversal matrix. The combined conditions allow one to eliminate $A^T_{\mu\nu}$, $A_\mu{}^A$, and A^P, so that[10]

$$\langle 0|\psi_\alpha(x)\bar{\psi}_\beta(0)|0\rangle = \delta_{\alpha\beta}A^S(x) + i(\gamma_\mu)_{\alpha\beta}A_\mu{}^V(x). \tag{10.28}$$

We next introduce the sum over the complete set of eigenstates of P_μ between ψ and $\bar{\psi}$ in Eq. (10.28). Thus

$$\langle 0|\psi_\alpha(x)\bar{\psi}_\beta(0)|0\rangle = \sum_n e^{i(p_n x)}\langle 0|\psi_\alpha(0)|n\rangle\langle n|\bar{\psi}_\beta(0)|0\rangle. \tag{10.29}$$

In this sum we may imagine fixing the four-momentum p and summing over the degenerate set of states for a given p. For example, for the one-nucleon state we would sum over the spins for a fixed p. We call[11]

$$\theta(-p^2)\theta(p_0)W_{\alpha\beta}(p) = (2\pi)^3 \sum_\lambda \langle 0|\psi_\alpha(0)|p\lambda\rangle\langle p\lambda|\bar{\psi}_\beta(0)|0\rangle, \tag{10.30}$$

where λ is any quantum number, such as spin, that labels the degenerate state, and the θ functions are introduced to restrict the four-momenta to physical values which, in our metric, correspond to $p^2 < 0$ and $p_0 > 0$. If we transform the sum over n into an integral over p, we find

$$\langle 0|\psi_\alpha(x)\bar{\psi}_\beta(0)|0\rangle = \frac{1}{(2\pi)^3}\int d^4p\, e^{i(px)}\theta(p_0)\theta(-p^2)W_{\alpha\beta}(p). \tag{10.31}$$

The symmetry arguments above show that

$$W_{\alpha\beta}(p) = \delta_{\alpha\beta}A^S(p) + i(\gamma_\mu)_{\alpha\beta}A_\mu{}^V(p). \tag{10.32}$$

From Lorentz invariance it follows that

$$A^S(p) = A_1(-p^2) \tag{10.33}$$

and

$$A_\mu{}^V(p) = p_\mu A_2(-p^2). \tag{10.34}$$

Thus, calling $p^2 = -m^2$, doing the θ-function integrals, and noting that from charge conjugation invariance $\langle 0|\psi_\alpha(x)\bar{\psi}_\beta(0)|0\rangle$ and $\langle 0|\bar{\psi}_\beta(0)\psi_\alpha(x)|0\rangle$ have essentially identical covariant expansions, we arrive at the equation

---

10. We have redefined $A_\mu{}^V$ by extracting an i so that A^S and $A_\mu{}^V$ are real by time-reversal invariance.
11. See Schweber, op. cit.

$$\mathbf{S}_F(p)_{\alpha\beta} = \int_0^\infty dm^2 \frac{1}{p^2 + m^2 + i\epsilon} (\delta_{\alpha\beta} A_1(-m^2) + i(\gamma p)_{\alpha\beta} A_2(-m^2)). \quad (10.35)$$

It is instructive to examine the contribution to A_1 and A_2 from the one-nucleon intermediate state in the sum, which would be the only contribution if the nucleons were free. We can find the contribution to A_1 by taking the trace of $\mathbf{S}_F(p)_{\alpha\beta}$ with respect to the spinor indices. Thus

$$A_1(-m^2) = \frac{(2\pi)^3}{4} \sum_{n,\alpha} \langle 0|\psi_\alpha(0)|n\rangle\langle n|\bar{\psi}_\alpha(0)|0\rangle, \quad (10.36)$$

where the sum extends over those $|n\rangle$ with $p_n{}^2 = -m^2$. We may now pick out the one-nucleon state with mass M from the sum, so that

$$A_1(-m^2) = \frac{(2\pi)^3}{4} \sum_{s,\alpha} \langle 0|\psi_\alpha(0)|\mathbf{p}\rangle_s\langle \mathbf{p}|_s\bar{\psi}_\alpha(0)|0\rangle\, \delta(m^2 - M^2) + A_1'(-m^2), \quad (10.37)$$

where A_1' represents the higher mass states. But, as we have seen in Chapter 7,

$$\langle 0|\psi_\alpha(0)|\mathbf{p}\rangle_s = \frac{\sqrt{Z_2}}{(2\pi)^{3/2}} \sqrt{2E}\, u_\alpha(\mathbf{p})_s, \quad (10.38)$$

where $E = \sqrt{\mathbf{p}^2 + M^2}$. We have normalized $\langle 0|\psi_\alpha(0)|\mathbf{p}\rangle_s$ in this special way to conform to the generally adopted definition of the "renormalization" constant Z_2. Thus[12]

$$A_1(-M^2) = \tfrac{1}{2} E Z_2 \operatorname{Tr}\left[\frac{-i(\gamma p) + M}{2E}\right] = Z_2 M, \quad (10.39)$$

so that

$$A_1(-m^2) = Z_2 M\, \delta(M^2 - m^2) + A_1'(-m^2). \quad (10.40)$$

Using[13]

$$\operatorname{Tr}(\gamma_4 i(\gamma p)) = -4E \quad (10.41)$$

and multiplying both sides of Eq. (10.35) by γ_4, we show that

$$A_2(-M^2) = -Z_2 \quad (10.42)$$

so that,

$$\mathbf{S}_F(p) = \frac{Z_2}{i(\gamma p) + M} + R_F(p), \quad (10.43)$$

where $R_F(p)$, since it receives contributions only from higher mass states, has the property that it remains finite when p is put on the one-nucleon mass shell.

---

12. With our spinor normalization

$$\sum_{s,\alpha} u_\alpha(p)_s \bar{u}_\alpha(p)_s = \operatorname{Tr}\left[\frac{-i(\gamma p) + M}{2E}\right].$$

13. $\operatorname{Tr}[\gamma_4\gamma_\mu] = 4\delta_{\mu 4}$.

On the shell, $\mathbf{S}_F(p)$ has a pole at the physical nucleon mass with residue Z_2. Thus, and this is the essential property that we need in what follows, as $i(\gamma p) \longrightarrow -M$ (the mass shell condition)

$$\mathbf{S}_F(p)^{-1} \longrightarrow \frac{1}{Z_2}(i(\gamma p) + M). \tag{10.44}$$

It is at this point that we may distinguish between renormalized and unrenormalized fields ψ. We call the renormalized field ψ_R, with the definition

$$\psi_R = \frac{1}{\sqrt{Z_2}}\psi. \tag{10.45}$$

Thus, on the shell,

$$\mathbf{S}_{FR}^{-1}(p) \longrightarrow i(\gamma p) + M, \tag{10.46}$$

provided we define

$$\mathbf{S}_{FR}(x - y) = \langle 0|(\psi(x)_R\bar{\psi}(y)_R)_+|0\rangle. \tag{10.47}$$

We shall next prove the so-called Ward[14] identity for β-decay and from it and the structure of $\mathbf{S}_F$ prove that $G_V/G = 1$. We recall (see Chapter 4) that the isotopic spin current is generated by the combined transformation

$$\begin{aligned} \psi &\longrightarrow \left(1 + i\boldsymbol{\Lambda}(x)\cdot\frac{\boldsymbol{\tau}}{2}\right)\psi, \\ \boldsymbol{\phi} &\longrightarrow \boldsymbol{\phi} + \boldsymbol{\phi}\times\boldsymbol{\Lambda}(x), \end{aligned} \tag{10.48}$$

where $\boldsymbol{\Lambda}(x)$ is an infinitesimal space-dependent gauge function. Under this transformation $\mathfrak{L}$, the *total* pion-nucleon Lagrangian, goes into

$$\mathfrak{L} \longrightarrow \mathfrak{L} - \mathbf{V}_\mu\cdot\partial_\mu\boldsymbol{\Lambda}, \tag{10.49}$$

where $\mathbf{V}_\mu$ is the conserved isotopic vector current. Under this transformation the Green's function, $\mathbf{S}_F(x - y)$, is also not invariant. Indeed, substituting $\psi(x) \longrightarrow \left(1 + i\boldsymbol{\Lambda}(x)\cdot\frac{\boldsymbol{\tau}}{2}\right)\psi(x)$ in $\langle 0|(\psi(x)\bar{\psi}(y))_+|0\rangle$ and keeping terms to first order in $\boldsymbol{\Lambda}$, we see that the change in $\mathbf{S}_F$, $\Delta\mathbf{S}_F$, is given by

$$\Delta\mathbf{S}_F(x - y) = i\boldsymbol{\Lambda}(x)\cdot\frac{\boldsymbol{\tau}}{2}\mathbf{S}_F(x - y) - i\mathbf{S}_F(x - y)\boldsymbol{\Lambda}(y)\cdot\frac{\boldsymbol{\tau}}{2}. \tag{10.50}$$

In momentum space this equation becomes

$$\Delta\mathbf{S}_F(p, p') = i\boldsymbol{\Lambda}(p)\cdot\frac{\boldsymbol{\tau}}{2}\mathbf{S}_F(p) - i\mathbf{S}_F(p')\boldsymbol{\Lambda}(p')\cdot\frac{\boldsymbol{\tau}}{2}. \tag{10.51}$$

However, we may compute $\Delta\mathbf{S}_F(p, p')$ in another way. Since $\boldsymbol{\Lambda} \ll 1$, the added term in $\mathfrak{L}$ in Eq. (10.49) can be taken as a perturbation to be treated to first

---

14. The original Ward identity was proved for electrodynamics by J. C. Ward, *Phys. Rev.* **77**:2931 (1950). The proof given here follows J. Bernstein et al., op. cit., and can be extended to any current, conserved or not, generated by phase transformations of the type discussed in Chapter 2. For another proof using the general definition of the vertex function, see Y. Takahashi, *Nuovo Cimento* **6**:371 (1957).

order in Λ. If we think in terms of Feynman diagrams then we may represent $\mathbf{S}_F(p)$ by a line

If we turn on the perturbation, the effect on the propagator (to lowest order) is to replace $\mathbf{S}_F(p)$ by

where x stands for the action of the perturbation. As a simple example, consider the effect of an added weak external electromagnetic field on the propagator of a free electron. Diagramatically this would correspond to

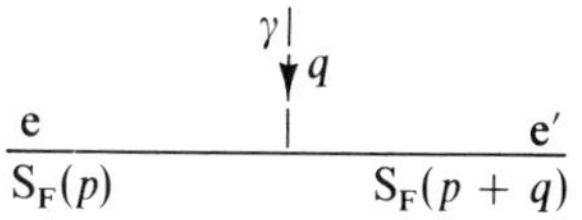

so that the change in the propagator due to this field would be, in first order,

$$\mathbf{S}_F(p)A_\mu(q)V_\mu(p, p')\mathbf{S}_F(p'),$$

where $V_\mu(p, p')$ is the vertex function corresponding to the absorption of the external photon. In our case we have[15]

$$\Delta\mathbf{S}_F(p, p') = -i\mathbf{S}_F(p)[(p_\mu - p_\mu')(\Lambda(p - p')\cdot \mathbf{V}_\mu(p', p))]\mathbf{S}_F(p'). \tag{10.52}$$

Since Λ is arbitrary we can, by equating the two methods of calculating ΔS_F, conclude that

$$i\frac{\tau}{2}\mathbf{S}_F(p) - i\mathbf{S}_F(p')\frac{\tau}{2} = -i(p - p')_\mu \mathbf{S}_F(p)\mathbf{V}_\mu(p, p')\mathbf{S}_F(p') \tag{10.53}$$

or that

$$i\frac{\tau}{2}\mathbf{S}_F^{-1}(p') - i\mathbf{S}_F^{-1}(p)\frac{\tau}{2} = -i(p - p')_\mu V_\mu(p, p'). \tag{10.54}$$

This is the celebrated Ward identity in its "integral" form. The "differential" Ward identity is obtained by letting $p \longrightarrow p'$ in Eq. (10.54). Thus

$$\frac{\tau}{2}\frac{\partial}{\partial p_\mu}\mathbf{S}_F^{-1}(p) = \mathbf{V}_\mu(p, p). \tag{10.55}$$

Since

$$\mathbf{S}_{FR}^{-1} = \mathbf{S}_F^{-1}Z_2, \tag{10.56}$$

15. This is the Fourier transform of

$$\int d^4u \int d^4v \int d^4z \mathbf{S}_F(x - u)\mathbf{V}_\mu(u, z, v)\frac{\partial \Lambda(z)}{\partial z_\mu}\mathbf{S}_F(v - y),$$

where $\mathbf{V}_\mu(u, z, v)$ is the vertex, or "three-point function," that describes the effect of the perturbation on the propagator line.

we also have

$$\frac{\tau}{2}\frac{\partial}{\partial p_\mu}\mathbf{S}_{FR}^{-1}(p) = Z_2\mathbf{V}_\mu(p, p). \tag{10.57}$$

But by general invariance arguments, now very familiar, we may write

$$\bar{u}(\mathbf{p})\mathbf{V}_\mu(p, p)u(\mathbf{p}) = \frac{i}{Z_1}\bar{u}(\mathbf{p})\gamma_\mu\frac{\tau}{2}u(\mathbf{p}), \tag{10.58}$$

where the $u(\mathbf{p})$ are the free, mass shell, nucleon spinors. The factor $1/Z_1$ has been introduced in this way to conform to the standard definition of the vertex renormalization constant. There is, of course, no loss of generality in this definition; we have simply called the constant factor in the matrix element at zero-momentum transfer $1/Z_1$. However, as we have seen, on the shell

$$\frac{\partial}{\partial p_\mu}\mathbf{S}_{FR}^{-1}(p) = i\gamma_\mu. \tag{10.59}$$

Thus

$$\bar{u}(\mathbf{p})\gamma_\mu\frac{\tau}{2}u(\mathbf{p}) = \frac{Z_2}{Z_1}\bar{u}(\mathbf{p})\gamma_\mu\frac{\tau}{2}u(\mathbf{p}), \tag{10.60}$$

so that

$$\frac{Z_2}{Z_1} = 1, \tag{10.61}$$

but

$$\frac{Z_2}{Z_1} = \frac{G_V{}^{\text{obs}}}{G}. \tag{10.62}$$

So we have shown, once again, that current conservation implies no renormalization of the vector constant. We shall make use of Ward-like identities later when we discuss nonconserved currents.

As pointed out in the last chapter, experimentally G_V/G is not exactly one but differs from one by, perhaps, two percent. We can look at this difference in, at least, two quite distinct ways.

1. When the electromagnetic interactions are turned on, the isotopic current is no longer conserved and it is this small failure in the current conservation, taken with the rest of the electromagnetic corrections, that accounts for the difference. From this point of view, in the *absence* of electromagnetism, the coupling of the vector current (to, say, leptons) would be exactly G.[16]

2. The difference is only partially accounted for by lack of current conservation and radiative corrections. From this point of view, and in the absence of electromagnetism, we would write the weak current in the form

---

16. It is worthwhile noting here that higher-order effects in the *weak* interactions can also renormalize G. Indeed, as we indicate, in the next chapter, in the presence of these couplings $\partial_\mu V_\mu \neq 0$. These corrections turn out to be very hard to compute, but they must have a small effect on G.
17. N. Cabibbo, *Phys. Rev. Letters* **10**:531 (1963).
18. These transitions are from spin-zero positive parity to spin-zero positive parity nucleon states.

$$\frac{G}{\sqrt{2}}(\cos\theta(V_\mu + A_\mu) + L_\mu), \tag{10.63}$$

where V_μ and A_μ are the β-decay currents and L_μ is the lepton current. We have written the factor expressing the departure of G from G^V_{obs} in the form $\cos\theta$, anticipating later work. The angle θ is known as the Cabibbo angle.[17] From the Cabibbo point of view $\theta \neq 0$, even if the electromagnetic couplings are turned off. The origin of this angle, and this is a point to which we shall return in detail later, must lie outside the framework of the Feynman-Gell-Mann theory. Its origin is not well understood, although, as we shall see, a good deal of experimental evidence appears to favor the existence of a Cabibbo angle and hence this point of view as opposed to the first.

We turn now from these rather abstract matters to the more concrete applications of the Feynman-Gell-Mann hypothesis. We begin by applying Eq. (10.19) to the prediction of the "*ft* values" in 0^+–0^+ transitions[18] in $T = 1$ nuclear isobars. The *ft* value of a β transition is essentially a measure of the square of the β-decay matrix element. The quantity f is[19] a dimensionless representation of the phase space of the decay, so that

$$\frac{1}{\tau} \propto |H_{WK}|^2 f. \tag{10.64}$$

If we deal with a heavy nucleus and treat the nuclear matrix element in the approximation in which we assume that the momentum transferred to the nucleus is negligible, then a 0^+–0^+ transition, which is pure allowed Fermi,[20] measures directly the charge associated with the vector current. However, in the Feynman-Gell-Mann theory, this charge is just T^+ (or T^-) for electron (or positron) decay. If we deal with a $T = 1$ isobar we see from Eq. (10.19) that, for all such nuclei, $|H_{WK}| = \sqrt{2}$, so that the *ft* values should be essentially constant over this entire group of nuclei. Indeed, a recent compilation[21] shows this to be true (Table I). Apart from a somewhat anomalous situation in Al^{26} (which appears not to be well understood), we see a remarkable uniformity in the *ft* values for these 0–0 transitions. While this is *consistent* with the Feynman-Gell-Mann theory, it is important to understand to what extent this result *confirms* the theory. As we pointed out at the very beginning of the discussion, Fermi's original β-decay theory, in modern isotopic notation, consisted of supposing that the β-decay current was simply

$$\mathbf{V}_\mu(x) = i\bar{\psi}(x)\gamma_\mu \frac{\tau}{2}\psi(x), \tag{10.65}$$

19. See, for example, *Nuclear Theory*, by R. Sachs, Addison-Wesley, Cambridge (1953), Chapter 11.
20. It occurs as an allowed transition of the operator $V_0(0)$ which, when taken between states of zero momentum, is the total weak vector charge.
21. See V. L. Telegdi, *Proceedings of the Argonne National Laboratory International Conference on Weak Interactions*, 1965, p. 352.

Table I

Decay	*ft Value (sec)*
$O^{14}(\beta^+)N^{14*}$	3127 ± 10
$Al^{26}(\beta^+)Mg^{26}$	3086 ± 12
$Cl^{34}(\beta^+)S^{34}$	3138 ± 19
$Sc^{42}(\beta^+)Ca^{42}$	3122 ± 9
$V^{46}(\beta^+)Ti^{46}$	$\begin{cases} 3138 \pm 25 \\ 3131 \pm 8 \end{cases}$
$Mn^{50}(\beta^+)Cr^{50}$	3125 ± 9
$Co^{54}(\beta^+)Fe^{54}$	3132 ± 17

where the ψ's are the nucleon fields. This would be the total isotopic current if there were no pions. With pions, as well as nucleons,

$$\mathbf{V}_\mu = i\bar{\psi}\gamma_\mu \frac{\boldsymbol{\tau}}{2}\psi - (\boldsymbol{\phi} \times \partial_\mu \boldsymbol{\phi}) \tag{10.66}$$

for the pseudoscalar theory, and

$$\mathbf{V}_\mu = i\bar{\psi}\gamma_\mu \frac{\boldsymbol{\tau}}{2}\psi - (\boldsymbol{\phi} \times \partial_\mu \boldsymbol{\phi}) + \frac{if}{m_\pi}\bar{\psi}(\gamma_5\gamma_\mu\boldsymbol{\tau} \times \boldsymbol{\phi})\psi \tag{10.67}$$

for the pseudovector theory. These currents are each conserved in their respective theories. But in neither theory is Fermi's $\mathbf{V}_\mu$, *without pions*, conserved. The difference between the Feynman-Gell-Mann $\mathbf{V}_\mu$ and Fermi's $\mathbf{V}_\mu$ is just in these pionic terms.[22] However, we might argue that, in nuclei, pionic contributions

22. We can see, incidentally, how these pionic terms lead to the nonrenormalization of G from the point of view of Feynman graphs. In the Fermi theory there is one nonpionic graph for neutron β-decay:

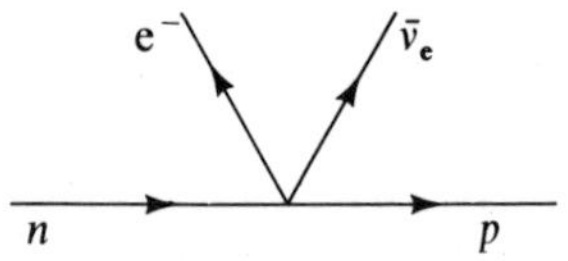

There are also pion corrections to this graph, of the form

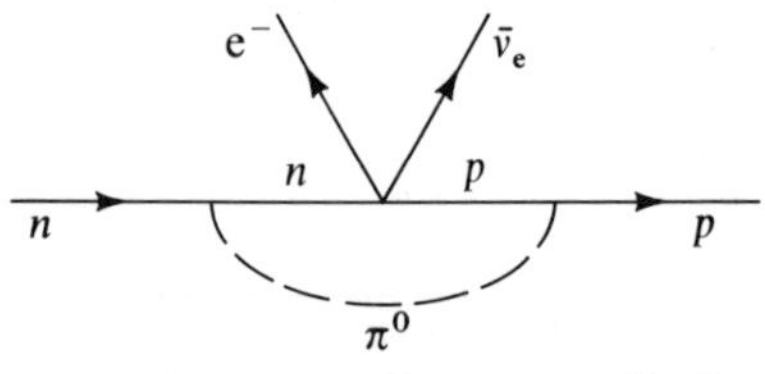

and these lead to a vertex renormalization. However, in the Feynman-Gell-Mann theory, to every such graph there is a "compensating" graph of the same order in the pion nucleon coupling constant; for example,

can be neglected to a good approximation. It is well known that an independent particle theory of nuclei gives, for many nuclei, a fairly reasonable description of, say, low-energy electromagnetic phenomena such as the magnetic moments, and we could imagine that pionic interactions might also be unimportant for weak processes such as nuclear β-decay. For β-decay, so far as meson corrections can be ignored, the allowed Fermi 0^+–0^+ transitions would be described by an effective nucleonic Hamiltonian of the form

$$H_{\text{eff}} = \frac{G}{\sqrt{2}} \sum_{i=1}^{N} \tau_{i+}\, \delta^3(\mathbf{r} - \mathbf{r}_i),$$

where the sum is over the nucleons in the nucleus. This H_{eff} corresponds to the picture in which β-decay occurs from the independent nucleons in the nucleus. It ignores corrections that might arise because the nucleons can communicate with each other by meson exchange. But the matrix elements of this Hamiltonian between the independent particle wave functions will simply be, once again, the matrix elements of T^+, the total isotopic spin raising operator (ignoring mesons). Hence this approximation will yield the same result as the Feynman-Gell-Mann theory even though its theoretical basis is totally different. It is therefore important to display tests of the theory that are more uniquely connected to it.

In this respect one of the most interesting processes is $\pi^- \longrightarrow e^- + \bar{\nu}_e + \pi^0$ (or $\pi^+ \longrightarrow e^+ + \nu_e + \pi^0$), which can take place, although with limited phase space, because of the small mass difference between $\pi^\mp$ and π^0 (about 4.5 MeV).

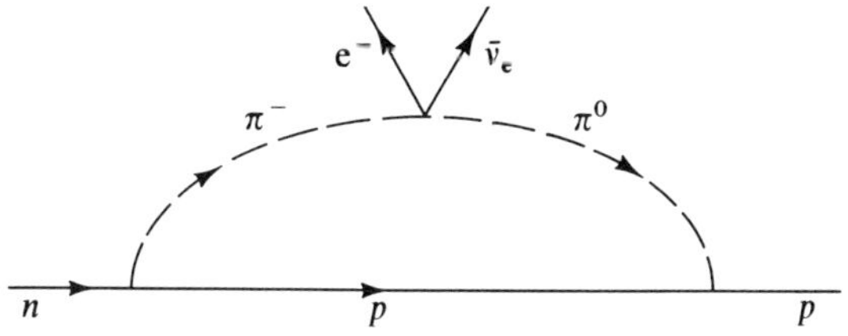

The conservation of $\mathbf{V}_\mu$ guarantees that, at zero lepton momentum transfer, these graphs cancel against the graphs in which the leptons come out of the nucleons. (One may, if one likes, verify this, graph by graph.) In the Fermi theory leptons can also be emitted from pions, as for example in

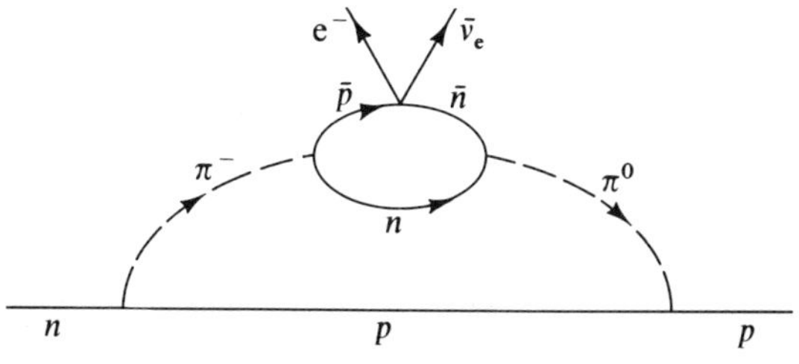

but these graphs are of higher order in the strong coupling constants and do not cancel against the pionic graphs in which the leptons emerge from the nucleons.

In the current-current theory the matrix element for this process can be written, say, for π^-, as

$$m = \frac{G}{\sqrt{2}}\langle\pi^0|J_\mu(0)|\pi^-\rangle\bar{u}_e\gamma_\mu(1 + \gamma_5)u_{\bar{\nu}_e},$$

with

$$J_\mu = V_\mu + A_\mu.$$

We now analyze the new quantity here, $\langle\pi^0|J_\mu(0)|\pi^-\rangle$. Even though parity is not conserved in the decay we have, since we are treating the weak coupling to lowest order, so that the pionic states $|\mathbf{p}^+\rangle$ and $|\mathbf{p}^0\rangle$ are eigenstates of the strong interaction alone, the parity conditions

$$\begin{aligned}\langle\mathbf{p}^0|\mathbf{V}(0)|\mathbf{p}^-\rangle &= -\langle-\mathbf{p}^0|\mathbf{V}(0)|-\mathbf{p}^-\rangle,\\ \langle\mathbf{p}^0|V_0(0)|\mathbf{p}^-\rangle &= \langle-\mathbf{p}^0|V_0(0)|-\mathbf{p}^-\rangle,\end{aligned} \tag{10.68}$$

and

$$\begin{aligned}\langle\mathbf{p}^0|\mathbf{A}(0)|\mathbf{p}^-\rangle &= \langle-\mathbf{p}^0|\mathbf{A}(0)|-\mathbf{p}^-\rangle,\\ \langle\mathbf{p}^0|A_0(0)|\mathbf{p}^-\rangle &= -\langle-\mathbf{p}^0|A_0(0)|-\mathbf{p}^-\rangle.\end{aligned} \tag{10.69}$$

But by Lorentz invariance, as in Chapter 4, we have[23]

$$(2\pi)^3\sqrt{4E_{\pi^-}E_{\pi^0}}\langle\mathbf{p}^0|J_\mu(0)|\mathbf{p}^-\rangle = (p_\mu^- + p_\mu^0)F_+(q_-^2) + (p_\mu^- - p_\mu^0)F_-(q_-^2), \tag{10.70}$$

with

$$q_-^2 = (p_\mu^- - p_\mu^0)^2. \tag{10.71}$$

It is then clear, from Eqs. (10.68) and (10.69), that the axial vector current cannot contribute to this transition; that is,

$$\langle\pi^0|J_\mu(0)|\pi^-\rangle = \langle\pi^0|V_\mu(0)|\pi^-\rangle. \tag{10.72}$$

In fact the $\pi^- - \pi^0$ decay is a prime example of an allowed Fermi 0^+–0^+ transition. The same argument that showed that V_μ cannot be exactly conserved in neutron β-decay unless $M_N = M_P$ applies here. Hence, to apply current conservation in this decay, we must make the approximation $m_{\pi^-} = m_{\pi^0}$ in the matrix element $\langle\pi^0|V_\mu(0)|\pi^-\rangle$. Given current conservation, and setting $m_{\pi^-} = m_{\pi^0}$, we eliminate $F_-(q^2)$ and write

$$(2\pi)^3\sqrt{4E_{\pi^-}E_{\pi^0}}\langle\pi^0|V_\mu(0)|\pi^-\rangle = (p_\mu^- + p_\mu^0)F_+(q^2). \tag{10.73}$$

Under the assumption that $\mathbf{V}(0)$ transforms like a time-reversal-symmetric current we have

$$\langle\mathbf{p}^0|\mathbf{V}(0)|\mathbf{p}^-\rangle = -\langle-\mathbf{p}^0|\mathbf{V}(0)|-\mathbf{p}^-\rangle^*, \tag{10.74}$$

---

23. The numerical factors on the left-hand side of Eq. 10.70 are chosen to simplify $F\pm$.

which can be satisfied if

$$F_+(q^2) = F_+^*(q^2). \tag{10.75}$$

From the kinematics of the decay it is easy to see, if we compute the decay in the frame of reference in which $\mathbf{p}^- = \mathbf{0}$, that

$$\frac{|\mathbf{p}^0|}{E(\mathbf{p}^-) + E(\mathbf{p}^0)} \ll 1. \tag{10.76}$$

Therefore it is a very good approximation to neglect the space components of the current in this process and to assume that the matrix element is given by

$$\begin{aligned}\langle \mathbf{0}|_0 V_0(0)|\mathbf{0}\rangle_-(2\pi)^3 2m_\pi &= 2m_\pi F_+(0)\\ &= \left\langle \mathbf{0}|_0 \int V_0(\mathbf{r})\,d\mathbf{r}|\mathbf{0}\right\rangle_- 2m_\pi = 2m_\pi \langle \mathbf{0}|_0 T_+|\mathbf{0}\rangle_- = 2m_\pi\sqrt{2}. \end{aligned}\tag{10.77}$$

Thus the matrix element for $\pi^- \longrightarrow \pi^0 + e^- + \bar{\nu}_e$ can be written in the explicit form[24]

$$M = G\bar{u}(\mathrm{e})\gamma_4(1 + \gamma_5)u(\bar{\nu}). \tag{10.78}$$

Using this matrix element it can be shown that the theoretical ratio of the rate $\pi^- \longrightarrow \pi^0 + e^- + \bar{\nu}_e$ to the dominant pion mode $\pi^- \longrightarrow \mu^- + \bar{\nu}_\mu$ (to be discussed in the next chapter) is $R_{\text{th}} = 1.07 \times 10^{-8}$; the rate for $\pi^- \longrightarrow \pi^0 + e^- + \nu_e$ itself, ignoring small corrections, is given by

$$\frac{1}{\tau} = \frac{G^2}{30\pi^3}(m_{\pi^-} - m_{\pi^0})^5. \tag{10.79}$$

The very small ratio is due to the small phase space. The three body pion-lepton phase space goes approximately as $(m_{\pi^-} - m_{\pi^0})^5$ and the small mass $\pi^- - \pi^0$ difference gives the small ratio. There are, by now, several experiments on this ratio[25] and the best value, at present, is

$$R_{\text{exp}} = (1.12 \pm 0.08) \times 10^{-8},$$

a number which is clearly in very good agreement with theory. However, we would be remiss in not pointing out that *any* β-decay theory would lead to *some* β-decay of the pion, for example, via graphs of the form

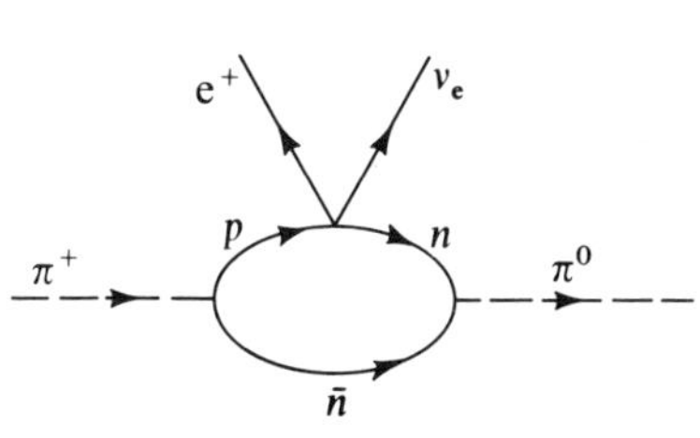

24. We ignore the small correction due to the Cabibbo angle.
25. See T. D. Lee and C. S. Wu, op. cit., for a complete survey of the references.

However, the point to emphasize is the precise numerical agreement of the experiment with the Feynman-Gell-Mann prediction. It would be a miracle if another theory, based on graphs like the one shown, gave the same numerical answer.

So far, our discussion of the predictions of the Feynman-Gell-Mann theory has been confined to phenomena that can be characterized as occurring at zero-momentum transfer. In other words, in matrix elements of the form $\langle \mathbf{a}|V_\mu(0)|\mathbf{b}\rangle$, we have been able to make the approximation $\mathbf{a} \simeq \mathbf{b} \simeq 0$, so that the matrix element reduces essentially to the matrix element of the total charge, which is simply T^+ or T^-. However, some of the most striking predictions of the theory occur for nonzero-momentum transfers. The simplest place to begin this discussion is in neutron β-decay, although the best experimental results are in heavy nuclei. We again consider

$$(2\pi)^3\langle \mathbf{p}'|_N V_\mu(0)|\mathbf{p}\rangle_P = i\bar{u}(\mathbf{p}')[\gamma_\mu f_V(q^2) + \sigma_{\mu\nu}q_\nu f_M(q^2)]u(\mathbf{p}). \tag{10.80}$$

The essential observation is the following: the electromagnetic current, in the good *i*-spin approximation, can be written as

$$J_\mu{}^\gamma = J_\mu{}^S + J_{\mu 3}{}^V, \tag{10.81}$$

where $J_\mu{}^S$ is the isotopic scalar current and $J_{\mu 3}{}^V$ is the third component of the isotopic vector current. But it is exactly the *same* $J_{\mu i}{}^V$, according to the Feynman-Gell-Mann theory, that enters the weak interactions. In brief, if the isotopic vector part of the electromagnetic current is $J_{\mu 3}{}^V$, then V_μ, the β-decay current, is $J_{\mu 1}{}^V + iJ_{\mu 2}{}^V$. Clearly this observation must allow us to relate some electromagnetic processes to some weak processes.

As a first example of how this works we shall exhibit a relation between neutron β-decay and electron-nucleon scattering (always in the good *i*-spin limit). To make the isotopic dependence as explicit as possible, we call $V_\mu = J_{\mu+}{}^V = J_{\mu 1}{}^V + iJ_{\mu 2}{}^V$. With

$$T_+ = \int d\mathbf{r}(J_0{}^V(\mathbf{r}, 0)_1 + iJ_0{}^V(\mathbf{r}, 0)_2), \tag{10.82}$$

we can show, from the isotopic commutation relations, that

$$J_{\mu+}{}^V = [J_{\mu 3}{}^V, T_+]_-. \tag{10.83}$$

Thus

$$\begin{aligned}\langle \mathbf{p}'|_P J_{\mu+}{}^V|\mathbf{p}\rangle_N = \langle \mathbf{p}'|_P[J_{\mu 3}{}^V, T_+]_-|\mathbf{p}\rangle_N &= \langle \mathbf{p}'|_P J_{\mu 3}{}^V|\mathbf{p}\rangle_P - \langle \mathbf{p}'|_N J_{\mu 3}{}^V|\mathbf{p}\rangle_N \\ &= \langle \mathbf{p}'|_P J_\mu|\mathbf{p}\rangle_P - \langle \mathbf{p}'|_N J_\mu|\mathbf{p}\rangle_N,\end{aligned} \tag{10.84}$$

where we have used the properties of T_+ as an isotopic raising operator and the relation given in Chapter 6 between the matrix elements of $J_{\mu 3}{}^V$ and the matrix elements of the total electric current $J_\mu{}^\gamma = J_\mu{}^S + J_{\mu 3}{}^V$ between neutron and proton. From this equation it follows that the form factors for the vector part of neutron β-decay are related to those of electron scattering of neutron and proton by the equation[26]

$$F_i(q^2)^+ = F_i(q^2)^P - F_i(q^2)^N, \tag{10.85}$$

where $i = 1, 2$. This relation enables us to make several predictions about β-decay and related processes. In the first place we can now estimate the variation of the weak form factors, as functions of q^2, by using the known electron scattering results. To get an idea of the scale of the variation we expand $F_i(q^2)^+$ in powers of q^2 using Eq. (10.85):

$$F_1(q^2)^+ = 1 - \frac{q^2}{6}(\langle \mathbf{r}^2\rangle_1{}^P - \langle \mathbf{r}^2\rangle_1{}^N) \simeq 1 - \frac{q^2}{6}\langle \mathbf{r}^2\rangle_1{}^P \simeq 1 - \frac{.03}{m_\pi{}^2}q^2. \tag{10.86}$$

For ordinary β-decay[27] $|q^2| \sim m_e{}^2$, so it is quite correct to ignore the q^2 dependence in the β-decay form factors. There is another weak reaction involving the same nucleon matrix elements, where the four-momentum transfer is much higher, so that we might expect variations in the vector form factors. This is muon capture by the proton: the process $\mu^- + p \longrightarrow n + \nu_\mu$. Here the nucleon matrix element (we discuss only the vector part) is given by

$$\begin{aligned}\langle \mathbf{p}'|_N J_\mu{}^V(0)_-|\mathbf{p}\rangle_P &= \langle \mathbf{p}'|_N[T_-, J_{\mu 3}{}^V]_-|\mathbf{p}\rangle_P \\ &= \langle \mathbf{p}'|_P J_\mu(0)|\mathbf{p}\rangle_P - \langle \mathbf{p}'|_N J_\mu(0)|\mathbf{p}\rangle_N. \end{aligned}\tag{10.87}$$

In this process we can take the initial proton and muon at rest so that

$$q^2 = (p - n)^2 = (\nu_\mu - \mu)^2 = -m_\mu{}^2 + 2\nu_{\mu 0}m_\mu. \tag{10.88}$$

But from energy conservation,

$$m_\mu + M_P = \nu_{\mu 0} + E_N \simeq \nu_{\mu 0} + M_P, \tag{10.89}$$

so that

$$q^2 \simeq m_\mu{}^2. \tag{10.90}$$

But

$$\left(\frac{m_\mu}{m_\pi}\right)^2 \simeq .56, \tag{10.91}$$

so that, again from Eq. (10.91), in muon capture we can neglect the q^2 dependence of the vector form factors.

There is a process, however, which has been studied experimentally, al-

---

26. In this notation, in terms of our previous definitions,

$$F_1(q^2)^+ = f_V(q^2),$$
$$F_2(q^2)^+ = f_M(q^2).$$

27. For β-decay, in our metric, $q^2 < 0$. Thus, strictly speaking, we should use the vector meson model or some equivalent analytical expression to make the extrapolation to small negative q^2. If we do, the conclusion remains the same.

though not with extremely high precision, in which the form factor dependence is *not* negligible. This is the reaction $\nu_\mu + n \longrightarrow \mu^- + p$ or $\bar{\nu}_\mu + p \longrightarrow \mu^+ + n$. (The processes $\bar{\nu}_\mu + n \longrightarrow \mu^- + p$ and $\nu_\mu + p \longrightarrow \mu^+ + n$ are forbidden by muon number conservation.) The muon neutrinos are obtained from the $\pi \longrightarrow \mu + \nu_\mu$ decay of very high-energy, charged pions, which are produced when protons, in the CERN or Brookhaven accelerators[28] (where the neutrino experiments were done), collide with, say, target protons, in a reaction such as $p + p \longrightarrow p + n + \pi^+$. Typical neutrino energies, and hence typical q^2, are in the BeV range in these experiments. The matrix element for, say, $\nu_\mu + n \longrightarrow \mu^- + p$ can be written in the form

$$\frac{G}{\sqrt{2}}\langle P|V_\mu(0) + A_\mu(0)|N\rangle i\bar{u}(p_\mu)\gamma_\mu(1 + \gamma_5)u(\nu_\mu), \tag{10.92}$$

so that both the vector and axial vector form factors are involved. Though the data are not precise[29] they are consistent with taking for $f_V(q^2)$

$$f_V(q^2) = \frac{1}{\left(1 - \dfrac{q^2}{M_V{}^2}\right)^2}, \tag{10.93}$$

where $M_V = .84$ BeV. This form is compatible in a general way with the fit to the electron scattering data given in Chapter 6, although the neutrino experiments are as yet too crude to allow a really precise comparison. (We defer a discussion of the axial vector form factors until the next chapter.)

The most impressive test of the Feynman-Gell-Mann theory involves what Gell-Mann has called "weak magnetism." [30] If we return, once again, to the neutron β-decay matrix element,

$$(2\pi)^3\langle \mathbf{p}'|_P V_\mu(0)|\mathbf{p}\rangle_N = i\bar{u}(\mathbf{p}')[\gamma_\mu f_V(q^2) + \sigma_{\mu\nu}q_\nu f_M(q^2)]u(\mathbf{p}), \tag{10.94}$$

we observe that the term proportional to f_M has the form of a Pauli magnetic moment. In any theory of β-decay there would be such a term. But it would arise from complicated, and unreliably estimated, meson corrections to the simple Fermi theory, which involves only the nucleon currents. However, in the Feynman-Gell-Mann theory, $f_M(q^2)$ is given directly from electron-nucleon scattering. In particular,

$$f_M(q^2) = f_2(q^2)^P - f_2(q^2)^N. \tag{10.95}$$

Hence[31]

---

28. See the proceedings of the Argonne Conference on Weak Interactions, op. cit., p. 257 and ff. for a recent summary of the data.
29. Typical cross sections, in the BeV range for $\nu_\mu + n \longrightarrow p + \mu^-$, are on the order of $.6 \times 10^{-38}$ cm^2, which means, in the CERN setup, that for a total day of running, corresponding to 1.7×10^{16} protons, there are only 15 events per *ton* of target material.
30. M. Gell-Mann, *Phys. Rev.* **111**:362 (1958).

$$f_M(0) = \frac{\mu_P - \mu_N}{2M} = \frac{1.79 + 1.91}{2M} = \frac{3.70}{2M}. \tag{10.96}$$

Since q/M is so small in neutron β-decay, it is all but impossible to confirm this remarkable prediction there. However, in nuclear β-decay the electron energies can run to well over 10 MeV, so that the "weak magnetic" terms can be observed as small corrections to the allowed, phase space, β-decay spectra. Indeed, Gell-Mann[32] has produced a beautiful system of nuclear levels suitable for studying weak magnetism. The figure below summarizes the situation in the A^{12} triad—that is, the three nuclei B^{12}, C^{12}, N^{12}.

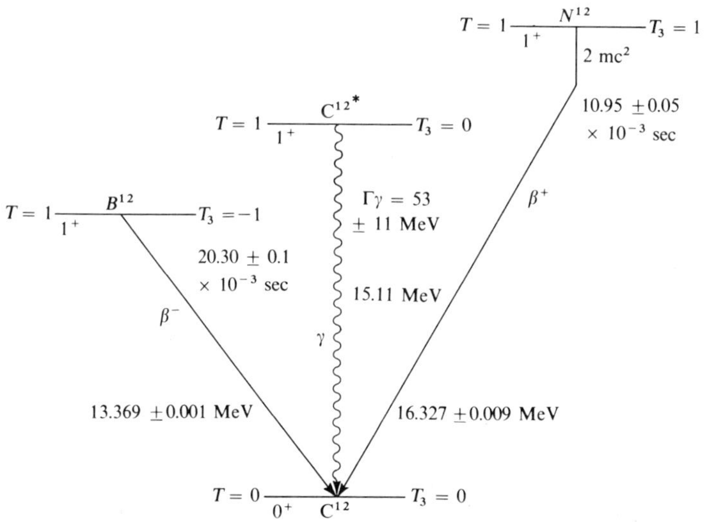

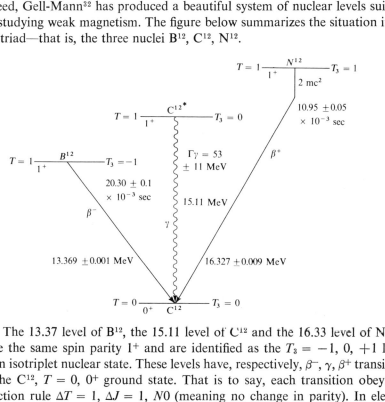

The 13.37 level of B^{12}, the 15.11 level of C^{12} and the 16.33 level of N^{12} all have the same spin parity 1^+ and are identified as the $T_3 = -1, 0, +1$ levels of an isotriplet nuclear state. These levels have, respectively, β^-, γ, β^+ transitions to the C^{12}, $T = 0$, 0^+ ground state. That is to say, each transition obeys the selection rule $\Delta T = 1$, $\Delta J = 1$, NO (meaning no change in parity). In electromagnetic parlance this is the selection rule of the isotopic vector part of the magnetic dipole operator. A magnetic dipole operator has the form $\mathbf{r} \times \mathbf{v}$ for a dipole due to a simple current $e\mathbf{v}$ (where $\mathbf{v}$ is the velocity of the charged particle) or $\boldsymbol{\sigma}$, if we are discussing spin magnetic moment transitions. In either case, the

31. To keep the dimensions straight we have written the Pauli moments explicitly in terms of nucleon Bohr magnetons. Of course, in going from electromagnetism to β-decay one must make the substitution, in the overall coupling constant multiplying the current,

$$\frac{G}{\sqrt{2}} \leftrightarrow e.$$

32. Gell-Mann, op. cit.

operator transforms like a pseudovector and hence leads to $\Delta J = 1$, $N0$ allowed transitions. This is also the selection rule for an allowed axial vector (Gamow-Teller) β-decay transition. We have already seen that for neutron β-decay

$$(2\pi)^3\langle\mathbf{0}|_P\mathbf{A}(0)|\mathbf{0}\rangle_N = \bar{u}(\mathbf{0})\boldsymbol{\sigma}u(\mathbf{0})g_A(0). \tag{10.97}$$

In the independent particle model the effective β-decay axial vector Hamiltonian for allowed Gamow-Teller transitions will have the form

$$H_{\text{eff}} = \frac{G}{\sqrt{2}} g_A \sum_{i=1}^{N} \delta^3(\mathbf{r} - \mathbf{r}_i)\tau_{i+}\boldsymbol{\sigma}_i\cdot\mathbf{L} \tag{10.98}$$

where $\mathbf{L}$ is the lepton current $i\bar{\psi}_e\boldsymbol{\gamma}(1 + \gamma_5)\psi_{\nu_e}$ and the sum runs over the nucleons in the nucleus in question. The nuclear matrix element of Eq. (10.98) is usually written simply as $\int \boldsymbol{\sigma}$. The B^{12} and N^{12} β-transitions are, predominantly, allowed axial and depend mostly on $\int \boldsymbol{\sigma}$. However, the vector current also contributes small "forbidden" corrections to this transition. In fact, the weak magnetic term, since it transforms like a magnetic dipole, obeys the same selection rule as $\int \boldsymbol{\sigma}$. The weak magnetic and allowed terms can interfere in the β-energy spectra. To display this explicitly, we first write the full, allowed Gamow-Teller matrix element, including the leptons. Thus

$$H_0 = \frac{G}{\sqrt{2}} g_A(0) \int \boldsymbol{\sigma}\cdot u^\dagger(\mathbf{e})\boldsymbol{\sigma}(1 + \gamma_5)u(\bar{\boldsymbol{\nu}}_e), \tag{10.99}$$

where we have used the identity

$$i\gamma_4\boldsymbol{\gamma}\gamma_5 = \boldsymbol{\sigma} \tag{10.100}$$

to rewrite the leptonic matrix element. We may now, following the original discussion of Gell-Mann, deduce the effective Hamiltonian for the weak magnetic corrections. To go from the electromagnetic isovector Hamiltonian to the β-decay Hamiltonian it suffices to take

$$H_{e_m}{}^V = eJ_{\mu 3}{}^V A_\mu \tag{10.101}$$

and to make the replacements

$$\begin{aligned} J_{\mu 3}{}^V &\longrightarrow J_{\mu +}{}^V, \\ A_\mu &\longrightarrow \frac{G}{e\sqrt{2}} i\bar{u}(\mathbf{e})\gamma_\mu(1 + \gamma_5)u(\bar{\boldsymbol{\nu}}_e). \end{aligned} \tag{10.102}$$

This replacement must hold also for all the moments of the current; the electric dipole, magnetic dipole, etc. Now the effective magnetic dipole Hamiltonian is of the form

$$H' = \boldsymbol{\mu}\cdot\frac{e}{2M}\boldsymbol{\nabla}\times\mathbf{A}, \tag{10.103}$$

where $\boldsymbol{\mu}$ is the matrix element of the magnetic dipole operator. In general, $\boldsymbol{\mu}$ will be an isoscalar plus an isovector, since this is how the current transforms. For our purposes, it will suffice to consider the isovector part alone. We now prove a simple lemma,[33] in which the states are labeled by T, T_3 and the subscripts on $\boldsymbol{\mu}$ are isotopic indices:

$$\langle 00|\boldsymbol{\mu}_+|1\;-1\rangle = \langle 00|[\boldsymbol{\mu}_3, T_+]_-|1\;-1\rangle = \sqrt{2}\,\langle 00|\boldsymbol{\mu}_3|10\rangle. \tag{10.104}$$

This lemma says that to transform the electromagnetic dipole nuclear matrix element between the state $T = 1$, $T_3 = 0$ and the state $T = 0$, $T_3 = 0$ into the *weak* magnetic nuclear matrix element connecting the $T = 1$, $T_3 = -1$ and $T = 0$, $T_3 = 0$ states, we merely multiply the former by $\sqrt{2}$. We may now derive the weak magnetic effective Hamiltonian for, say, the B^{12} transition shown in Fig. 1. Using Eqs. (10.103) and (10.102) we have

$$H' = \frac{G}{\sqrt{2}}\frac{\boldsymbol{\mu}}{\sqrt{2}\,M}\cdot\boldsymbol{\nabla}\times\bar{u}(\mathrm{e})i\boldsymbol{\gamma}(1+\gamma_5)u(\bar{\nu}_e), \tag{10.105}$$

where the $\boldsymbol{\mu}$ here is the *same* number as the magnetic dipole matrix element that governs the C^{12} γ-ray transition. The full[34] effective Hamiltonian, H_{eff}, can then be written

$$H_{\text{eff}} = \frac{G}{\sqrt{2}}\Big[g_A(0)\int\boldsymbol{\sigma}\cdot u^{\dagger}(\mathrm{e})\boldsymbol{\sigma}(1+\gamma_5)u(\bar{\nu}_e)$$
$$-\,i\frac{\boldsymbol{\mu}}{\sqrt{2}\,M}\cdot\mathbf{q}\times\bar{u}(\mathrm{e})i\boldsymbol{\gamma}(1+\gamma_5)u(\bar{\nu}_e)\Big], \tag{10.106}$$

where we have replaced $\boldsymbol{\nabla}$ by $-i\mathbf{q}$, the sum of the lepton momenta. The essential parameter that measures the size of the weak magnetic correction relative to the Gamow-Teller terms is

$$a \cdot\!\equiv\!\cdot \frac{|\mu_3|}{\sqrt{2}\,M}\,\frac{1}{|g_A(0)|\left|\int\boldsymbol{\sigma}\right|}. \tag{10.107}$$

In terms of this parameter the correction to the phase space spectrum for the β^-, B^{12} decay, with E the electron energy, turns out to be simply

$$1 + \tfrac{8}{3}aE, \tag{10.108}$$

a linear correction with slope $\frac{8}{3}a$. As Gell-Mann pointed out, a can be determined either directly from experiment (at least up to a sign) or from a simple theoretical consideration. The quantity $|\mu_3|/\sqrt{2}\,M$ is the weak magnetic coefficient. But for a single neutron[35] the matrix element is effectively

$$\frac{\mu_P-\mu_N}{2M}\,i\boldsymbol{\sigma}\cdot\mathbf{q}\times\mathbf{L}, \tag{10.109}$$

---

33. Note that for the isoscalar part of $\boldsymbol{\mu}$; $\boldsymbol{\mu}_S$

$$\langle 00|\boldsymbol{\mu}_S|10\rangle = \frac{1}{\sqrt{2}}\langle 00|\boldsymbol{\mu}_S T_+|1\;-1\rangle = \langle 00|T_+ + \boldsymbol{\mu}_S|1\;-1\rangle = 0,$$

so that we do not have to consider $\boldsymbol{\mu}_S$ in isospin changing transitions.

34. For simplicity we have dropped small additional correction terms representing other forbidden effects; see Gell-Mann, op. cit.
35. See Eq. (10.96).

where **L** is the space part of the leptonic current. Hence, in the independent particle model,

$$a = \frac{\mu_P - \mu_N}{2M|g_A(0)|}. \tag{10.110}$$

Here the $|\int \boldsymbol{\sigma}|$ term, which is common to the magnetic and weak magnetic matrix elements, has canceled out. Thus in this model, numerically,

$$a \simeq \frac{2}{M} > 0. \tag{10.111}$$

But $|a|$ can also be determined *directly* from the measured rate of the $C^{12*} \longrightarrow C^{12} + \gamma$ transition. The experimental number[36] for $|a|$, $2.40 \pm .25/M$, is in such good agreement with the independent particle model calculation of a that one may predict, with confidence, that the slope of the correction to the β spectrum is positive. Several experiments have been done to measure this slope. The most recent[37] and most accurate is in essentially perfect agreement with the sign and magnitude of a as given by Gell-Mann's theory. Moreover, there is a second prediction of the theory that is also in perfect agreement with the experiment of Lee et al. If we pass from the electron decay of B^{12} to the positron decay of N^{12} we go from the lepton current L_μ to its charge conjugate current and from the isotopic current $J_{\mu+}{}^V$ to $J_{\mu-}{}^V$. It is easy to see, from the work of the last chapter, that this simply corresponds to letting $a \longrightarrow -a$, since a is an interference term between vector and axial vector contributions to the β-decay and these transform with opposite signs when we charge conjugate L_μ. Thus the N^{12} β^+ spectrum correction should have the form

$$1 - \tfrac{8}{3}aE. \tag{10.112}$$

The N^{12} spectrum was also measured by Lee et al. and it also has the correct

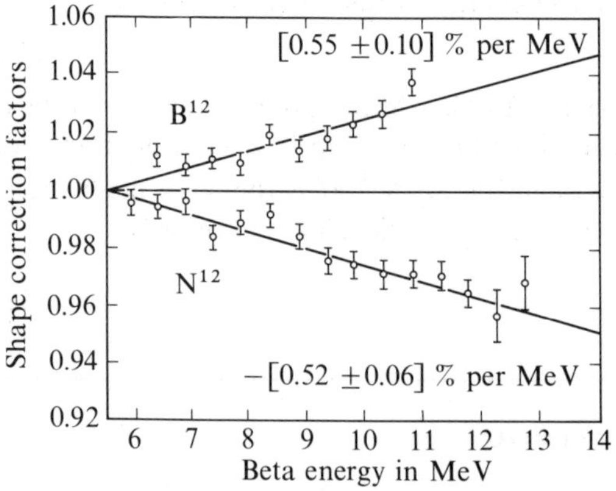

36. The rate is given in terms of μ^2 by the formula $\Gamma_\gamma = \mu^2\omega^3/3(137)M^2$, where ω is the photon energy. Since $\Gamma_\gamma = 53 \pm 11$ eV, μ^2 and hence $|a|$ can be found. See Gell-Mann, op. cit., for a detailed discussion and for references to the experimental literature.
37. Y. K. Lee, L. W. Mo, and C. S. Wu, *Phys. Rev. Letters* **10**:253 (1963).

magnitude and sign predicted by the theory.[38] There is little doubt that the V_μ of β decay is identical to the isotopic spin current. This same current couples to the photon and, as we have seen earlier, there is reason to suppose that it also couples to the ρ meson. If the W meson exists, then we can summarize this situation by stating that the vector mesons, ρ, W, and A, the photon, all seem to couple to the conserved isotopic spin current. It is an extraordinary fact, not understood on its deepest level, that the isotopic spin current coupling plays such a universal rule in strong, electromagnetic, and weak couplings. We now leave the vector current and, in the next chapter, discuss its axial vector counterpart.

38. There are other, more complicated, tests of the theory in nuclear β-decay; see, for example, J. Bernstein and R. R. Lewis, *Phys. Rev.* **112**:232 (1958), and C. S. Wu, *Rev. Mod. Phys.* **36**:618 (1964); these, too, are in good agreement.

11

The Axial Vector I

The burden of the last chapter was to show that the lack of renormalization of the vector coupling constant, expressed by the experimental observation that $f_V(0) \simeq 1$, has a natural explanation if it is assumed that $\partial_\mu V_\mu = 0$. As we have seen this cannot be an exact statement. However, it can hold in the good *i*-spin limit in which the masses of the particles in a given isotopic spin multiplet are degenerate. We have, up to now, not had any occasion to discuss the weak interactions in anything but lowest order in the weak coupling constant, $G/\sqrt{2}$, in the current-current theory, or g^2 in the *W* theory. While none of the experiments done so far on reactions involving leptons require, in their explanation, higher-order weak effects, these effects involve, in any case, some important matters of principle. In the first place, it is easy to see that the first Born approximation cannot be valid for leptonic reactions at very high energies. From the leptonic Lagrangian

$$H_0 = \frac{G}{\sqrt{2}} [L_\mu L_\mu^*], \tag{11.1}$$

with

$$L_\mu = \bar{\psi}_e \gamma_\mu (1 + \gamma_5) \psi_{\nu_e} + \bar{\psi}_\mu \gamma_\mu (1 + \gamma_5) \psi_{\nu_\mu}, \tag{11.2}$$

we may compute the total cross section for the process $\nu_\mu + e^- \longrightarrow \nu_e + \mu^-$, which must occur as one of the "crossed" reactions to muon decay. In terms of the center of mass neutrino momentum, ν_μ, the Born approximation cross section is given by

$$\frac{d\sigma}{d\Omega} = \frac{G^2}{\pi^2} \nu_\mu^2, \tag{11.3}$$

corresponding to the diagram

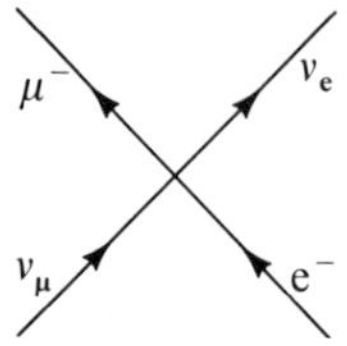

It is easy to see, from the angular dependence of the cross section, that the scattering, in the center of mass, is pure *S*-wave, a circumstance that can be traced back to the local character of the $L_\mu L_\mu^*$ coupling; the currents interact at a point. It is well-known that every angular momentum scattering channel has an upper bound dictated by the conservation of probability—a "unitarity limit." In this case, since only the S-wave contributes, the unitarity condition gives

$$\sigma_{l=0} < \frac{\pi}{2}\frac{1}{\nu_\mu^2}, \tag{11.4}$$

which implies that unitarity breaks down for $\nu_\mu \sim 300$ BeV. This means that the Born approximation, in the weak interactions, cannot be valid at such very large energies. It turns out that if the same computation is done in the W meson theory with the graph

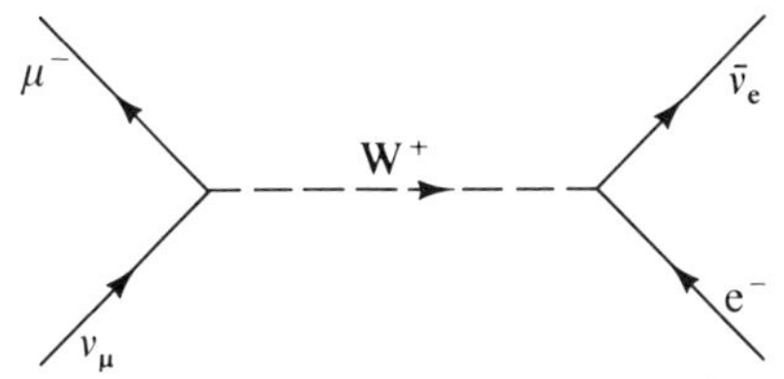

the unitarity catastrophe is postponed to higher energies,[1] but it is still there. The moral: to make a consistent theory of the weak interactions we must be able to compute to higher orders than the first Born approximation. The problem: both the current-current theory and the W meson theory are very singular, so that it has turned out to be very difficult to attach a meaning to any graphs[2] except those of the first Born approximation. Since the experiments, done at low energies, are well accounted for by these graphs, we have as yet not been able to obtain experimental guidance toward modifying the theory. The point that concerns us, at the moment, is that, even in the absence of electromagnetism, the current V_μ will not be conserved, if we include higher-order weak correction terms. We can see this by remembering that the weak Lagrangian will contain terms like $(G/\sqrt{2})L_\mu V_\mu^*$. Now, V_μ transforms like an isovector, but L_μ does not. The weak interactions violate isotopic spin conservation. Therefore V_μ cannot be conserved if we include effects of the weak dynamics on the fields such as ψ_N and π, out of which V_μ is composed. In all of the applications of the conserved vector theory, discussed in the last chapter, we have ignored the effect of the weak interactions on V_μ itself and have treated it, in the approximation, in

---

1. The unitarity condition becomes

$$g^4 \ln^2\left(\frac{2\nu_\mu^2 + M_B^2}{M_B^2}\right) < 1,$$

where M_B is the boson mass.

2. For one such attempt, see G. Feinberg and A. Pais, *Phys. Rev.* **131**:2724 (1963), and G. Feinberg and A. Pais, *Phys. Rev.* **133**:B 477 (1964).

which the fields in V_μ obey *only* the strong interaction dynamics, which *are* isotopic spin–conserving.

At first sight it might be tempting to try to demand that A_μ, the axial current, be conserved, perhaps in the good *i*-spin limit, a point of view perhaps encouraged by the experimental observation that

$$\left|\frac{g_A(0)}{f_V(0)}\right| = 1.18 \pm .02,$$

so that $|g_A(0)| \simeq 1$. We open our discussion of A_μ by showing that it cannot be conserved, even in the good *i*-spin limit. We argue in two ways as follows:

1. Since[3], with $q = p' - p$,

$$(2\pi)^3\langle \mathbf{p}'|_P A_\mu(0)|\mathbf{p}\rangle_N$$
$$= i\bar{u}(\mathbf{p}')\gamma_5[\gamma_\mu g_A(q^2) + i(p' - p)_\mu g_P(q^2) + i(p + p')_\mu g_E(q^2)]u(\mathbf{p}), \quad (11.5)$$

we have, if we assume that $\partial_\mu A_\mu = 0$ and we work in the good *i*-spin limit, so that $(p - p')(p + p') = -M^2 + M^2 = 0$,

$$0 = \bar{u}(\mathbf{p}')\gamma_5[i((p - p')\gamma)g_A(q^2) + (p - p')^2 g_P(q^2)]u(\mathbf{p}) \quad (11.6)$$

or, using the Dirac equation,

$$0 = \bar{u}(\mathbf{p}')\gamma_5 u(\mathbf{p})[2Mg_A(q^2) - (p - p')^2 g_P(q^2)]. \quad (11.7)$$

This equation leads (by an analysis similar to the one used in eliminating f_S for an absolutely conserved vector current) to a relation between g_A and g_P that is in contradiction with the experiments on the β-decay of the neutron. We must be a little more careful here because

$$\bar{u}(\mathbf{0})\gamma_5 u(\mathbf{0}) = 0. \quad (11.8)$$

However, in neutron β-decay, $q^2 < 0$, so that we can conclude that if Eq. (11.7) were true, then

$$\frac{2m_e M|g_A(q^2)|}{|q^2|} = m_e|g_P(q^2)|. \quad (11.9)$$

Therefore the effective pseudoscalar coupling constant would be

$$m_e|g_P(q^2)| \sim \frac{M}{m_e} \sim 2000, \quad (11.10)$$

which contradicts experiment. We might imagine a hypothetical world in which the mass of the nucleon were allowed to vanish. In this fictitious limit we could, if we wanted to, enforce the conservation of the axial vector current. We return in the next chapter to the question of whether this would necessarily lead to the nonrenormalization of g_A.

2. The second argument against the absolute conservation of A_μ comes from

3. For the moment we do not assume charge symmetry for A_μ. We again use capital M for the nucleon mass.

pion decay. It is well known that the $\pi^\pm$ mesons have as their principal decay modes

$$\pi^\pm \longrightarrow \mu^\pm + \nu_\mu \text{ (or } \bar{\nu}_\mu).$$

Experimentally, $\pi^\pm$ are known to have the same lifetimes (as they must by *CPT* invariance; see Chapter 9). The common lifetime is

$$\tau_{\text{exp}} = (2.55 \pm .02) \times 10^{-8} \text{ sec.}$$

Although we have not yet specified the form of A_μ we can nonetheless say a good deal about π-decay, assuming only that it is described as a first-order current-current weak interaction. Indeed, we may write the matrix element for, say π^--decay as

$$\frac{G}{\sqrt{2}} \langle \mu^- \bar{\nu}_\mu | J_\mu J_\mu{}^* | \pi^- \rangle \simeq \frac{G}{\sqrt{2}} \langle \mu^- \bar{\nu}_\mu | L_\mu{}^*(0) | 0 \rangle \langle 0 | J_\mu(0) | \pi^- \rangle$$

$$= \frac{G}{\sqrt{2}} i\bar{u}(p_\mu)\gamma_\mu(1 + \gamma_5)u(\bar{p}_\nu)\langle 0 | J_\mu(0) | \pi^- \rangle. \quad (11.11)$$

In deriving this equation we have inserted a complete sum of states in between J_μ and $J_\mu{}^*$ in the first term on the left side of Eq. (11.11). The leptonic current has first-order matrix elements between the vacuum and the $\mu^-\bar{\nu}_\mu$ state. Any other state either gives no contribution because of some absolute conservation law, such as conservation of baryon or lepton number, or gives a contribution to a higher order in G. The form of the matrix element, $\langle \bar{\nu}_\mu \mu^- | L_\mu(0) | 0 \rangle$, can be obtained most simply by assuming that $\bar{\nu}_\mu$ and μ^- are free, which is correct to the order in G to which we work, and by applying the creation and annihilation operators in the decomposition of the fields making up L_μ. This way we are led to the right side of Eq. (11.11). We shall next derive the general form of $\langle 0 | J_\mu(0) | \mathbf{p}_\pi \rangle$. Using the decomposition of J_μ, into vector and axial vector currents, and the fact that the π meson is a pseudoscalar, relative to the vacuum, which means, in terms of the parity operator P, that[4]

$$P|\mathbf{p}_\pi\rangle = -|-\mathbf{p}_\pi\rangle, \quad (11.12)$$

we have

$$\langle 0 | \mathbf{J}(0) | \mathbf{p}_\pi \rangle = \langle 0 | \mathbf{V}(0) + \mathbf{A}(0) | \mathbf{p}_\pi \rangle = -\langle 0 | -\mathbf{V}(0) + \mathbf{A}(0) | -\mathbf{p}_\pi \rangle. \quad (11.13)$$

There is, however, an additional fact about this matrix element that follows from Lorentz covariance and the spin-zero character of the pion. The quantity $\langle 0 | J_\mu(0) | \mathbf{p}_\pi \rangle$ must transform like a Lorentz vector (or pseudovector). But there is only one Lorentz vector that can be associated with the spin-zero pion; that is its four-momentum. (There is no pseudovector.) Thus we may define

4. We ignore, in the spirit of lowest-order perturbation theory, the effect of parity nonconservation on the fields in J_μ and on the states so that, in particular, $P|0\rangle = |0\rangle$.

$$\frac{G}{\sqrt{2}}\langle 0|J_\mu(0)|\mathbf{p}_\pi\rangle = \frac{ip_{\pi\mu}}{\sqrt{2E_\pi}}\, g_\pi(p_\pi{}^2). \tag{11.14}$$

The factor of i and $1/\sqrt{2E_\pi}$ have been introduced to conform with the definition of $g_\pi(p_\pi{}^2)$ most often found in the literature. The $g_\pi(p_\pi{}^2)$ is a Lorentz scalar form factor associated with this matrix element. However, for the decay of the physical pion, $p_\pi{}^2 = -m_\pi{}^2$, so that g_π is just a number. Equations (11.13) and (11.14) taken together imply that

$$\langle 0|\mathbf{V}(0)|\mathbf{p}_\pi\rangle = 0, \tag{11.15}$$

and from Eq. (11.14) and the fact that[5]

$$\langle 0|V_0(0)|\mathbf{p}_\pi\rangle = -\langle 0|V_0(0)|-\mathbf{p}_\pi\rangle, \tag{11.16}$$

we conclude that

$$\langle 0|V_\mu(0)|\mathbf{p}_\pi\rangle = 0. \tag{11.17}$$

Thus

$$\frac{G}{\sqrt{2}}\langle 0|J_\mu(0)|\mathbf{p}_\pi\rangle = \frac{G}{\sqrt{2}}\langle 0|A_\mu(0)|\mathbf{p}_\pi\rangle = \frac{ip_{\pi\mu}}{\sqrt{2E_\pi}}\, g_\pi(-m\pi^2), \tag{11.18}$$

so that pion decay is a Gamow-Teller transition. From T invariance we have the condition

$$\langle 0|\mathbf{A}(0)|\mathbf{p}_\pi\rangle = \langle 0|\mathbf{A}(0)|-\mathbf{p}_\pi\rangle^*, \tag{11.19}$$

which means, with our definitions, that $g_\pi(-m_\pi{}^2)$ is real.[6] It now follows that A_μ cannot be conserved; if it were, then

$$0 = p_{\pi\mu}\langle 0|A_\mu(0)|\mathbf{p}_\pi\rangle = \frac{-im_\pi{}^2}{\sqrt{2E_\pi}}\, g_\pi(-m_\pi{}^2), \tag{11.20}$$

so that

$$g_\pi(-m_\pi{}^2) = 0, \tag{11.21}$$

which means that the pion decay would be forbidden altogether. However, $g_\pi(-m_\pi{}^2)$ can be found directly from the pion lifetime. Indeed, using energy

---

5. We use the parity operator on $V_0(0)$ and $|\mathbf{p}_\pi\rangle$.
6. From CPT it follows that

$$\langle 0|\mathbf{A}(0)|\mathbf{p}_{\pi^-}\rangle = -\langle 0|\mathbf{A}^\dagger(0)|-\mathbf{p}_{\pi^+}\rangle^*.$$

We see that this condition is sufficient to guarantee that the π^+ and π^- lifetimes are identical. In terms of our representation for the matrix elements of A_μ it means that

$$g_{\pi^-} = g^*{}_{\pi^+},$$

which means that, even not assuming time-reversal invariance,

$$|g_{\pi^-}|^2 = |g_{\pi^+}|^2.$$

Assuming time reversal, these form factors are real, and so the matrix elements would be identical. There is a subtle point here that we have slurred over: we should in fact apply

momentum conservation, $p_\pi = p_\mu + p_\nu$, we have for the $\pi^+ \longrightarrow \mu^+ + \nu_\mu$ matrix element $V - A$ (with the pion at rest)

$$\frac{g_\pi(-m_\pi^2)}{\sqrt{2m_\pi}}\, i(p_{\mu\alpha} + p_{\nu\alpha}) i u(p_\mu)^C \gamma_\alpha(1 + \gamma_5)u(p_\nu)$$
$$= \frac{-m_\mu}{\sqrt{2m_\pi}}\, g_\pi(-m_\pi^2) i u^C(p_\mu)(1 + \gamma_5)u(p_\nu). \quad (11.22)$$

(We have chosen to write the matrix element of the decay $\pi^+ \longrightarrow \mu^+ + \nu_\mu$, so that the muon wave function, u^C, is the charge conjugate wave function to the μ^- wave function u. $u^C(p)$ obeys the Dirac equation $u^C(p)(i\gamma p + m_\mu) = 0$.)

This is a very interesting result for several reasons. In the first place we see that the matrix element vanishes in the limit in which $m_\mu \longrightarrow 0$. This result could have been anticipated by the following line of reasoning. If we choose the frame of reference in which the pion is at rest, it then follows that, in this frame,

$$\mathbf{p}_\mu + \mathbf{p}_\nu = 0, \quad (11.23)$$

so that the π^+ decay, for example, can be pictured as

$$\overset{\mu^+}{\longleftarrow} \pi^+ \overset{\nu_\mu}{\longrightarrow}.$$

Now, suppose ν_μ has a definite helicity, say left-handed, in analogy to ν_e.[7] We may then argue that, in π-decay, this helicity will be forced on the μ^+ by angular momentum conservation. Indeed, if we redraw the momentum picture, but now with *both* the spins and momenta, we see that since the pion has no spin we must have the spin of the μ^+ and ν_μ emitted in the decay pointing in opposite directions:

$$\underset{\mu^+}{\overset{\Rightarrow}{\longleftarrow}} \pi^+ \underset{\nu_\mu}{\overset{\Leftarrow}{\longrightarrow}}$$

This is to say that the muon must be emitted with its momentum antiparallel to its spin. (If we consider the charge conjugate decay, $\pi^- \longrightarrow \mu^- + \bar{\nu}_\mu$, then here the spins of $\bar{\nu}_\mu$ and μ^- have the opposite orientation relative to their respective momenta.) We might, at first sight, worry that the orbital angular momentum of ν_μ and μ complicates this discussion. But along the direction of the momenta, say the z direction, there is no orbital angular momentum. Thus the

---

CPT to the full weak Hamiltonian, which is to say, in this case, we should take the *CPT* of $A_\mu L_\mu{}^*$. We know from previous work that under C the vector and axial vector currents transform oppositely, so that taking *CPT* of $L_\mu{}^*$ will change the sign of the γ_μ term relative to the $\gamma_\mu\gamma_5$ term. This change of sign will not affect the rate (or lifetime) since this is a quantity that is insensitive to the nonconservation of parity. Thus the fact that $|g_{\pi^+}| = |g_{\pi^-}|$ is enough to guarantee lifetime equality. Parity-*violating* effects in the decay will, as we shall see, differ for π^+ and π^-.

7. Since we are talking about the muon neutrino, it is important to emphasize that the β-decay experiments on polarized nuclei that involve electron neutrinos and show that this neutrino is left-handed (see Lee and Wu, op. cit., p. 438) do *not* imply that the muon neutrino is left-handed, although this is the assumption made when one uses muon-electron symmetry in constructing the muonic part of L_μ. The left-handedness of the muon neutrino is confirmed directly by measuring the μ helicity in π-decay.

π-decay is a perfect polarizer of muons and it is the experimental observation of this muon polarization that confirms the left-handedness of ν_μ and hence this aspect of muon-electron symmetry. Any Dirac spinor may be decomposed as follows:

$$u(p) = \frac{(1+\gamma_5)}{2}u(p) + \frac{(1-\gamma_5)}{2}u(p) = u_+(p) + u_-(p), \tag{11.24}$$

where

$$\gamma_5 u_\pm(p) = \pm u_\pm(p). \tag{11.25}$$

However, the leptonic coupling in $\pi^+ \longrightarrow \mu^+ + \nu_\mu$ has the form

$$iu^C(p_\mu)\gamma_\mu(1+\gamma_5)u(p_\nu) = iu^C(p_\mu)(1-\gamma_5)\gamma_\mu u(p_\nu). \tag{11.26}$$

This means that if we write

$$u^C(p_\mu) = u^C(p_\mu)\frac{(1+\gamma_5)}{2} + u^C(p_\mu)\frac{(1-\gamma_5)}{2}, \tag{11.27}$$

the coupling picks out only the term[8]

$$u^C(p_\mu)\frac{(1-\gamma_5)}{2}.$$

We may now ask what the helicity is of the state represented by $u^C(p_\mu)_- = u^C(p_\mu)(1-\gamma_5)/2$, which has the property that

$$u^C(p_\mu)_-\gamma_5 = -u^C(p_\mu)_-. \tag{11.28}$$

We recall that for a mass zero neutrino the eigenstates of γ_5 were also eigenstates of $\boldsymbol{\sigma}\cdot\hat{\mathbf{p}}$, the helicity. However, here $m_\mu \neq 0$ and

$$\begin{aligned} u^C(p_\mu)\boldsymbol{\sigma}\cdot\hat{\mathbf{p}}_\mu &= u^C(\mathbf{p}_\mu)_- i\gamma_4\gamma_5\boldsymbol{\gamma}\cdot\hat{\mathbf{p}}_\mu = iu^C(\mathbf{p}_\mu)_-(\boldsymbol{\gamma}\cdot\hat{\mathbf{p}}_\mu)\gamma_4 \\ &= \frac{E}{|\mathbf{p}_\mu|}u^C(\mathbf{p}_\mu)_- - u^C(\mathbf{p}_\mu)_+\gamma_4\frac{m_\mu}{|\mathbf{p}_\mu|}. \end{aligned} \tag{11.29}$$

In the limit $m_\mu = 0$, Eq. (11.29) shows that $u^C(\mathbf{p}_\mu)_-$ would have positive helicity, so that angular momentum conservation would forbid the decay. The decay diagram would have to look like

$$\underset{\mu^+}{\overset{\Leftarrow}{\longleftarrow}}\ \pi^+\ \underset{\nu_\mu}{\overset{\Leftarrow}{\longrightarrow}},$$

which represents a state of spin 1, while π^+ has spin zero. However, a state of finite lepton mass, as Eq. (11.29) indicates, even if it is an eigenstate of γ_5, is *not* an eigenstate of helicity. There is a mixing of helicities and it is this mixing that allows the decay to take place and accounts for the fact that its matrix element is proportional to the leptonic mass.

8. Clearly, $(1+\gamma_5)(1-\gamma_5) = 0$.

The fact that the decay rate of pions into leptons depends on the lepton mass allows an especially convincing experimental confirmation of muon-electron symmetry. According to the *V-A* theory, with muon-electron symmetry, the decay rate for $\pi^+ \longrightarrow e^+ + \nu_e$ should be deducible from the rate for $\pi^+ \longrightarrow \mu^+ + \nu_\mu$ simply by making the replacements $\mu \longrightarrow e$ and $\nu_\mu \longrightarrow \nu_e$ in Eq. (11.22). Using Eq. (11.22), an entirely straightforward calculation shows that the rate for $\pi \longrightarrow \mu + \nu_\mu$ is given by[9]

$$R_{\pi\to\mu+\nu_\mu} = \frac{(g_\pi(-m_\pi^2))^2}{4\pi m_\pi^3} m_\mu^2(m_\pi^2 - m_\mu^2)^2. \tag{11.30}$$

We cannot use this equation to predict the pion lifetime since g_π is unknown. In Feynman graph language, g_π would be determined by graphs of the form

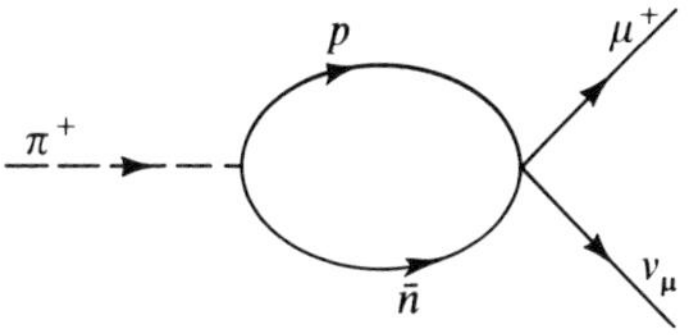

which involve the details of the strong interactions. We can of course use the experimental lifetime to determine g_π. With our conventions, g_π has the dimensions of $\left[\frac{1}{M}\right]$. It is conventional to use the lifetime to fix the dimensionless quantity $g_\pi^2 m_\pi^2$, and it turns out to be

$$(g_\pi m_\pi)^2 = (2.2 \pm .02) \times 10^{-14}.$$

If we assume μ-e symmetry, then the *V-A* theory makes a unique prediction for the ratio of the rates of the decays $\pi^+ \longrightarrow \mu^+ + \nu_\mu$ and $\pi^+ \longrightarrow e^+ + \nu_e$, since in this ratio the unknown $g_\pi(-m_\pi^2)$, which must, according to μ-e symmetry, be common to both matrix elements, cancels out. Indeed, this ratio (with μ-e symmetry) is

$$R = \frac{R_{\pi\to e+\nu_e}}{R_{\mu\to\mu+\nu_\mu}} = \frac{m_e^2\,(m_\pi^2 - m_e^2)^2}{m_\mu^2\,(m_\pi^2 - m_\mu^2)^2} \times 0.965. \tag{11.31}$$

The factor of .965 is a radiative correction[10] that differs slightly in the two cases because of the difference in the mass of the muon and electron. Numerically,

$$R_{\text{th}} = 1.23 \times 10^{-4}.$$

The reason why R is so small is, as we have seen, related to angular momentum conservation and the fact that the leptonic current has the special structure

9. The weak constant $G/\sqrt{2}$ does not appear here since it has been absorbed in the definition of g_π.
10. See Lee and Wu, op. cit., p. 469.

$i\bar{\psi}_e\gamma_\mu(1 + \gamma_5)\psi_{\nu_l}$, which means that only the parts of the ψ's with opposite symmetries under γ_5 are connected. The phase space ratio, assuming that the matrix element were just a constant, is given by

$$R \text{ phase space} = \frac{(m_\pi^2 - m_e^2)^2(m_\pi^2 + m_e^2)}{(m_\pi^2 - m_\mu^2)^2(m_\pi^2 + m_\mu^2)} = 3.5. \qquad (11.32)$$

It is the ratio $(m_e/m_\mu)^2$, which enters because of the helicity argument, that cuts R down from 3.5 to $\sim 10^{-4}$. Experimentally,[11]

$$R_{\text{exp}} = (1.21 \pm 0.07) \times 10^{-4}, \qquad (11.33)$$

which is in excellent agreement with the theoretical value and gives strong confirmation of the whole set of assumptions that go into the theory.

If we review the discussion up to this point, we find that we have learned next to nothing about the functional form of A_μ or even of $\partial_\mu A_\mu$, except that the latter cannot be zero unless m_π or M, or both, equal zero. Furthermore, it is not obvious that even if we were to consider a theory of zero mass pions interacting with nucleons, and impose current conservation, that this would imply that $g_A(0) = 1$. To see the sort of thing that can happen (we come back to this question in more detail later) let us study the matrix element of the axial vector charge between states of single physical nucleons. We call

$$Q_5^+(t) = \int d\mathbf{r}\, A_0(\mathbf{r}, t) \qquad (11.34)$$

recalling that A_μ is defined to raise electric charge and that Q_5^+ will be time-dependent if $\partial_\mu A_\mu \neq 0$. Let us assume (to see what happens) that $\partial_\mu A_\mu = 0$, so that Q_5^+ is time-independent. This means, extending the arguments of Chapter 2, that it transforms like a pseudoscalar. Thus

$$\langle \mathbf{0}|_P Q_5^+ |\mathbf{0}\rangle_N = Z_5\bar{u}(\mathbf{0})\gamma_5 u(\mathbf{0}) = 0, \qquad (11.35)$$

since $\bar{u}(\mathbf{0})\gamma_5 u(\mathbf{0}) = 0$. Hence current conservation teaches us nothing about the value of the "axial charge," Z_5. The fact that A_μ is an axial vector, and not a vector, makes a profound difference in the renormalization discussion and this argument using Q_5^+ is just a simple example of the sort of thing that can happen.

In the old Fermi theory the axial vector current was simply[12] assumed to be

$$A_\mu = i\bar{\psi}\gamma_5\gamma_\mu\tau_+\psi. \qquad (11.36)$$

If the nucleon fields in this current are taken to be free, then

$$\partial_\mu A_\mu = -2iM\bar{\psi}\gamma_5\tau_+\psi, \qquad (11.37)$$

an expression that has no especially transparent interpretation. Of course if the

---

11. H. L. Anderson et al., *Phys. Rev.* **119**:2050 (1960).
12. It is customary to write $\gamma_5\gamma_\mu$ in A_μ (rather than $\gamma_\mu\gamma_5$) and in the matrix elements of A_μ. Thus a current such as the leptonic current, which is written as $L_\mu = i\bar{\psi}_e\gamma_\mu(1 + \gamma_5)\psi_\nu$, represents a V-A coupling since γ_5 appears here to the right of γ_μ. With this convention, $g_A(0)$ is negative relative to $f_V(0)$.

fields are taken free, then $\langle 0|i\bar{\psi}(0)\gamma_5\gamma_\mu\tau_+\psi(0)|\pi^-\rangle = 0$, so that the Fermi current with free fields cannot account for pion decay. On the other hand, if the pion dynamics are included, so that A_μ has a nonvanishing matrix element for pion decay, then there is no reason to suppose that the Fermi current alone constitutes all of A_μ. For example, $\partial_\mu\pi^+$ is a perfectly good pseudovector candidate for part of A_μ. In brief, it is not especially clear at the outset where, if anywhere, the simplifying features of A_μ lie.

This was essentially where the situation stood until 1958, when Goldberger and Treiman[13] discovered a remarkable connection between pionic phenomena and $g_A(0)$. The Goldberger-Treiman formula is expressed as follows:[14]

$$Gg_A(0) \simeq \frac{f_\pi g_\pi}{M_P}. \tag{11.38}$$

Here, M_P is the mass of the proton; g_π, $g_A(0)$, and G have been defined before. The quantity f_π is the renormalized coupling constant of neutral pi mesons to nucleons with pseudoscalar coupling. By isotopic spin invariance the coupling of *charged* pions to nucleons is $\sqrt{2}f_\pi$.[15] We will have occasion to return to the definition of f_π in the sequel. However, let us make a few preliminary remarks about it based on the electromagnetic field as an analogy. By taking matrix elements of Maxwell's equation

$$\Box^2 A_\mu = -J_\mu{}^\gamma \tag{11.39}$$

between the one-proton states and using

$$\partial_\nu[\partial_\nu A_\mu] = \frac{1}{i}[P_\nu, \partial_\nu A_\mu]_-, \tag{11.40}$$

we derive the relation

$$q^2\langle \mathbf{p}'|A_\mu(0)|\mathbf{p}\rangle = \langle \mathbf{p}'|J_\mu{}^\gamma(0)|\mathbf{p}\rangle. \tag{11.41}$$

Hence we see that the matrix element $\langle \mathbf{p}'|A_\mu(0)|\mathbf{p}\rangle$ has a *pole* at $q^2 = 0$. Furthermore, the residue of this pole is related to the observed proton charge, $eF_1{}^P(0)$, as it is actually measured in electron-proton scattering. This suggests that if we define $\mathbf{j}_\pi$, an isotopic vector, by the field equation

$$(-\Box^2 + m_\pi{}^2)\boldsymbol{\phi}_\pi = \mathbf{j}_\pi, \tag{11.42}$$

and take matrix elements of the neutral pion component between proton states, then

$$(q^2 + m_\pi{}^2)\langle \mathbf{p}'|\pi^0(0)|\mathbf{p}\rangle = \langle \mathbf{p}'|j_\pi(0)_3|\mathbf{p}\rangle. \tag{11.43}$$

Hence the matrix element $\langle \mathbf{p}'|\pi^0(0)|\mathbf{p}\rangle$ has a pole at $q^2 = -m_\pi{}^2$ whose residue

13. M. L. Goldberger and S. B. Treiman, *Phys. Rev.* **110**:1178 (1958).
14. The reason for the $\simeq$ symbol in Eq. (11.38) will become clear in the sequel.
15. Recall that, if $\boldsymbol{\tau}$ is the nucleon isotopic spin matrix, then

$$\boldsymbol{\tau}\cdot\boldsymbol{\phi}_\pi = \tau_3\pi_0 + \sqrt{2}\,\pi^+\tau_- + \sqrt{2}\,\pi^-\tau_+.$$

we shall identify with the pion nucleon coupling constant.[16] To make this identification more precise we shall analyze, in detail, the structure of the matrix element $\langle \mathbf{p}'|\pi^0(0)|\mathbf{p}\rangle$ or more generally $\langle \mathbf{p}'|O_5(0)|\mathbf{p}\rangle$, where $O_5(0)$ is *any* pseudoscalar operator. The most general form that $\langle \mathbf{p}'|O_5(0)|\mathbf{p}\rangle$ can have, in a parity conserving theory, is

$$\langle \mathbf{p}'|O_5(0)|\mathbf{p}\rangle = i\bar{u}(\mathbf{p}')\gamma_5 u(\mathbf{p})F_5(q^2). \tag{11.44}$$

If we assume that $O_5(\mathbf{r}, t)$ transforms under time reversal like the neutral pion field (see Chapter 5) so that

$$TO_5(\mathbf{r}, t)T^{-1} = -O_5(\mathbf{r}, -t), \tag{11.45}$$

we then have the condition

$$\langle \mathbf{p}'|_{s'}O_5(0)|\mathbf{p}\rangle_s = -\langle -\mathbf{p}'|_{-s'}O_5(0)|-\mathbf{p}\rangle_{-s}{}^*. \tag{11.46}$$

Thus

$$i\bar{u}(\mathbf{p}')_{s'}\gamma_5 u(\mathbf{p})_s F_5(q^2) = F_5{}^*(q^2) i\bar{u}(\mathbf{p}')_{s'}(t\gamma_5 t^\dagger)^* u(\mathbf{p})_s, \tag{11.47}$$

which implies that

$$F_5(q^2)^* = F_5(q^2). \tag{11.48}$$

If we define $f_\pi(q^2)$ by the equation[17]

$$\langle \mathbf{p}'|\pi^0(0)|\mathbf{p}\rangle = f_\pi(q^2) i\bar{u}(\mathbf{p}')\gamma_5 u(\mathbf{p}), \tag{11.49}$$

then the above analysis shows that we can write

$$f_\pi(q^2) = \frac{F_\pi(-m_\pi{}^2)D(q^2)}{q^2 + m_\pi{}^2}, \tag{11.50}$$

where, from this definition of the pion-nucleon coupling constant,

$$D(-m_\pi{}^2) = 1. \tag{11.51}$$

One may well wonder how this coupling constant is to be measured. Again electron-proton scattering suggests a method. In Born approximation, electron-proton scattering is given by the one-photon exchange graph

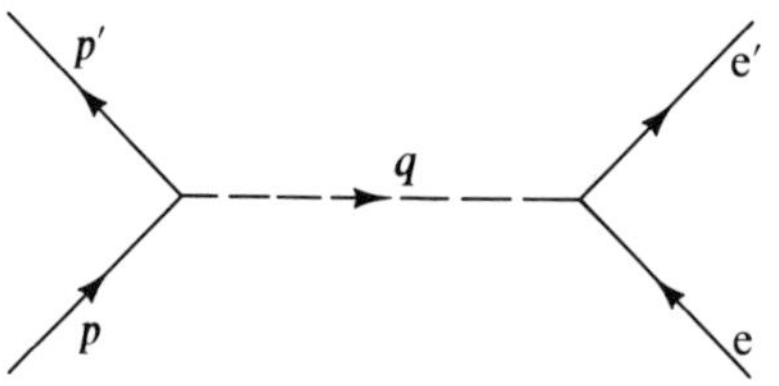

where each vertex is expressed as the appropriate one-particle matrix element

16. The point $q^2 = 0$ does not have any special significance in this matrix element and there is no reason to define the coupling constant with respect to it.
17. It will be real by the time reversal discussion just above.

of the electric current operator. In the limit $q^2 = 0$ the scattering amplitude is, as we have seen, proportional to the product of the electron and proton charges. In the proton-proton elastic scattering $p + p \longrightarrow p + p$, there will be a one-pion exchange contribution (called by the pp scattering experts *OPEC*) represented by the graph

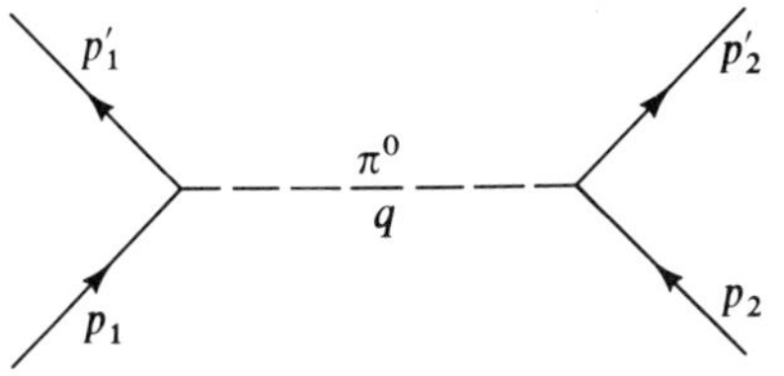

The virtual pion emission is given by the pion nucleon vertex $\langle \mathbf{p}'|\mathbf{j}_\pi(0)|\mathbf{p}\rangle$ just as the virtual photon emission is given by the vertex $\langle \mathbf{p}'|J_\mu{}^\gamma(0)|\mathbf{p}\rangle$. At $q^2 = -m_\pi{}^2$ this vertex gives a direct measure of the pion nucleon coupling constant, at least as it is defined in this way. There are at least two important problems connected with actually using this graph to find the coupling constant (or its square, since the vertex enters twice). In the first place, $q^2 = -m_\pi{}^2$ is not a point that is actually reached in physical pp scattering, where $0 \leq q^2 \leq \infty$. In the second place, since the pion is strongly coupled there is no reason to expect that the *OPEC* gives anything like the complete theory of pp scattering. For example, multipion exchange graphs, exchanges of vector mesons, and the like, must also contribute strongly to the scattering in the physical region. However, the pion is the longest range particle (smallest mass) that the protons can exchange, leaving out the photon which produces complicating electromagnetic effects that also enter the analysis of pp scattering. Hence, despite the complications, the *OPEC* must make an important, perhaps decisive, contribution to the longest range part of the pp nuclear force, which is to say, the part contributing to high angular momentum channels in the pp center of mass.[18] Using this philosophy, several groups[19] have analyzed pp scattering and find that

$$\frac{f_\pi{}^2}{4\pi} \simeq 14.8.$$

As a technical point we note that we could also have written

$$\langle \mathbf{p}'|\pi^0(0)|\mathbf{p}\rangle = i\bar{u}(\mathbf{p}')i\gamma_5(\gamma q)u(\mathbf{p})f_\pi{}'(q^2). \tag{11.52}$$

Hence, if we had used this equivalent "pseudovector" representation of $\langle \mathbf{p}'|\pi^0(0)|\mathbf{p}\rangle$, we would have defined another form factor $f_\pi{}'(q^2)$ with the dimensions of an inverse mass. It is customary to take this mass to be the pion mass

18. Since $L \sim rp$, for fixed p, the long-range collisions involve the largest angular momenta.
19. See, for example, 1962 International Conference on High-Energy Physics at CERN, published by CERN, Geneva, p. 131 and ff.

m_π so that the "pseudovector" and "pseudoscalar" coupling constants are related by

$$\frac{m_\pi}{2M_P} f_{\pi P.S.} = f_{\pi P.V.}. \tag{11.53}$$

As we shall shortly see, the *OPEC* is not the only method of defining and measuring f_π. It appears as if the best consensus of the different methods of defining and measuring f_π gives the value[20]

$$\frac{f_\pi^{\,2}}{4\pi} = 14.8 \pm .3.$$

Using these numbers,[21] we can evaluate the two sides of the Goldberger-Treiman (*G–T*) relation written in the form

$$Gg_A(0)M_P^2 \simeq f_\pi g_\pi M_P. \tag{11.54}$$

Thus

$$Gg_A(0)M_P^2 = (1.19 \pm .02) \times 10^{-5}, \tag{11.55}$$

and

$$f_\pi g_\pi M_P \simeq 1.3 \times 10^{-5}. \tag{11.56}$$

Since the derivations of the *G–T* relation contain several approximations, this numerical confirmation represents remarkably good agreement.

Before turning to these derivations we make a few remarks about the form factors g_A and g_P that enter the matrix element $\langle \mathbf{p}'|A_\mu(0)|\mathbf{p}\rangle$. A direct dispersion treatment of these form factors, done in the same spirit as in electron-proton scattering (Chapter 7), shows, as we shall see below, that the lowest mass states, involving familiar particles, that contribute to g_A and g_P in the dispersion integral are nucleon-antinucleon states with $J = 1$, to g_A, and, the one-pion state, to g_P. Diagramatically,

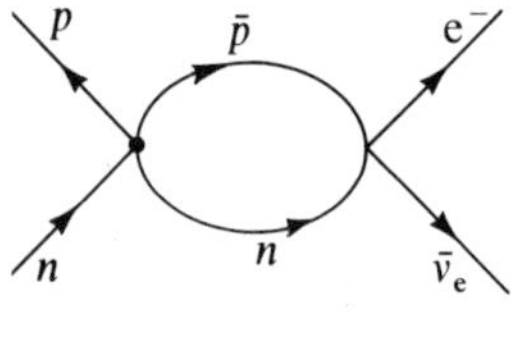

g_A

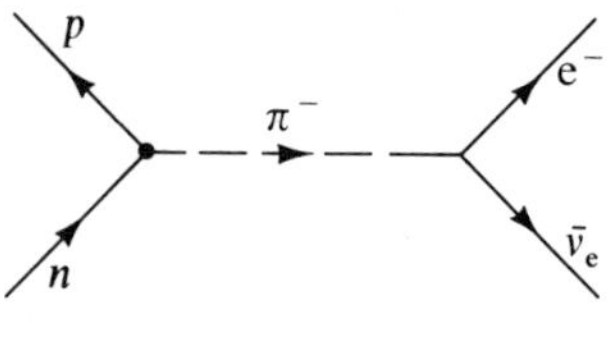

g_P

20. See, for example, V. K. Samaranagake and W. S. Woolcock, *Phys. Rev. Letters* **15**:936 (1965). Henceforth by f_π we shall always mean $f_{\pi P.S.}$.
21. Remember that $M_P/m_\pi = 6.7$.
22. See A. H. Rosenfeld et al., *Rev. Mod. Phys.* **39**:1 (1967).
23. As we have seen from the dispersion treatment of electron-scattering the intermediate states that contribute to the imaginary parts of the dispersion integral are on their mass shells, so that this diagram is not the conventional Feynman diagram where the nucleons would be off their shells. It is merely a simple mnemonic for keeping track of the intermediate states that contribute to the imaginary parts.

If there were an axial vector meson (or mesons), it (they) would, of course, contribute to g_A. Assuming that the axial vector current is first-class Weinberg (see Chapter 9), such mesons would have to have negative G parity, like the pion, so they would be likely to decay into three pions, or into a ρ and a pion. There is at least one tentative candidate for such a meson,[22] the so-called A_1 particle, with a mass of about 1080 Mev and a width of about 125 Mev. It appears to decay almost entirely into ρ and π. If further experiments continue to support the existence of the A_1, it could play a crucial role in the structure of $g_A(q^2)$. However, because of the complexity and uncertainty of these states, it is not profitable to make a direct frontal attack on $\langle \mathbf{p}' | A_\mu(0) | \mathbf{p} \rangle$. Indeed, in their original derivation of the G–T relation, Goldberger and Treiman started instead from the exact amplitude for π-decay. In the dispersion theoretic treatment of this matrix element they made the approximation that the π-decay proceeds largely through the ${}^1S_0$, n, $\bar{n}$ channel.[23]

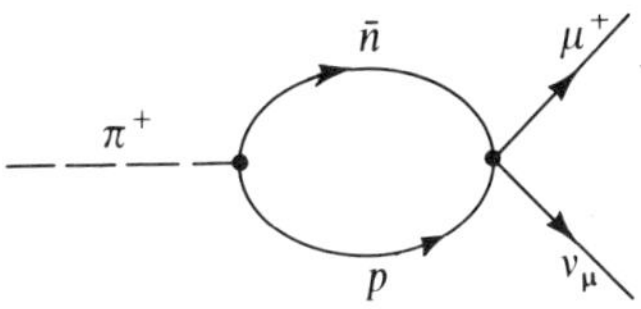

Hence they inevitably became involved in the complexities of the strongly interacting $n\bar{n}$ system. In fact, the formula they actually ended up with can be written (in our notation) in the form

$$g_\pi = M_P G g_A(0) \frac{f_\pi J}{1 + f_\pi^2 J}, \tag{11.57}$$

where J contains all the complicated details of the $n\bar{n}$ scattering.[24] Using explicit models for $n\bar{n}$ scattering they were able to argue, plausibly, that $f_\pi^2 J \gg 1$. Hence they could neglect the 1 in the denominator of *Eq.* (11.57) and end up with the G–T relation in its familiar form, which is independent of J, and of the $n\bar{n}$ dynamics. The final formula was so simple and its agreement with experiment so striking that it seemed plausible that there must be a simpler way to derive it. In fact, it has turned out to be possible to give several alternate derivations,[25] which at least appear to be more straightforward. We now turn to a treatment of them.

---

24. The reader who is interested in the exact definition of J is referred to M. L. Goldberger and S. B. Treiman, *Phys. Rev.* **110**:1178 (1958).
25. See, for example, K. C. Chou, *J. Exptl. Theoret. Phys.* (USSR) **39**:703 (1960); Y. Nambu, *Phys. Rev. Letters* **4**:380 (1960); M. Gell-Mann and M. Lévy, *Nuovo Cimento* **16**:705 (1960); J. Bernstein, S. Fubini, M. Gell-Mann, and W. Thirring, *Nuovo Cimento* **17**:757 (1960); R. P. Feynman, *Symmetries in Elementary Particle Physics*, edited by A. Zichichi, Academic Press (1965), p. 111 and ff.

All of these derivations have in common the idea that even though $\partial_\mu A_\mu$ cannot be zero in the real physical world, it may, anyway, be a "simple" operator. (The meaning of "simple" will become clearer as we proceed with the derivations.) To begin with, we discuss the general symmetry features of $\partial_\mu A_\mu$. Let (just to give it a name)

$$\partial_\mu A_\mu = O_5{}^+(\mathbf{r}, t) \tag{11.58}$$

The superscript $+$ is in honor of the fact that A_μ raises the electric charge by one unit, and so then must $O_5{}^+$. By integrating Eq. (11.58) over space and assuming that $\mathbf{A}(\mathbf{r}, t)$ is sufficiently localized, so that

$$\int \nabla \cdot \mathbf{A}(\mathbf{r}, t)\, d\mathbf{r} = 0, \tag{11.59}$$

we see that

$$\frac{\partial}{\partial t} Q_5{}^+(t) = O_5{}^+(t), \tag{11.60}$$

where $Q_5{}^+(t)$, the axial "charge," is defined by

$$\int A_0(\mathbf{r}, t)\, d\mathbf{r} = Q_5{}^+(t) \tag{11.61}$$

and

$$O_5{}^+(t) = \int O_5{}^+(\mathbf{r}, t)\, d\mathbf{r}. \tag{11.62}$$

From the properties of A_μ as a first-class Weinberg, charge-symmetric, pseudovector current, we can draw the following conclusions.

1. $O_5{}^+(\mathbf{r}, t)$ is a pseudoscalar operator,[26] $PO_5{}^+(\mathbf{r}, t)P^{-1} = -O_5{}^+(-\mathbf{r}, t)$
2. It transforms under time reversal as

   $$TO_5{}^+(\mathbf{r}, t)T^{-1} = -O_5{}^+(\mathbf{r}, -t).$$

3. Because of the charge symmetry of A_μ,

   $$e^{i\pi T_2} O_5{}^+ e^{-i\pi T_2} = -O_5{}^-.$$

4. Since A_μ is first-class Weinberg,

   $$GO_5{}^+G^{-1} = -O_5{}^+.$$

5. Since $G = Ce^{i\pi T_2}$ (or directly from the C character of A_μ),

   $$CO_5{}^+C^{-1} = O_5{}^-.$$

6. $CPT\, O_5{}^+(\mathbf{r}, t)(CPT)^{-1} = O_5{}^-(-\mathbf{r}, -t)$.

---

26. $P\left[\nabla_{\mathbf{r}} \cdot \mathbf{A}(\mathbf{r}, t) + \frac{\partial}{\partial t} A_0(\mathbf{r}, t)\right] P^{-1} = -\nabla_{-\mathbf{r}} \cdot \mathbf{A}(-\mathbf{r}, t) - \frac{\partial}{\partial t} A_0(-\mathbf{r}, t).$

It will not have escaped the attention of the alert reader that $O_5^+(\mathbf{r}, t)$ has, so far as these symmetries go, all of the properties of the pion field $\pi^+(\mathbf{r}, t)$.

We next show that the matrix element $\langle 0|O_5^+(0)|\pi^-\rangle$ can be determined numerically from the decay rate of the physical π^-. This follows directly from the string of equations

$$\frac{G}{\sqrt{2}}\langle 0|O_5^+|\pi^-\rangle = \frac{G}{\sqrt{2}}\langle 0|\partial_\mu A_\mu|\pi^-\rangle = \frac{1}{i}\frac{G}{\sqrt{2}}\langle 0|[P_\mu, A_\mu]_-|\pi^-\rangle$$

$$= \frac{G}{\sqrt{2}}\, ip_{\pi\mu}\langle 0|A_\mu|\pi^-\rangle = \frac{m_\pi^2}{\sqrt{2E_\pi}}\, g_\pi(-m_\pi^2), \tag{11.63}$$

where we have, in the last stage, used Eq. (11.14). Thus we have fixed $\langle 0|O_5^+(0)|\pi^-\rangle$ in terms of the pion lifetime. Next, we consider O_5^+ between states of the physical neutron and proton. Once again,

$$\frac{G}{\sqrt{2}}\langle \mathbf{p}'|O_5^+(0)|\mathbf{p}\rangle = \frac{1}{i}(p'-p)_\mu\langle \mathbf{p}'|A_\mu(0)|\mathbf{p}\rangle\frac{G}{\sqrt{2}}$$

$$= \frac{1}{i}\frac{(p'-p)_\mu}{(2\pi)^3}\frac{G}{\sqrt{2}}\, i[\bar{u}(p')\gamma_5[\gamma_\mu g_A(q^2) + i(p'-p)_\mu g_P(q^2)]u(p)]$$

$$= -i\frac{G}{\sqrt{2}}\frac{[\bar{u}(p')\gamma_5 u(p)]}{(2\pi)^3}[2M_P g_A(q^2) - q^2 g_P(q^2)]. \tag{11.64}$$

Hence, defining a function $H(q^2)$ with

$$H(q^2) = -(2M_P g_A(q^2) - q^2 g_P(q^2))\frac{G}{\sqrt{2}(2\pi)^3}, \tag{11.65}$$

we have

$$\langle \mathbf{p}'|O_5^+(0)|\mathbf{p}\rangle = i\bar{u}(p')\gamma_5 u(p)H(q^2), \tag{11.66}$$

where, from the reality of $g_A(q^2)$ and $g_P(q^2)$ for $q^2 \geq 0$ (which follows from the symmetry character of A_μ), it must be that $H(q^2)$ is real, at least in the interval $0 \leq q^2 \leq \infty$.[27] So far no approximations or special assumptions about O_5^+ or $H(q^2)$ have been made. We now make one. We *assume* that $\lim_{q^2\to\infty} H(q^2) = 0$, or that, in particular, $H(q^2)$ obeys an *unsubtracted* dispersion relation. (We return later to the question of how strongly the argument depends on this assumption.) Thus we may write

$$H(q^2) = \frac{1}{\pi}\int_{-\infty}^{0}\frac{\operatorname{Im} H(q'^2)}{q'^2 - q^2}\, dq'^2. \tag{11.67}$$

We do not have to worry about taking the principal value here since q^2 is in the physical region for neutron β-decay and q'^2 is in the unphysical region, where

---

27. If we were to take into account the n, p mass difference, the time-reversal argument actually implies that $H(q^2)$ is real in the interval

$$-(M_N - M_P)^2 \leq q^2 \leq \infty.$$

Im $H(q'^2) \neq 0$. Thus the denominator cannot vanish and we can consider the integral as an ordinary integral.

We may now repeat the entire discussion of Chapter 7, with the reduction formula, to find an expression for Im $H(q^2)$. Using the fact that $O_5{}^+(0)$ (because it is a pseudoscalar) satisfies

$$\langle 0|O_5{}^+(0)|0\rangle = 0, \tag{11.68}$$

where $|0\rangle$ is the vacuum state, we arrive at the identity (P is here the time-ordering symbol)

$$\langle \mathbf{p}'|O_5{}^+(0)|\mathbf{p}\rangle = \int d^4x\, \langle \mathbf{p}'|P[O_5{}^+(0)\bar{j}_N(x)]|0\rangle f_{\mathbf{p}}(x)$$
$$- \int d^4x\, \langle \mathbf{p}'|[O_5{}^+(0), \bar{\psi}(x)]_-|0\rangle f_{\mathbf{p}}(x)\, \delta(t), \tag{11.69}$$

with, as in Chapter 7,

$$f_{\mathbf{p}}(x) = \frac{e^{i(px)}u(\mathbf{p})}{(2\pi)^{3/2}} \tag{11.70}$$

and

$$\bar{j}_N(x) = \bar{\psi}(x)(-\overleftarrow{\partial}_\mu\gamma_\mu + M_p) \tag{11.71}$$

We must now, as in Chapter 7, make some further assumption about the equal time commutator,

$$[O_5{}^+(0), \bar{\psi}(\mathbf{r}, 0)]_-.$$

In all the theories in which specific models for A_μ have been constructed (We shall exhibit several below.) this commutator has the property that[28]

$$[O_5{}^+(0), \bar{\psi}(\mathbf{r}, 0)]_- = \delta^3(\mathbf{r})P(\mathbf{r}), \tag{11.72}$$

where $P(\mathbf{r})$ is some function of the fields ψ and π. Thus

$$\int d^4x\, \langle \mathbf{p}'|[O_5{}^+(0), \bar{\psi}(\mathbf{r}, 0)]_-|0\rangle f_{\mathbf{p}}(x)\, \delta(t)$$
$$= \int d^4x\, \delta^4(x)\, \langle \mathbf{p}'|P(\mathbf{r})|0\rangle u(\mathbf{p})\, \frac{e^{i(px)}}{(2\pi)^{3/2}} = \langle \mathbf{p}'|P(0)|0\rangle \frac{u(\mathbf{p})}{(2\pi)^{3/2}}. \tag{11.73}$$

By invariance,[29]

$$\langle \mathbf{p}'|P(0)|0\rangle = i\bar{u}(p')\gamma_5 N, \tag{11.74}$$

where N is a constant. Thus

$$\langle \mathbf{p}'|O_5{}^+(0)|\mathbf{p}\rangle = iN\bar{u}(p')\gamma_5 u(p) + \int d^4x\, \langle \mathbf{p}'|P[O_5{}^+(0)\bar{j}_N(x)]|0\rangle f_{\mathbf{p}}(x). \tag{11.75}$$

However, we have assumed that

---

28. To give the simplest example, if $A_\mu = i\bar{\psi}\gamma_5\gamma_\mu\psi$, then, for free fields, $O_5{}^+(0) = -2iM\bar{\psi}\gamma_5\psi$. and this, when commuted with $\bar{\psi}$, clearly has the local property assumed in Eq. (11.72).
29. $\bar{\psi}$ is a spinor and $O_5{}^+$ is a pseudoscalar.

$$\langle \mathbf{p}'|O_5{}^+(0)|\mathbf{p}\rangle = i\bar{u}(\mathbf{p}')\gamma_5 u(\mathbf{p})H(q^2) \tag{11.76}$$

with

$$\lim_{q^2\to\infty} H(q^2) \longrightarrow 0. \tag{11.77}$$

The simplest assumption to make consistent with this condition is that

$$N = 0. \tag{11.78}$$

(Otherwise we must suppose, to be consistent with the large q^2 limit of $H(q^2)$, that the second term on the right side cancels out the N for large q^2.) Though this set of assumptions appears at first glance quite reasonable and, as we shall see below, there is even a class of theories for which

$$[O_5{}^+(0), \bar{\psi}(\mathbf{r}, 0)]_- = 0, \tag{11.79}$$

it is not without somewhat paradoxical-looking implications. In particular, suppose we pass to the limit in which the pion-nucleon coupling goes to zero. If this limiting process were a straightforward one, then, since when there is no pion coupling, $\bar{j}_N(x) = 0$, we could apparently conclude that

$$\langle \mathbf{p}'|O_5{}^+(0)|\mathbf{p}\rangle = 0. \tag{11.80}$$

This result certainly contradicts the notion that, with no pion couplings present, the current A_μ reduces simply to $i\bar{\psi}\gamma_5\gamma_\mu\tau\psi$, since this current is not conserved unless $M_p = 0$. Hence we should not be surprised if $O_5(0)$ has a singular behaviour for $f_\pi \longrightarrow 0$.

In any event we have now arrived at the identity

$$\langle \mathbf{p}'|O_5{}^+(0)|\mathbf{p}\rangle = \int d^4x\, \langle \mathbf{p}'|P[O_5{}^+(0)\bar{j}_N(x)]|0\rangle f_\mathbf{p}(x). \tag{11.81}$$

From this point on, the argument proceeds without incident, as in Chapter 7, and we show here, as there, that

$$2i\bar{u}(\mathbf{p}')\gamma_5 u(\mathbf{p})\,\mathrm{Im}\,H(q^2) = (2\pi)^{5/2}\sum_n \delta^4(p_n - q)\langle \mathbf{p}'|\bar{j}_N(0)|n\rangle\langle n|O_5{}^+(0)|0\rangle u(\mathbf{p}), \tag{11.82}$$

where the sum is over all the eigenstates (the complete set) of P_μ. We now make the next crucial approximation. If we analyze the set of states that $O_5{}^+(0)$ can connect to the vacuum, we see from the symmetries of $O_5{}^+(0)$ they must have the following properties.

1. $J = 0$.
2. $P = -1$.
3. $G = -1$.
4. $Q = 1$.

The candidates, among others, are the states π^+, $\pi^+\pi^0\pi^0$, $p\bar{n}$, and so on. We assume that in this sum *only* the state π^+ contributes—in other words, that the operator $O_5{}^+(0)$ acts like a π^+ field to the extent that it tends to connect the vacuum with the one-pion state to the exclusion of the many-pion or nucleon-antinucleon states. This approximation cannot really be justified a priori. It has sometimes been called "partial current conservation," PCAC,[30] and we shall return to it often in the sequel. The only real justification for it is that it leads to simple results like the *G-T* formula that agree with experiment.

Having made this approximation we can now study the combination

$$(2\pi)^{5/2}\,\delta^4(p_{\pi^+} - q)\langle p'|\bar{j}_N(0)|p_{\pi^+}\rangle\langle p_{\pi^+}|O_5^+(0)|0\rangle \tag{11.83}$$

with

$$q = p' - p \tag{11.84}$$

and

$$q^2 = -m_\pi^2.$$

There is no difficulty in interpreting the expression $\langle p_{\pi^+}|O_5^+(0)|0\rangle$. Using the symmetries of O_5^+ and Eq. (11.63),

$$\frac{G}{\sqrt{2}}\langle p_{\pi^+}|O_5^+(0)|0\rangle = -\langle 0|O_5^-(0)|p_{\pi^+}\rangle^*\frac{G}{\sqrt{2}} = -\frac{m_\pi^2}{\sqrt{2E_\pi}}\,g_\pi(-m_\pi^2), \tag{11.85}$$

since $\langle p_{\pi^+}|$ is, by construction, the physical one-pion state. The quantity $\langle p'|\bar{j}_N(0)|p_{\pi^+}\rangle$ clearly has to do only with the *strong* interactions. We are treating the weak interactions to first order and can thus ignore their effect on $\langle p'|$, $\langle p_{\pi^+}|$, and $\bar{j}_N(0)$. In fact, it is clear from our previous work that $\langle p'|\bar{j}_N(0)|p_{\pi^+}\rangle$ must represent the vertex for the transition

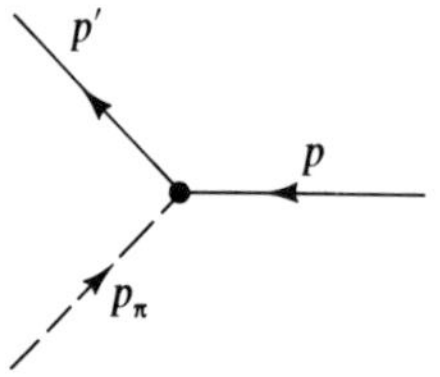

$$p_\pi = p' - p$$

where the heavy lines are nucleons and the dotted line a pion. We know that by construction p_π and p' are on their respective mass shells. What about p? Consider

$$p^2 = (p' - p_\pi)^2. \tag{11.86}$$

We can always evaluate this invariant expression in the frame in which, say, the π^+ is at rest. Thus

$$p^2 = -M_P^2 - m_\pi^2 + 2E(p')m_\pi \neq M_P^2. \tag{11.87}$$

Therefore this vertex is the vertex for the absorption of a *virtual* nucleon by a

---

30. Some authors restrict the use of the name PCAC to that class of theories in which O_5^+ actually *is* the pion field. The whole point of the discussion given here is to derive the *G-T* relation without assuming PCAC in this narrow sense.

charged physical pion to yield a *physical* nucleon. It is part of the one-nucleon exchange diagram in pion-nucleon scattering:

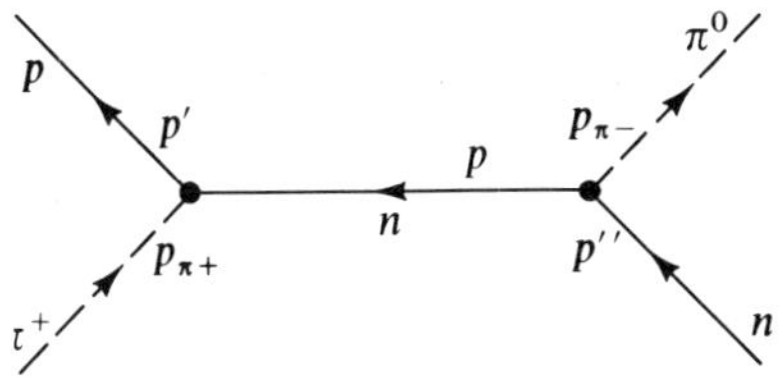

We may express $\langle p'|\bar{j}_N(0)|p_\pi\rangle$ in covariant form,

$$(2\pi)^3\langle p'|\bar{j}_N(0)|p_\pi\rangle = i\bar{u}(p')\gamma_5\frac{f_\pi((p'-p_\pi)^2)\sqrt{2E_\pi}\sqrt{2}}{(2\pi)^{3/2}}, \tag{11.88}$$

with $f_\pi((p'-p_\pi)^2)$ the real form factor for this vertex. The factor $(2\pi)^{-3/2}(2E_\pi)^{1/2}$ is put in to conform to the standard definition of the dimensionless pion-nucleon coupling constant. The factor of $\sqrt{2}$ comes in because the pion-nucleon vertex here involves the coupling of charged pions. The one-nucleon pole in pion-nucleon scattering plays the same role as the one-pion pole in nucleon-nucleon scattering in determining the pion-nucleon coupling constant. The smallest possible value for p^2 in the physical region for pion-nucleon scattering is, from Eq. (11.87), $-(M_P - m_\pi)^2$. Hence the extrapolation to the nucleon pole, in pion-nucleon scattering, requires the same sorts of continuations in momenta as the extrapolation to the pion pole in nucleon-nucleon scattering. The quantity $f_\pi(-M_P{}^2)$ is identified as the pion-nucleon coupling constant as measured in pion-nucleon scattering. This coupling constant is, experimentally, about the same as the pion coupling constant that is measured in nucleon-nucleon scattering. We can now put all the pieces together and write, in the one-pion approximation,

$$2i\bar{u}(p')\gamma_5 u(p)\,\mathrm{Im}\,H(q^2) = -\delta^4(p_\pi - q)i\bar{u}(p')\gamma_5 u(p)\frac{f_\pi(p^2)\sqrt{2}m_\pi{}^2 g_\pi(-m_\pi{}^2)}{(2\pi)^2}. \tag{11.89}$$

Thus, evaluating the dispersion integral at the one-pion pole, we have

$$\frac{G}{\sqrt{2}}(2M_P g_A(q^2) - q^2 g_P(q^2)) \simeq m_\pi{}^2\frac{g_\pi(-m_\pi{}^2)f_\pi(-M_P{}^2)\sqrt{2}}{q^2 + m_\pi{}^2}. \tag{11.90}$$

In addition to the one-pion pole approximation we have also assumed that $f_\pi(q^2)$ is given, approximately, by its value at the one-nucleon pole in pion-nucleon scattering; that is, we replace $f_\pi(q^2)$ by the pion-nucleon coupling constant. As stated in the beginning, in the absence of a detailed dynamical theory of pions and nucleons it is essentially impossible to give a reliable estimate of the errors involved in making these approximations. We can only test them retrospectively by examining what they lead to. The first thing that Eq. (11.90) leads to is the

G-T relation. If $g_P(q^2)$ is finite at $q^2 = 0$ (We come back to this question later.), then, setting $q^2 = 0$ in Eq. (11.70), we have

$$M_P G g_A(0) \simeq g_\pi(-m_\pi^2) f_\pi(-M_P^2). \tag{11.91}$$

It might appear, from the way this equation is written, that the relative algebraic sign of g_π and f_π has been fixed. This is not true, since we have obtained g_π by taking, essentially, the square root of the expression for the pion decay rate, and hence our derivation only fixes the magnitudes of the quantities involved. This derivation of the *G-T* relation has the virtue that it involves the fewest conditions on the explicit form of $\partial_\mu A_\mu$. We see also that had we not assumed that $H(q^2) \longrightarrow 0$ at $q^2 = \infty$ we would have introduced unknown subtraction constants into the problem and, hence, would have sacrificed the prospect of relating $g_A(0)$ to known quantities.

We can use Eq. (11.70) to go one step farther. We may ask where the pion pole in $2M_P g_A(q^2) - q^2 g_P(q^2)$ occurs—in $g_A(q^2)$, in $g_P(q^2)$, or in both. To study this matter we once again return to the covariant representation of $\langle \mathbf{p}'|_P A_\mu(0)|\mathbf{p}\rangle_N$:

$$(2\pi)^3 \langle \mathbf{p}'|_P A_\mu(0)|\mathbf{p}\rangle_N = i\bar{u}(\mathbf{p}')\gamma_5[\gamma_\mu g_A(q^2) + i(p' - p)_\mu g_P(q^2)]u(\mathbf{p}). \tag{11.92}$$

We can now imagine repeating the arguments of Chapter 7 to derive the dispersion relations for g_A and g_P in terms of integrals over their imaginary parts. We know—and this is the point that concerns us here—that these imaginary parts involve sums over states of the form

$$\sum_n \langle p'|\bar{j}_N(0)|n\rangle\langle n|A_\mu(0)|0\rangle\, \delta^4(p_n - q)u(p). \tag{11.93}$$

Now the one-pion term will be of the form

$$\langle p'|\bar{j}_N(0)|\pi\rangle\langle \pi|A_\mu(0)|0\rangle. \tag{11.94}$$

The whole Lorentz vector character is contained in $\langle \pi|A_\mu(0)|0\rangle$. But, as we have seen before, this factor must be proportional to q_μ, the pion momentum. There is no other Lorentz four-vector available. Thus from Eq. (11.72) we see that the one-pion state can contribute *only* to $g_P(q^2)$. Nucleon-antinucleon states, axial vector mesons, if they exist, can contribute to $g_A(q^2)$, but the one-pion state cannot. This means that we must be able to write

$$\frac{G}{\sqrt{2}}\, g_P(q^2) = \frac{f(q^2)}{q^2 + m_\pi^2}, \tag{11.95}$$

where $f(-m_\pi^2)$ is the residue at the pion pole. Thus, near the pole, where we can drop the $g_A(q^2)$ term in Eq. (11.70), we have

$$m_\pi^2 f(-m_\pi^2) = m_\pi^2 g(-m_\pi^2) f_\pi(-M_P^2)\sqrt{2}. \tag{11.96}$$

If we assume that

31. Remember that $m_e g_P(0)$ is the effective pseudoscalar constant for β-decay.

$$Gg_P(0) \simeq \frac{f(-m_\pi^2)}{m_\pi^2} = \frac{g(-m_\pi^2)f_\pi(-M_P^2)2}{m_\pi^2}, \tag{11.97}$$

then we have[31]

$$M_P^2 G m_e g_P(0) \simeq \left(\frac{M_P}{m_\pi}\right)^2 2g(-m_\pi^2)f_\pi(-M_P^2)m_e. \tag{11.98}$$

We can evaluate this term directly or, somewhat more simply and inexactly, by using the *G-T* relation, which gives

$$M_P^2 G m_e g_P(0) \sim (M_P^2 G) g_A(0) 2 \frac{M_P m_e}{m_\pi^2} \tag{11.99}$$

or

$$m_e g_P(0) \sim 4.8 \times 10^{-2} g_A(0). \tag{11.100}$$

Hence the induced pseudoscalar in ordinary β-decay is certainly too small to play any role there. However, muon capture is another story. For the basic process $\mu^- + p \longrightarrow n + \nu_\mu$, we have the kinematical equations (the μ and p are initially at rest)

$$q^2 = (\nu - \mu)^2 = -m_\mu^2 + 2\nu m_\mu = m_\mu^2 \left[\frac{2\nu}{m_\mu} - 1\right] \tag{11.101}$$

and

$$M_\mu + M_P = \nu + \sqrt{\nu^2 + M_P^2}, \tag{11.102}$$

from which we conclude that, in m_μ capture,

$$q^2 = .88\, m_\mu^2. \tag{11.103}$$

Thus, the set of steps above, that determines g_P, yields, for the effective pseudoscalar coupling in μ capture, $m_\mu g_P$,

$$\begin{aligned} m_\mu g_P(.88 m_\mu^2) &\simeq \frac{2g_A(0) M_P m_\mu}{.88 m_\mu^2 + m_\pi^2} \\ &= 2g_A(0)\frac{M_P}{m_\mu}\frac{1}{.88 + \left(\dfrac{m_\pi}{m_\mu}\right)^2} \simeq 7g_A(0). \end{aligned} \tag{11.104}$$

The detailed theory of muon capture, especially in complex nuclei, is an extensive exercise in nuclear physics and beyond the scope of this book.[32] In principle, the induced pseudoscalar plays a role in total capture rates, rates for 0^+–0^- transitions between particular nuclear levels, the form of the angular distribution of the neutron emitted following the nuclear capture of the muon, the shape of the energy spectrum in the radiative capture process $\mu^- + p \longrightarrow n + \nu_\mu + \gamma$, and the like. Experiments on all of these quantities have been done, but their interpretation is clouded by the details of the nuclear physics. It appears

---

32. See Lee and Wu, op. cit., p. 458 and ff. for an up-to-date review. Many references to the literature are contained in this article.

that there is evidence for the existence of a g_P term but we cannot yet say whether its sign and magnitude, relative to g_A, are consistent with the *G-T* prediction.

As concerns the relative sign of g_A and g_P, it is worth the following comment. Even before the *G-T* relation was discovered the existence of g_P, and its magnitude, had been conjectured.[33] This was done on the basis of the simple perturbation theory diagram

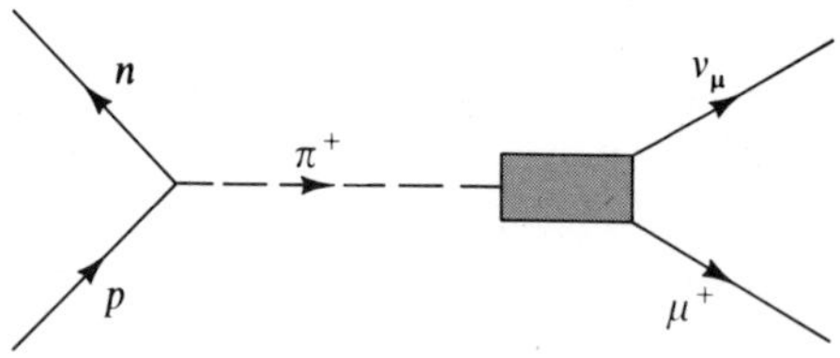

which leads to a value of g_P of about the same order of magnitude as the *G-T* relation gives, providing that the vertex

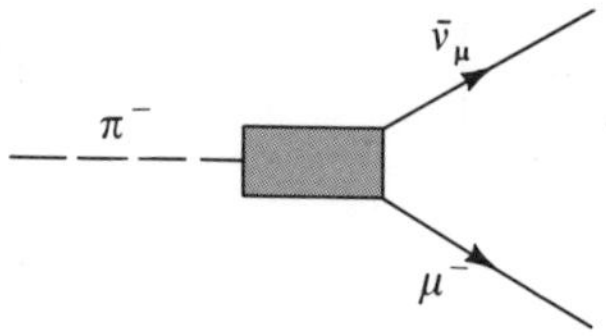

is estimated by using the experimental pion lifetime. However, this derivation does not lead to a statement about the relative sign of g_A and g_P, since there is a sign ambiguity that arises when we take (essentially) the square root of the lifetime to get the matrix element. The *G-T* relation resolves this ambiguity (see Eq. (11.84)) and predicts that g_A and g_P have the *same* sign.

We shall conclude this chapter by presenting two alternative derivations of the *G-T* relation, both of which serve to introduce some new ideas. The first one is due to Y. Nambu.[34] He argues as follows: if we imagine a fictitious world in which the pion mass is zero, the axial vector current could be conserved. Thus, in this world, we would have (Eq. (11.7))

$$g_P(q^2) = \frac{2M_P g_A(q^2)}{q^2}, \qquad (11.105)$$

which is to say that $g_P(q^2)$ would have a pole at $q^2 = 0$ with residue $2M_P g_A(0)$. We may now imagine turning on the pion mass. If we do this, $g_P(q^2)$ acquires a

33. L. Wolfenstein, *Nuovo Cimento*, **8**:882 (1958).
34. Y. Nambu, *Phys. Rev. Letters* **4**:380 (1960).
35. The fact that the position of the pole in $g_P(q^2)$ is correlated to the pion mass allows us to argue that $g_P(0)$ is finite, a fact that we have used in our previous derivation of the *G-T* relation.

pole at $q^2 = -m_\pi{}^2$ with a residue fixed by the factors in the diagram (say for μ capture)

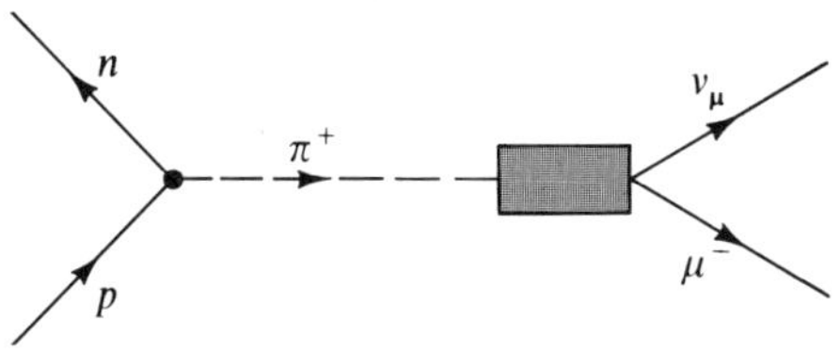

$$\frac{\sqrt{2}f_\pi g_\pi}{\dfrac{G}{\sqrt{2}}} = \lim_{q^2 \to -m_\pi{}^2} g_P(q^2)(q^2 + m_\pi{}^2). \tag{11.106}$$

If turning on the mass is a "gentle" process, then we might imagine that what happens is that in Eq. (11.85):

$$g_P(q^2) = \frac{2M_P g_A(q^2)}{q^2} \longrightarrow \frac{2M_P g_A(q^2)}{q^2 + m_\pi{}^2}. \tag{11.107}$$

In other words, though the position of the pole in $g_P(q^2)$ is shifted,[35] the value of its residue is not much changed. Equating the two methods of determining the residue, we have

$$2M_P g_A(q^2) \simeq \frac{2f_\pi g_\pi}{G}, \tag{11.108}$$

which is again the *G-T* relation. Sometimes this point of view is called the "partially conserved" or "almost conserved" axial current, since if

$$g_P(q^2) \simeq \frac{2M_P g_A(q^2)}{q^2 + m_\pi{}^2}$$

the matrix element looks as nearly as possible (in view of the nonzero pion mass) like that of the conserved current. This derivation has suggested to many people that for $q^2 \gg m_\pi{}^2$ the axial vector current becomes *effectively* conserved, in the sense that its matrix elements are to a very good approximation identical to those of a theory with zero mass pions and a truly conserved axial current. Such a point of view has led to specific predictions about certain pionic processes, but the reader should refer to the literature for the discussion.[36]

The second type of derivations of the *G-T* relation makes use of the existence of a class of theories with the property that, as a consequence of the dynamical equations for the fields,

---

36. S. L. Adler, *Phys. Rev.* **137B**:1022 (1965); S. L. Adler, *Phys. Rev.* **139B**:1638 (1965); Y. Nambu and D. Lurié, *Phys. Rev.* **125**:1429 (1962); Y. Nambu and E. Shrauner, *Phys. Rev.* **128**:862 (1962).

$$\partial_\mu A_\mu(\mathbf{r}, t) = K\pi^+(\mathbf{r}, t), \tag{11.109}$$

where $\pi^+(\mathbf{r}, t)$ is the actual pion field operator in the theory in question and K is a constant characteristic of the theory. A few explicit theories of this kind have actually been constructed, and we shall return to them later in the book. The only theory involving pions and nucleons, *alone*, with interactions of the conventional Yukawa type, that is known to produce a current of this kind, is the pseudovector theory characterized by the Lagrangian

$$\mathfrak{L} = -\frac{1}{2}\left\{\frac{\partial \boldsymbol{\phi}}{\partial x_\nu} \cdot \frac{\partial \boldsymbol{\phi}}{\partial x_\nu} + m_\pi^2 \boldsymbol{\phi}\cdot\boldsymbol{\phi}\right\} - \{\bar{\psi}(\gamma_\mu\partial_\mu + M_P)\psi\} - \frac{if}{m_\pi}\frac{\partial \boldsymbol{\phi}}{\partial x_\mu} \cdot \bar{\psi}\gamma_5\gamma_\mu\boldsymbol{\tau}\psi. \tag{11.110}$$

Although this theory has many nice features, it is well known that it contains infinities in its perturbation expansion and these infinities cannot be suppressed by the usual renormalization techniques. To make any sense out of the higher-order Feynman graphs in the theory, we must cut off the momentum integrals in some more or less arbitrary manner. Fortunately, this important technical difficulty will not effect the discussion we wish to give.[37]

In contrast to the electrodynamic case we do not have an obvious principle here, such as the principle of minimal electromagnetic coupling, that allows us to generate A_μ in a unique fashion (by, say, gauge transformations) from this Lagrangian. Indeed, there are several gauge transformations that generate different A_μ's. For example, we might use

$$\psi \longrightarrow (1 + i\boldsymbol{\Lambda}(x)\cdot\boldsymbol{\tau}\gamma_5)\psi \tag{11.111}$$

or

$$\boldsymbol{\phi} \longrightarrow \boldsymbol{\phi} + a\boldsymbol{\Lambda}(x), \tag{11.112}$$

as well as combinations of these and others. We simply do not know a priori how to select from among these generators the simplest or most fundamental one. As it happens, the transformation

$$\psi \longrightarrow \psi, \ \boldsymbol{\phi} \longrightarrow \boldsymbol{\phi} + a\boldsymbol{\Lambda}(x) \tag{11.113}$$

produces an interesting A_μ in the pseudovector theory. Indeed,[38] using Eq. (11.115) on Eq. (11.110) we generate

$$\mathbf{A}_\mu = \frac{-if}{m_\pi} a\bar{\psi}\gamma_5\gamma_\mu\boldsymbol{\tau}\psi - a\partial_\mu\boldsymbol{\phi}, \tag{11.114}$$

which we write more simply by choosing

$$a = \frac{-m_\pi}{f}, \tag{11.115}$$

37. Note that in these theories $[\partial_\mu A_\mu(0), \bar{\psi}(\mathbf{r}, 0)]_- = 0$, so that the constant N in Eq. (11.75) vanishes.

so that

$$A_\mu = i\bar{\psi}\gamma_5\gamma_\mu\boldsymbol{\tau}\psi + \frac{m_\pi}{f}\,\partial_\mu\boldsymbol{\phi}. \tag{11.116}$$

Apart from anything else, this current has the striking property that the nucleonic part is negligible compared to the pionic part in the limit of small f, which is not what one might expect naively.[39] Moreover, from the relation

$$\partial_\mu \mathbf{A}_\mu = \frac{\delta\mathcal{L}}{\delta\Lambda} = \frac{m_\pi^{\,3}}{f}\,\boldsymbol{\phi}, \tag{11.117}$$

we see that this A_μ has a divergence proportional to the pion field. In fact

$$\partial_\mu \mathbf{A}_\mu = i\,\partial_\mu(\bar{\psi}\gamma_5\gamma_\mu\boldsymbol{\tau}\psi) + \frac{m_\pi}{f}\,\Box^2\boldsymbol{\phi}, \tag{11.118}$$

so that from Eq. (11.97)

$$\frac{f}{m_\pi}\,i\,\partial_\mu(\bar{\psi}\gamma_5\gamma_\mu\boldsymbol{\tau}\psi) = (-\Box^2 + m_\pi^{\,2})\boldsymbol{\phi}, \tag{11.119}$$

which is just the equation of motion of $\boldsymbol{\phi}$ in the pseudovector theory. It is also clear from Eq. (11.117) that in this theory, if m_π were zero, A_μ would be conserved, a fact that can be traced back to the noninvariance of the term $m_\pi^{\,2}\boldsymbol{\phi}\cdot\boldsymbol{\phi}$ under the transformation of Eq. (11.113). We can now carry out the proof the *G-T* relation in this theory. We consider

$$\begin{aligned}
(2\pi)^3\frac{G}{\sqrt{2}}\,\langle\mathbf{p}'|_P\partial_\mu A_\mu|\mathbf{p}\rangle_N &= -\frac{iG}{\sqrt{2}}\,[\bar{u}(p')\gamma_5 u(p)][2M_P g_A(q^2) - q^2 g_P(q^2)]\\
&= (2\pi)^3\,\frac{m_\pi^{\,3}}{f}\,\langle\mathbf{p}'|_P\phi_\pi^{\,+}|\mathbf{p}\rangle_N\\
&= \frac{im_\pi^{\,3}}{f}\,\frac{\sqrt{2}f_\pi D(q^2)}{q^2 + m_\pi^{\prime 2}}\,\bar{u}(\mathbf{p}')\gamma_5 u(\mathbf{p}).
\end{aligned} \tag{11.120}$$

In making the last step we have used the connection, discussed earlier in the chapter, between the matrix element of the pion field taken between physical nucleon states and the observed pion-nucleon coupling constant. With the factors chosen here, $D(-m_\pi^{\,2}) = 1$. There is an important point contained in this formula. The numbers f and m_π are the mass and coupling constant parameters that enter the pion-nucleon Lagrangian. There is no reason why, except in lowest-order perturbation theory, that these should coincide with the position of the pole, in $\langle\mathbf{p}'|_P\phi_\pi^{\,+}|\mathbf{p}\rangle_N$ and its residue, which represent the mass and coupling constant of the *physical* particles. We have no principle in pion physics, as we have in electrodynamics, that guarantees that these parameters are unchanged by interactions. The quantities f_π and m_π' are the *measured* coupling constant

38. Use the method of Chapter 2 to compute $\delta\mathcal{L}/\delta\Lambda,\ \mu$.
39. See the discussion following Eq. (11.79).

and pion mass and can be very different from f and m_π. Fortunately we have a second equation in this theory that enables us to eliminate the unphysical quantities m_π and f. It reads:

$$\frac{G}{\sqrt{2}}\langle 0|\partial_\mu A_\mu|\pi^-\rangle = \frac{m_\pi{}^3}{f}\langle 0|\phi_\pi{}^+|\pi^-\rangle = \frac{m_\pi{}^3}{f\sqrt{2E_\pi}}$$

$$= \frac{-m_\pi{}'^2 g_\pi(-m_\pi{}^2)}{\sqrt{2E_\pi}}. \tag{11.121}$$

Making the substitutions indicated and setting $q^2 = 0$ in Eq. (11.120), it is clear that we arrive once again at the *G-T* relation.

While the *G-T* relation exhibits a very interesting correlation between strong and weak interactions, it does not appear to take us to the root of the structure of the axial vector current. As we have seen, the *G-T* relation can on the one hand be derived on the basis of rather general analyticity assumptions that do not appear to be connected with any special symmetries of the weak interactions, and on the other hand from rather special strong interaction models. What is lost in all of these derivations is a unification of A_μ and V_μ. The remarkable fact is that $|f_V(0)| \simeq |g_A(0)| \simeq 1$, which seems to indicate a connection between V_μ and A_μ something like the connection among the axial and vector leptonic currents expressed by the symmetric interaction $\gamma_\mu(1 + \gamma_5)$. The *G-T* relation, as interesting as it is, does not give us any hints on how V_μ and A_μ are tied together. We turn, in the next chapters, to some of the more recent attempts at unifying V_μ and A_μ.

12
The Axial Vector II

In the last chapter we saw a specific example of a theory, the pseudovector theory, in which $\partial_\mu A_\mu = 0$ in the limit $m_\pi \longrightarrow 0$. In analogy with electrodynamics we might try to carry out the following program.

1. Turn off the pion mass so that $\partial_\mu A_\mu = 0$.
2. Prove that, in this limit, $|g_A(0)| = 1$.
3. Argue that, when the pion mass is turned back on, this is a "gentle" operation so that $|g_A(0)| \simeq 1$ in agreement with experiment.

In this program we are stopped at step 2. It is simply not true that a conserved axial vector current necessarily implies no renormalization of g_A. We have already seen, in the last chapter, what goes wrong with the nonrenormalization argument in which the expectation value of the total "conserved" axial charge, Q_5, with

$$Q_5 = \int d^3\mathbf{r}\, A_0(\mathbf{r}, t) \tag{12.1}$$

is taken between one particle nucleon states—namely,

$$\langle \mathbf{0} | Q_5(0) | \mathbf{0} \rangle = 0. \tag{12.2}$$

Therefore no conclusion about $g_A(0)$ is possible from the conservation of charge. However, we might imagine that we could construct a Ward-like identity involving $\Gamma_{\mu 5}$, the axial vertex, and the nucleon propagators and use it, in the spirit of the work at the beginning of Chapter 10, to argue that $g_A(0) = 1$. In fact, it is quite simple to construct the Ward identities[1] that correspond to any current, conserved or not, that are generated by a gauge transformation on the nucleon and pion fields. As an example of how this works we construct the Ward identity corresponding to the transformation

$$\phi \longrightarrow \phi - \frac{m_\pi}{f}\Lambda, \qquad \psi \longrightarrow \psi, \tag{12.3}$$

which generates the A_μ of the pseudovector theory. (Here, ψ and ϕ are the nucleon and meson fields, respectively.) Under this transformation,

1. See J. Bernstein, M. Gell-Mann, and L. Michel, *Nuovo Cimento* **16**:560 (1960), for more details.

$$\mathfrak{L} \longrightarrow \mathfrak{L} + \partial_\mu \boldsymbol{\Lambda} \cdot \mathbf{A}_\mu - \frac{m_\pi^{\,3}}{f} \boldsymbol{\Lambda} \cdot \boldsymbol{\phi}. \tag{12.4}$$

The last term in Eq. (12.4) represents the nonconservation of **A** arising from the nonvanishing of the pion mass. Following the discussion of Chapter 10, we perform this gauge transformation on the nucleon propagator and compute the change in the nucleon Green's function $\langle 0|(\psi(x)\bar{\psi}(y))_+|0\rangle$. In fact, since $\psi \longrightarrow \psi$, under this transformation

$$\Delta S_F = 0. \tag{12.5}$$

However, as we saw in Chapter 10, ΔS_F can be computed by treating the change in $\mathfrak{L}$ as a perturbation. Thus, $q = (p' - p)$, and

$$0 = iq_\mu \Gamma_{\mu 5}{}^{+}(p, p') + \frac{m_\pi^{\,3}}{f} \frac{\sqrt{2} f_\pi}{q^2 + m_\pi^{\,2}} \Gamma_5(p, p'). \tag{12.6}$$

The reason for the first term in Eq. (12.6) is clear from the discussion of Chapter 10. The second term is a consequence of the appearance of the expression $(m_\pi^{\,3}/f)\boldsymbol{\Lambda}\cdot\boldsymbol{\pi}$, in the change in $\mathfrak{L}$. In computing ΔS_F, this term corresponds to the emission of a pion by two off mass shell nucleons:

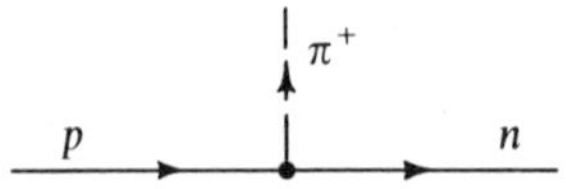

We have written the vertex function in such a way so as to make the pion pole and its residue explicit. If we put Eq. (12.6) between free nucleon spinors, we are simply led back to the *G-T* relation. If $m_\pi = 0$, corresponding to current conservation, then

$$q_\mu \Gamma_{\mu 5}{}^{+}(p, p') = 0. \tag{12.7}$$

Using general invariance arguments it is possible to construct, along the lines of the discussion given in electron scattering, the most general form that $\Gamma_{\mu 5}{}^{+}(p, p')$ can have. This turns out to involve seven[2] independent pseudovector quantities. Weinberg[3] has shown that the *G* transformation properties of $\mathbf{A}_\mu$—that is, whether $\mathbf{A}_\mu$ is a first-class or second-class Weinberg current in the sense discussed in Chapter 9—play an important role in restricting the number of form factors that can occur in $\Gamma_{\mu 5}{}^{+}(p, p')$. Since this discussion illustrates some important matters of principle we outline it here. It can be shown by general field

2. In J. Bernstein, M. Gell-Mann, and L. Michel, op. cit.; there is a form factor missing in Eq. (22). I am grateful to T. D. Lee for pointing this out.
3. S. Weinberg, *Phys. Rev.* **112**:1375 (1958).
4. See, for example, F. Low, *Phys. Rev.* **97**:1392 (1955).
5. When we write A_μ we refer to the axial current that *raises* the charge. When we write $\mathbf{A}_\mu$ we refer to the vector in isotopic space.

theoretic arguments[4] that[5] A_μ and $\Gamma_{\mu 5}{}^+$ are related by the following equation which, for our purposes, we can take as the definition of $\Gamma_{\mu 5}{}^+$.

$$\langle 0|(\bar{\psi}(x)_P\psi(x')_N A_\mu(y))_+|0\rangle$$
$$= -\int d^4u\, d^4u'\, \mathbf{S}_{FP}(x-u)\Gamma_{\mu 5}{}^+(u-y, y-u')\mathbf{S}_{FN}(u'-x'). \quad (12.8)$$

Here ψ_N is the Heisenberg field of a nucleon, and the $\mathbf{S}_F$'s are the nucleon Green's functions defined in Chapter 10.[6] We may then use the known G transformation properties of ψ_N and ψ_P and A_μ, $(GA_\mu G^{-1} = -A_\mu)$, to deduce the restrictions on $\Gamma_{\mu 5}{}^+(p, p')$. Indeed, as Weinberg shows, if A_μ is first class, with k_P and k_N the nucleon four-momenta, then

$$\Gamma_{\mu 5}{}^+(k_P, k_N) = c^{-1}\Gamma_{\mu 5}{}^{+T}(-k_N, -k_P)c, \quad (12.9)$$

where c is the unitary 4×4 charge conjugation matrix defined so that

$$-\gamma_\mu{}^T = c\gamma_\mu c^{-1} \quad (12.10)$$

and

$$c^T = -c. \quad (12.11)$$

We may, following the discussion of Chapter 5, write $\Gamma_{\mu 5}{}^+(k_P, k_N)$ as a function of twelve pseudovectors:

1. $\gamma_5 p_{+\mu}$.
2. $\gamma_5 p_{-\mu}$.
3. $\gamma_5\gamma_\mu$.
4. $\gamma_5\sigma_{\mu\nu}p_{+\nu}$.
5. $\gamma_5\sigma_{\mu\nu}p_{-\nu}$.
6. $\gamma_5(\gamma p_+)p_{-\mu}$.
7. $\gamma_5(\gamma p_-)p_{+\mu}$.
8. $\gamma_5(\gamma p_+)p_{+\mu}$.
9. $\gamma_5(\gamma p_-)p_{-\mu}$.
10. $\epsilon_{\mu\nu\lambda\sigma}\gamma_\nu p_{+\lambda}p_{-\sigma}$.
11. $\gamma_5(\gamma p_+)(\gamma p_-)p_{+\mu}$.
12. $\gamma_5(\gamma p_+)(\gamma p_-)p_{-\mu}$,

with $p_\pm = k_P \pm k_N$.

If the nucleons are off their mass shells, the twelve associated from factors will be functions of k_P and k_N separately; $F_{i5}(k_P, k_N)$, $i = 1, \ldots, 12$. Hence the G conditions will not restrict the number of F_{i5} but will simply yield symmetry conditions in k_P and k_N, for each of them. However, if we take the nucleons on

6. We note that by differentiating this equation with respect to y and evaluating the equal time commutators that arise from differentiating the θ functions defining the $+$ operation, and, then, Fourier transforming the result, we would be led back to the Ward identities. See Y. Takahashi, *Nuovo Cimento*, **6**:371 (1957), and J. Bernstein, M. Gell-Mann, and L Michel, op. cit. (Appendix).

their shells, so that $k_P^2 = k_N^2 = -M^2$, the F_{i5}'s become functions simply of $(k_P - k_N)^2$ and we can use the conditions to restrict their total number. In fact, on the shell,

$$\Gamma_{\mu 5}{}^{+} = F_1(q^2)\gamma_5\gamma_\mu + F_2(q^2)\gamma_5 p_{-\mu} + F_3(q^2)\gamma_5\sigma_{\mu\nu}p_{+\nu} + F_4(q^2)\gamma_5(\gamma p_+)p_{+\mu}$$
$$+ F_5(q^2)\gamma_5(\gamma p_-)p_{-\mu} + F_6(q^2)\epsilon_{\mu\nu\lambda\sigma}\gamma_\nu p_{+\lambda}p_{-\sigma}$$
$$+ F_7(q^2)\gamma_5(\gamma p_+)(\gamma p_-)p_{-\mu}. \tag{12.12}$$

The Ward identity states that[7] (with A_μ conserved)

$$p_{-\mu}\Gamma_{\mu 5}{}^{+} = 0 = F_1(q^2)\gamma_5(\gamma p_-) + F_2(q^2)\gamma_5 p_-{}^2 + F_3(q^2)\gamma_5\sigma_{\mu\nu}p_{-\mu}p_{+\nu}$$
$$+ F_5(q^2)\gamma_5(\gamma p_-)p_-{}^2 + F_7(q^2)\gamma_5(\gamma p_+)(\gamma p_-)p_-{}^2. \tag{12.13}$$

Taking this expression between spinors produces relations among the F's but does *not* show that there is no renormalization of the axial coupling constant.

This discussion shows that the usual arguments (that connect current conservation to the nonrenormalization of the charge) break down for axial currents. Of course, this does not imply that some other argument may not exist that establishes the connection. Several authors have tried to show by explicit counterexamples that no such general argument can exist in principle.[8] Though none of these discussions is fully rigorous, they all tend to show that axial current conservation does not imply that $g_A(0) = 1$. Hence we are left with a puzzle, which we can state as follows: in the leptonic current V and A are connected together in the very symmetric form

$$L_\mu = i\bar{u}_e\gamma_\mu(1 + \gamma_5)\nu_e + i\bar{u}_\mu\gamma_\mu(1 + \gamma_5)\nu_\mu,$$

and in low-energy β-decay phenomena, despite the effect of the strong interactions, V and A are connected together in the almost symmetric form

$$J_\mu \simeq i\bar{P}\gamma_\mu(1 + 1.18 \pm .02\gamma_5)N.$$

Why should these two currents look so nearly alike? Put in another way, is there a symmetry principle of some kind that is shared by these currents?

This is essentially where things stood in 1960. At this time Gell-Mann made a profound series of observations about the leptonic currents that eventually led to the formulation of SU_3 and the whole modern theory of current algebras.[9]

We begin the discussion by writing down the full leptonic Lagrangian including weak and electromagnetic interactions:

---

7. Keep in mind that $(p_+p_-) = (k_P{}^2 - k_N{}^2) = 0$.
8. J. Bernstein, M. Gell-Mann, and L. Michel, op. cit.; R. J. Blin-Stoyle, *Nuovo Cimento* **10**:132 (1958); S. B. Treiman (1964), unpublished.
9. The references in this field have grown exponentially since 1960. Here we give only those papers that bear directly on the question of the algebra of leptonic currents; later, when discussing other currents, we shall give additional references.

M. Gell-Mann, California Institute of Technology Synchrotron Laboratory Report CTSL-20 (1961), unpublished.

$$\mathfrak{L} = -[\bar{\psi}_\mu\gamma_\mu\partial_\mu\psi_\mu + \bar{\psi}_e\gamma_\mu\partial_\mu\psi_e + \bar{\psi}_{\nu_e}\gamma_\mu\partial_\mu\psi_{\nu_e} + \bar{\psi}_{\nu_\mu}\gamma_\mu\partial_\mu\psi_{\nu_\mu}] - ieA_\mu[\bar{\psi}_e\gamma_\mu\psi_e + \bar{\psi}_\mu\gamma_\mu\psi_\mu]$$
$$-(m_e\bar{\psi}_e\psi_e + m_\mu\bar{\psi}_\mu\psi_\mu) - \frac{G}{\sqrt{2}}[L_\mu L_\mu^*]. \quad (12.14)$$

The terms in this Lagrangian are by now all familiar. However, we wish to comment on the distinction between $L_\mu^\dagger$ and L_μ^* introduced in Chapter 9. To illustrate the point, consider

$$(i\bar{\psi}_e\gamma_\mu(1 + \gamma_5)\psi_{\nu_e})^\dagger = i\bar{\psi}_{\nu_e}[-2\,\delta_{\mu 4}\,\gamma_4 + \gamma_\mu](1 + \gamma_5)\psi_e. \quad (12.15)$$

From this expression it is clear that ordinary Hermitian conjugation introduces a change in sign in the fourth component of L_μ. If we define, as in Chapter 9,

$$\begin{aligned} L_\mu^* &= L_\mu^\dagger, \qquad & \mu &= 1, 2, 3, \\ L_\mu^* &= -L_\mu^\dagger, \qquad & \mu &= 4, \end{aligned} \quad (12.16)$$

this sign change is compensated. Since $L_\mu L_\mu^*$ is Hermitian, we do not have to add the Hermitian conjugate to make H_{WK} Hermitian. This change from $L_\mu^\dagger$ to L_μ^* is actually necessary if we wish the *fourth* components of the currents to transform like the *space* components under C and G. In Chapter 9 we have been careful to use the space components in arguments involving C and G. Note, in particular, that in contrast to Eq. (12.15), we have for L_μ^*

$$(i\bar{\psi}_e\gamma_\mu(1 + \gamma_5)\psi_{\nu_e})^* = i\bar{\psi}_{\nu_e}\gamma_\mu(1 + \gamma_5)\psi_e. \quad (12.17)$$

Thus

$$L_\mu = i[\bar{\psi}_e\gamma_\mu(1 + \gamma_5)\psi_{\nu_e} + \bar{\psi}_\mu\gamma_\mu(1 + \gamma_5)\psi_{\nu_\mu}] \quad (12.18)$$

and

$$L_\mu^* = i[\bar{\psi}_{\nu_e}\gamma_\mu(1 + \gamma_5)\psi_e + \bar{\psi}_{\nu_\mu}\gamma_\mu(1 + \gamma_5)\psi_\mu]. \quad (12.19)$$

The Lagrangian $\mathfrak{L}$, as it stands, is gauge invariant and leads to equations of motion that conserve the electromagnetic and leptonic currents. However, new symmetries appear if we make additional approximations. Although these symmetries are not rigorously true, they lead to important results. First we set $m_\mu = m_e = 0$. We then define ψ_e and ψ_ν by

$$\psi_l = \begin{pmatrix} \psi_e \\ \psi_\mu \end{pmatrix} \quad (12.20)$$

M. Gell-Mann, *Phys. Rev.* **125**:1067 (1962).

M. Gell-Mann, Proceedings of the 1960 Annual International Conference on High Energy Physics at Rochester, pp. 508–513. Interscience, New York (1960).

M. Gell-Mann, *Physics*, Vol. 1, No. 1, 63 (1964).

F. Gürsey and G. Feinberg, *Phys. Rev.* **128**:378 (1962).

T. D. Lee, *Nuovo Cimento* **35**:945 (1965).

and

$$\psi_\nu = \begin{pmatrix} \psi_{\nu_e} \\ \psi_{\nu_\mu} \end{pmatrix}. \tag{12.21}$$

In an obvious 2×2 notation,

$$\mathcal{L} = -[\bar{\psi}_l \partial_\mu \gamma_\mu \psi_l + \bar{\psi}_\nu \gamma_\mu \partial_\mu \psi_\nu] - ieA_\mu[\bar{\psi}_l \gamma_\mu \psi_l] - \frac{G}{\sqrt{2}}[L_\mu L_\mu^*], \tag{12.22}$$

with

$$L_\mu = i\bar{\psi}_l \gamma_\mu (1 + \gamma_5)\psi_\nu. \tag{12.23}$$

Written this way, it is clear that $\mathcal{L}$ is invariant under the full unitary group in two dimensions, U_2; that is, if u is any 2×2 unitary matrix[10] (we do not require a unit determinant), then the transformation

$$\begin{aligned} \psi_l' &= u\psi_l, \\ \psi_\nu' &= u\psi_\nu \end{aligned} \tag{12.24}$$

leaves $\mathcal{L}$ invariant. Moreover, we can enlarge the invariance group by decomposing[11] ψ_l as follows:

$$\psi_l = \frac{(1 + \gamma_5)}{2}\psi_l + \frac{(1 - \gamma_5)}{2}\psi_l = \psi_+ + \psi_-. \tag{12.25}$$

It is then easy to see that unitary transformations that take $\psi_+ \longrightarrow \psi_+$ and $\psi_- \longrightarrow \psi_-$ also leave $\mathcal{L}$ invariant, so that the full invariance group of $\mathcal{L}$ is the direct product $U_2 \otimes U_2$. Lee[12] has shown that this symmetry predicts relations among leptonic reactions at high energy, where the masses can be neglected. If $m_\mu \neq m_e \neq 0$, only the group of leptonic number phase transformations remains an invariance; this group is the product of two one-dimensional unitary groups $U_1 \otimes U_1$.

In the presence of the weak coupling the leptonic current, L_μ, is not conserved, even if the leptonic masses are set equal to zero. This is related to the fact that interchanging leptons and neutrinos in $\mathcal{L}$ does not leave it invariant. We can express this lack of conservation by introducing an "isotopic spin" as follows: we define a new object ψ with

$$\psi = \begin{pmatrix} \psi_l \\ \psi_\nu \end{pmatrix} \tag{12.26}$$

and three isotopic spin matrices:

---

10. $u^\dagger = (u^T)^* = u^{-1}$.
11. The neutrinos are left-handed to begin with.
12. T. D. Lee, *Nuovo Cimento* **35**:945 (1965).

$$\tau_+ = \begin{pmatrix} 0 & 0 & 1 & 0 \\ 0 & 0 & 0 & 1 \\ 0 & 0 & 0 & 0 \\ 0 & 0 & 0 & 0 \end{pmatrix},$$

$$\tau_- = \begin{pmatrix} 0 & 0 & 0 & 0 \\ 0 & 0 & 0 & 0 \\ 1 & 0 & 0 & 0 \\ 0 & 1 & 0 & 0 \end{pmatrix}, \tag{12.27}$$

$$\tau_3 = \begin{pmatrix} 1 & 0 & 0 & 0 \\ 0 & 1 & 0 & 0 \\ 0 & 0 & -1 & 0 \\ 0 & 0 & 0 & -1 \end{pmatrix}.$$

Setting $m_e = m_\mu = 0$,

$$\mathfrak{L} = -\bar{\psi}\gamma_\mu\partial_\mu\psi - ie\left[\bar{\psi}\gamma_\mu\frac{(1+\tau_3)}{2}\psi\right]A_\mu$$
$$- \frac{G}{\sqrt{2}}\,[(i\bar{\psi}\gamma_\mu(1+\gamma_5)\tau_+\psi)(i\bar{\psi}\gamma_\mu(1+\gamma_5)\tau_-\psi) + \text{h.c.}]. \tag{12.28}$$

With $G \neq 0$ and $e \neq 0$, $\mathfrak{L}$ is *not* invariant under the isotopic rotation groups defined by

$$U_5(\boldsymbol{\Lambda}) = e^{-i\gamma_5\boldsymbol{\Lambda}\cdot\boldsymbol{\tau}/2},$$
$$U(\boldsymbol{\Lambda}) = e^{-i\boldsymbol{\Lambda}\cdot\boldsymbol{\tau}/2}, \tag{12.29}$$

which means that the current

$$L_\mu = i\bar{\psi}\gamma_\mu(1+\gamma_5)\frac{\boldsymbol{\tau}}{2}\psi \tag{12.30}$$

is not conserved. However, if we turn off G and e, then evidently $U(\boldsymbol{\Lambda})$ and $U_5(\boldsymbol{\Lambda})$ are invariances of the theory and the currents

$$\mathbf{V}_{l\mu} = i\bar{\psi}\gamma_\mu\frac{\boldsymbol{\tau}}{2}\psi \tag{12.31}$$

and

$$\mathbf{A}_{l\mu} = i\bar{\psi}\gamma_5\gamma_\mu\frac{\boldsymbol{\tau}}{2}\psi \tag{12.32}$$

are separately conserved. Thus the Hermitian charges

$$\mathbf{Q}_l = \int \psi^\dagger\frac{\boldsymbol{\tau}}{2}\psi\, d\mathbf{r} \tag{12.33}$$

and

$$\mathbf{Q}_{5l} = \int \psi^\dagger\gamma_5\frac{\boldsymbol{\tau}}{2}\psi\, d\mathbf{r} \tag{12.34}$$

are constants of the motion. Moreover, employing techniques familiar from Chapter 3,

$$[Q_{il}, Q_{jl}]_- = i\epsilon_{ijk}Q_{kl}, \tag{12.35a}$$

$$[Q_{5il}, Q_{5jl}]_- = i\epsilon_{ijk}Q_{kl}, \tag{12.35b}$$

$$[Q_{il}, Q_{5jl}]_- = i\epsilon_{ijk}Q_{5kl}. \tag{12.35c}$$

In the highly artificial situation considered so far, all of the Q's are constants of the motion, so that these commutation relations hold at arbitrary times. However, even with $e \neq G \neq 0$, the Q's will still have the *same* commutation relations providing that we evaluate the commutators for the nonconserved Q's at equal times. In this sense, the ϵ_{ijk}, which are the "structure constants" of the algebra of SU_2, defined by the commutators, retain a significance even when the group disappears as an exact invariance of the theory. What Gell-Mann went on to assume was that the "universality" between leptons and hadrons in the weak interactions is expressed by the assumption that the "charges" associated with the hadron currents $\mathbf{V}_\mu$ and $\mathbf{A}_\mu$,

$$\begin{aligned} Q_i(t) &= \int d\mathbf{r}\, V_{0i}(\mathbf{r}, t), \\ Q_{5i}(t) &= \int d\mathbf{r}\, A_{0i}(\mathbf{r}, t), \end{aligned} \tag{12.36}$$

obey the same equal time commutation relations as the leptonic "charges" do (see Eq. (12.35)).

For the vector current this is not a new assumption, but rather a restatement of the hypothesis that this current is to be identified with that of the total isotopic spin group whose generators, of course, obey the relation

$$[T_i, T_j]_- = i\epsilon_{ijk}T_k. \tag{12.37}$$

The new developments start from the study of the consequences of the commutation relations involving Q_{5i}.

We begin the work by noting that the axial commutation relations are far from trivial in the sense that they are by no means fulfilled in all theories. If the axial vector current were simply

$$\mathbf{A}_\mu = i\bar{\psi}\gamma_5\gamma_\mu \frac{\boldsymbol{\tau}}{2}\psi, \tag{12.38}$$

then, even in the presence of pionic couplings, providing that the equal time commutation relations have the canonical form

$$[\psi(\mathbf{r}, t), \psi^\dagger(\mathbf{r}', t)]_+ = \delta^3(\mathbf{r} - \mathbf{r}'), \tag{12.39}$$

direct computation shows that

$$[Q_{5i}(t), Q_{5j}(t)]_- = i\epsilon_{ijk}Q_k^N(t), \tag{12.40}$$

where

$$Q_k^N(t) = \int \psi^\dagger \frac{\tau_k}{2}\psi\, d\mathbf{r}. \tag{12.41}$$

But $Q_k^N(t)$ is the "charge" associated with the Fermi current

$$\mathbf{V}_\mu^N = i\bar{\psi}\gamma_\mu \frac{\boldsymbol{\tau}}{2}\psi, \tag{12.42}$$

and as we have argued in detail in the last chapters, this nonconserved current is, according to the Feynman-Gell-Mann theory, just part of the vector current of β-decay. Clearly we must add the pionic terms to Q_{5i} if we hope to see them emerge in the commutator. However, adding pionic terms to A_μ is neither simple nor unique, and it is by no means clear at the outset that it can be done in such a way as to guarantee the Gell-Mann commutation relations. For example, from the work of the last chapter we found for the pseudovector theory an A_μ given by

$$\mathbf{A}_\mu = i\bar{\psi}\gamma_5\gamma_\mu\boldsymbol{\tau}\psi + \frac{m_\pi}{f}\,\partial_\mu\boldsymbol{\phi}, \tag{12.43}$$

with

$$\partial_\mu\mathbf{A}_\mu = \frac{m_\pi^{\ 3}}{f}\,\boldsymbol{\phi}. \tag{12.44}$$

Hence, in this theory,

$$\mathbf{A}_0 = -i\mathbf{A}_4 = \bar{\psi}\gamma_5\gamma_4\boldsymbol{\tau}\psi - \frac{m_\pi}{f}\,\dot{\boldsymbol{\phi}}. \tag{12.45}$$

However, the Lagrangian for this theory is

$$\mathfrak{L} = -\frac{1}{2}\left\{\frac{\partial\boldsymbol{\phi}}{\partial x_\nu}\cdot\frac{\partial\boldsymbol{\phi}}{\partial x_\nu} + m_\pi^{2}\boldsymbol{\phi}\cdot\boldsymbol{\phi}\right\} - \{\bar{\psi}(\gamma_\mu\partial_\mu + M)\psi\}$$
$$- \frac{if}{m_\pi}\frac{\partial\boldsymbol{\phi}}{\partial x_\mu}\cdot\bar{\psi}\gamma_5\gamma_\mu\boldsymbol{\tau}\psi, \tag{12.46}$$

which means that the canonical pion momentum $\delta\mathfrak{L}/\delta\dot{\boldsymbol{\phi}}$ is, in this theory,

$$\boldsymbol{\pi} = \dot{\boldsymbol{\phi}} - \frac{f}{m_\pi}\,\bar{\psi}\gamma_5\gamma_4\boldsymbol{\tau}\psi. \tag{12.47}$$

But

$$[\pi_i(\mathbf{r}, t), \pi_j(\mathbf{r}, t)]_- = 0. \tag{12.48}$$

Comparing Eqs. (12.48) and (12.45) we see that, in this theory,

$$[Q_{5i}(t), Q_{5j}(t)]_- = 0 \tag{12.49}$$

instead of Eq. (12.35)b. We shall return later to the question of whether there exist theories in which Eq. (12.35)b is valid for hadrons.

Equation (12.35)c appears to be on a somewhat different footing. Since in the Feynman-Gell-Mann theory Q_i is the isotopic charge, Eq. (12.35)c expresses the fact that $\mathbf{Q}_5$ transforms like an isotopic vector. We would expect that this commutation relation would hold for essentially any $\mathbf{A}_\mu$. For example, it is easy to see from the canonical commutation relations, Eq. (12.39) (which are valid when there are no derivatives in the pion-nucleon coupling) that it holds for the Fermi A_μ; $i\bar{\psi}\gamma_5\gamma_\mu\,(\tau/2)\psi$.

For the rest of this chapter we will adopt the attitude that Gell-Mann's commutation relations are valid and that indeed their validity may be a method of sorting out the different dynamical models of $\mathbf{A}_\mu$. We now turn to the tests and applications of Gell-Mann's commutator algebra.

The most celebrated test of the commutation relations is the so-called Adler-Weisberger sum rule.[13] It can be written in the form

$$1 - \frac{1}{g_A{}^2} = \left(\frac{2M_P}{f_{\pi N}}\right)^2 \frac{1}{2\pi} \int \frac{dq_0}{q_0}\,(\sigma_+{}^P(q_0) - \sigma_-{}^P(q_0)). \tag{12.50}$$

All of the quantities in Eq. (12.50) have been defined previously except those occurring in the integral. The objects $\sigma_\pm{}^P(q_0)$ are the *total* scattering cross sections for *zero-mass* $\pi^\pm$ particles at lab energy q_0, where the $\pi^\pm$ are incident on protons at rest. We shall see below why these somewhat unphysical expressions enter the sum rule and how they are to be evaluated, approximately, in terms of measurable quantities. We here remark, parenthetically, that if we set $M_P = 0$ and assume that nothing singular happens to the integral in this limit, then according to Eq. (12.50) $|g_A| = 1$. If A_μ were the simple Fermi current, $i\bar{\psi}\gamma_5\gamma_\mu\tau\psi$, and the fields ψ were free, then this limit would correspond to the limit in which $\mathbf{A}_\mu$ is conserved, since in this case $\partial_\mu\mathbf{A}_\mu$ is proportional to M_P. With $M_P = 0$, we can choose the nucleon states to be eigenstates of γ_5, so that the distinction between vectors and axial vectors is lost. In this limit, $M_P = 0$, we would expect that axial current conservation would imply no renormalization of g_A. This is what Eq. (12.50) seems to be saying. More generally, we have seen that $\partial_\mu\mathbf{A}_\mu = 0$ only if $m_\pi = 0$. But the right side of Eq. (12.50) is evaluated for $m_\pi = 0$. Hence Eq. (12.50) appears to be consistent with the sort of behavior that one might expect for g_A as a function of m_π and M_P.

Quantum mechanical sum rules are as old as the quantum theory itself.[14] All of these sum rules make use of three basic ideas.

1. The quantum mechanical commutation relations

$$[x, P]_- = i,$$
$$[x, H]_- = i\frac{P}{m}. \tag{12.51}$$

2. The existence of complete sets of states $|n'\rangle$, which enable one to evaluate the products of the operators between states $|n\rangle$ and $|m\rangle$ in terms of sums over

---

13. S. L. Adler, *Phys. Rev. Letters* **14**:1051 (1965) and *Phys. Rev.* **140B**:736 (1965); W. I. Weisberger, *Phys. Rev. Letters* **14**:1047 (1965). See also S. Fubini and G. Furlan, *Physics* **1**:229 (1965).
14. See, for example, H. A. Bethe and E. E. Salpeter, *Quantum Mechanics of One- and Two-Electron Atoms*, Academic Press, 1957, for a review of the electromagnetic sum rules of atomic physics.

matrix elements involving single operators; that is, if A and B are *any* two operators and $|n'\rangle$ is a member of *any* complete set, then

$$\langle m|AB|n\rangle = \sum_{n'} \langle m|A|n'\rangle\langle n'|B|n\rangle. \tag{12.52}$$

3. A connection (or connections) between these matrix elements and observable physical quantities.

In atomic physics the form of the interactions are known and, because of the smallness of e, matrix elements can be computed in perturbation theory. This simplifies step 3 above since the identification of matrix elements with physical processes can be made by inspection. For example, we know that the electric dipole transition operator is simply $\boldsymbol{\epsilon}\cdot\mathbf{r}$, where $\mathbf{r}$ is the radial coordinate associated with the charge and $\boldsymbol{\epsilon}$ is the photon polarization. Knowing the operator enables us to compute the transition matrix elements between any two states and to identify the total cross section for dipole absorption in terms of these matrix elements. Using this identification and the commutation relations, we can derive sum rules for dipole absorption, such as

$$\int \sigma_d(\omega)\, d\omega = 2\pi^2 r_0, \tag{12.53}$$

where the integral is over the total dipole cross section for the absorption of a photon of energy ω and r_0 is the "classical electron radius,"

$$r_0 = \frac{e^2}{m} = 2.82 \times 10^{-13} \text{ cm}. \tag{12.54}$$

Equation (12.53) is a connection between the coupling constant e^2 and a measurable quantity, $\int \sigma_d(\omega)\, d\omega$, of the type embodied in the Adler-Weisberger sum rule.

As we shall see, there is no problem in extending the first two steps into the domain of weak interactions and current algebras. The technical complications arise when we attempt to interpret the resulting matrix elements in terms of physically observable quantities.

The commutation relation that is used in deriving the Adler-Weisberger sum rule is obtained by considering

$$Q_{5\pm} = Q_{51} \pm iQ_{52}. \tag{12.55}$$

From Eq. (12.35)[b] it follows that[15]

$$[Q_{5+}, Q_{5-}]_- = 2Q_3, \tag{12.56}$$

where Q_3 is the isotopic vector charge. We can evaluate this relation between

15. All commutation relations, from now on, unless otherwise specified, are taken at equal times. It may seem strange that the commutator of Q_{5+} and Q_{5-}, both of which depend on time, is equal to a quantity, Q_3, which is time-independent. But this reflects a similar property of the canonical field commutators.

any two states and, to begin with, we evaluate it between the states of two physical protons of momenta **p** and **p**′. Since

$$Q_3 = \int d\mathbf{r}\, V_{03}(\mathbf{r}, 0), \tag{12.57}$$

it cannot transfer three-momenta. For any operator

$$O(\mathbf{r}) = e^{-i\mathbf{P}\cdot\mathbf{r}} O(0) e^{i\mathbf{P}\cdot\mathbf{r}}, \tag{12.58}$$

we find

$$\langle \mathbf{p}' | Q_3 | \mathbf{p} \rangle = \tfrac{1}{2}\, \delta^3(\mathbf{p} - \mathbf{p}'), \tag{12.59}$$

where we have made use of the normalization

$$\langle \mathbf{p}' | \mathbf{p} \rangle = \delta^2\, (\mathbf{p} - \mathbf{p}'). \tag{12.60}$$

In evaluating this matrix element we have also made use of the fact that Q_3 is the *total* isotopic charge; $|\mathbf{p}\rangle$ is, by construction, the physical nucleon state of isotopic spin $\frac{1}{2}$ including its mesonic cloud. We could, of course, maintain that Q_3 and $Q_3{}^N$, the nucleonic contribution to Q_3, have about the same matrix elements between physical protons and so Eq. (12.56) is approximately true even if Eq. (12.56) holds only for $Q_3{}^N$, as in the Fermi theory. In the absence of a dynamical theory this would be just a coincidence, and it is much more elegant to try to maintain Eq. (12.56) as long as possible.

From Eqs. (12.56) and (12.59) we now have the identity

$$\begin{aligned} \delta^3(\mathbf{p} - \mathbf{p}') &= \langle \mathbf{p}' | [Q_{5+}, Q_{5-}]_- | \mathbf{p} \rangle \\ &= \sum_n \left[\langle \mathbf{p}' | Q_{5+} | n \rangle \langle n | Q_{5-} | \mathbf{p} \rangle - \langle \mathbf{p}' | Q_{5-} | n \rangle \langle n | Q_{5+} | \mathbf{p} \rangle \right] \end{aligned} \tag{12.61}$$

when $|n\rangle$ is a member of the complete set of eigenstates of P_μ; that is, all of the strongly interacting particles with their physical masses. We can now separate from this sum the one-nucleon contribution. Since Q_{5-} lowers electric charge by one unit (Q_{5+} raises charge by one unit), only the intermediate neutron state can contribute. Thus, if we call **k** the momentum of this state and s the spin we have

$$\delta^3(\mathbf{p} - \mathbf{p}') = \sum_{s'} \int d\mathbf{k}\, \langle \mathbf{p}' |_s Q_{5+} | \mathbf{k} \rangle_{s'}\; {}_{s'}\langle \mathbf{k} | Q_{5-} | \mathbf{p} \rangle_s + R(\mathbf{p}', \mathbf{p})_{\mathbf{p}'=\mathbf{p}}. \tag{12.62}$$

Here, $R(\mathbf{p}', \mathbf{p})$ is the contribution from the rest of the states, to be evaluated below, and we have summed over spins, s, of the neutron and integrated over momenta, **k**. As we have seen in the last chapters, assuming that the axial current is first-class Weinberg,

$$(2\pi)^3 \langle \mathbf{p}' |_{s'} A_\mu(0) | \mathbf{p} \rangle_s = i\bar{u}(\mathbf{p}')_{s'} \gamma_5 [\gamma_\mu g_A(q^2) + i q_\mu g_P(q^2)] u(\mathbf{p})_s, \tag{12.63}$$

with

$$q_\mu = (p' - p)_\mu. \tag{12.64}$$

Since

$$Q_{5+}(t) = \int d\mathbf{r}\, A_0(\mathbf{r}, t), \tag{12.65}$$

we have, anticipating the evaluation of the one-nucleon contribution to Eq. (12.62),

$$\begin{aligned}\langle \mathbf{p}'|_{s'} Q_{5+} |\mathbf{k}\rangle_s &= (2\pi)^3 \delta^3(\mathbf{k} - \mathbf{p}')\langle \mathbf{p}'|_{s'} A_0(0)|\mathbf{k}\rangle_s \\ &= \delta^3(\mathbf{k} - \mathbf{p}')\bar{u}(\mathbf{p}')_{s'}\gamma_5\gamma_4 g_A(0) u(\mathbf{k})_s.\end{aligned} \tag{12.66}$$

We have used the δ^3 function condition, which tells us that the four-vector[16] $p' - k = 0$. Using the fact that

$$Q_{5+} = Q_{5-}{}^{\dagger},$$

we can write, doing the $\mathbf{k}$ integration over the δ functions:

$$\delta^3(\mathbf{p} - \mathbf{p}') = \delta^3(\mathbf{p} - \mathbf{p}') \sum_{s'} |\langle \mathbf{p}'|_{s'} A_0(0)|\mathbf{p}\rangle_s|^2 + R(\mathbf{p}', \mathbf{p})_{\mathbf{p}=\mathbf{p}'} \tag{12.67}$$

It is very instructive[17] to evaluate the one-nucleon contribution by taking explicit representations of $u(\mathbf{p})_s$ and γ_5. With our choice of γ's,

$$\gamma_5 = -\begin{pmatrix} 0 & I \\ I & 0 \end{pmatrix}. \tag{12.68}$$

It is clear the Eq. (12.67) is not a Lorentz scalar equation. We choose to evaluate it by taking the initial proton to have momentum $\mathbf{p}$ in, say, the z direction, and we suppose its spin is also in this direction. In general, with our normalization

$$\bar{u}(\mathbf{p})u(\mathbf{p}) = \frac{M}{E(\mathbf{p})}, \tag{12.69}$$

$u(\mathbf{p})$ can be written in terms of the two component spin functions, u, in the form

$$u(\mathbf{p}) = \begin{pmatrix} u \\ \dfrac{\boldsymbol{\sigma}\cdot\mathbf{p}}{E(\mathbf{p}) + M} u \end{pmatrix} \sqrt{\frac{E(\mathbf{p}) + \mathrm{M}}{2E(\mathbf{p})}}. \tag{12.70}$$

Thus, with spin up and $\mathbf{p} = (0, 0, |\mathbf{p}|)$,

$$u(\mathbf{p}) = \begin{pmatrix} 1 \\ 0 \\ \dfrac{|\mathbf{p}|}{E(\mathbf{p}) + M} \\ 0 \end{pmatrix} \sqrt{\frac{E(\mathbf{p}) + M}{2E(\mathbf{p})}}. \tag{12.71}$$

Since $\mathbf{p} = \mathbf{p}'$ in Eq. (12.66), a trivial calculation shows that[18]

---

16. With $\mathbf{p}' = \mathbf{k}$, remember that $E(\mathbf{p}') = E(\mathbf{k})$, since we neglect the N, P mass difference.
17. We follow here an extremely lucid discussion of the sum rules and their implications given by J. S. Bell, *Proceedings of the 1966 CERN School of Physics*, Vol. I, CERN 66–29.
18. It is easy to see that fixing the initial spin s, as indicated above, fixes $s' = s$, since $A_0(0)$ does not flip spin.

$$\langle \mathbf{p}|_s A_0(0)|\mathbf{p}\rangle_s = -g_A(0)\frac{|\mathbf{p}|}{E(\mathbf{p})} = -vg_A(0),$$

where v is the speed of the initial nucleon. Thus the sum rule, Eq. (12.67), takes the form

$$1 - v^2 g_A{}^2(0) = \sum_n |\langle n|A_0(0)|\mathbf{p}\rangle|^2(2\pi)^3\delta^3(\mathbf{p}-\mathbf{p}_n) - \sum_n |\langle n|A_0{}^\dagger(0)|\mathbf{p}\rangle|^2(2\pi)^3\delta^3(\mathbf{p}-\mathbf{p}_n), \quad (12.72)$$

where we have used, to derive the right side of Eq. (12.72), the fact that $Q_{5\pm}$ cannot transfer three-momenta. For each value of v we get a different sum rule. The v-dependence of the right side of Eq. (12.72) is, of course, reflected in the state vector $|\mathbf{p}\rangle$. We notice that the right side of Eq. (12.72) is the difference between two positive quantities, so that a priori it is not clear whether this calculation will give $g_A(0)$ a value which is larger or smaller than one. Moreover, it will, in any case, predict only the magnitude of $g_A(0)$, so that it cannot distinguish between, say, a $V - A$ and a $V + A$ theory.

To estimate the right side of Eq. (12.72) it is now customary to take the following steps. We note that

$$|\langle n|A_0(0)|\mathbf{p}\rangle| = \left|\frac{\left\langle n\middle|\frac{\partial}{\partial t}A_0\middle|\mathbf{p}\right\rangle}{(E(\mathbf{p})-E_n)}\right|, \quad (12.73)$$

since

$$i\frac{\partial}{\partial t}A_0 = [A_0, H]_-. \quad (12.74)$$

It is important to remark that this expression, when inserted in Eq. (12.72), is only well defined if $\partial_\mu A_\mu \neq 0$. If $\partial_\mu A_\mu = 0$, then Q_{5+} is time-independent, which means that Q_{5+} can transfer neither energy nor three-momenta. Thus in Eq. (12.72)

$$E(\mathbf{p}) - E_n = 0 = \mathbf{p} - \mathbf{p}_n. \quad (12.75)$$

Moreover,

$$\langle \mathbf{n}|\nabla\cdot\mathbf{A}|\mathbf{p}\rangle = i\langle \mathbf{n}|[A_i, P_i]_-|\mathbf{p}\rangle = i(\mathbf{p}_n - \mathbf{p})\cdot\langle \mathbf{n}|\mathbf{A}|\mathbf{p}\rangle. \quad (12.76)$$

If we use the condition $\delta^3(\mathbf{p}-\mathbf{p}_n)$ in Eq. (12.72), we see that in this equation[19]

$$\left\langle \mathbf{n}\middle|\frac{\partial}{\partial t}A_0\middle|\mathbf{p}\right\rangle = \langle \mathbf{n}|\partial_\mu A_\mu|\mathbf{p}\rangle, \quad (12.77)$$

so that, in case $\partial_\mu A_\mu = 0$, both the numerator and denominator of Eq. (12.73) would vanish when inserted in Eq. (12.72) and the equation is then not well defined. If $\partial_\mu A_\mu \neq 0$, we can then write, using Eqs. (12.73) and (12.77), the sum rule in the form

19. Remember that in our metric

$$\partial_\mu A_\mu = \nabla\cdot\mathbf{A} + \partial_4 A_4 = \nabla\cdot\mathbf{A} + \frac{\partial}{\partial t}A_0.$$

$$1 - v^2 g_A{}^2(0) = \sum_n |\langle \mathbf{n}|A_0|\mathbf{p}\rangle|^2 (2\pi)^3 \delta^3(\mathbf{p} - \mathbf{p}_n) - \sum_n |\langle \mathbf{n}|A_0{}^\dagger|\mathbf{p}\rangle|^2 (2\pi)^3 \delta^3(\mathbf{p} - \mathbf{p}_n)$$

$$= \int \frac{d(\Delta E)}{2\pi(\Delta E)^2} \sum_n |\langle \mathbf{n}|\dot{A}_0(0)|\mathbf{p}\rangle|^2 \delta^4(p_n - p - q)(2\pi)^4 - (A_0 \leftrightarrow A_0{}^\dagger), \quad (12.78)$$

where

$$q = (p_n - p) = (0, 0, 0, i(E(\mathbf{p}_n) - E(\mathbf{p}))). \quad (12.79)$$

Therefore, using the Eq. (12.77), we can write the sum rule as

$$1 - v^2 g_A{}^2(0) = \int \frac{d(\Delta E)}{2\pi(\Delta E)^2} \sum_n |\langle \mathbf{n}|\partial_\mu A_\mu|\mathbf{p}\rangle|^2 \delta^4(p_n - p - q)(2\pi)^4 - (A_\mu \leftrightarrow A_\mu{}^\dagger). \quad (12.80)$$

It is convenient for the applications of the sum rule to evaluate the matrix elements occurring on the right side of Eq. (12.80) in the frame of reference in which the nucleon is initially at rest. The quantity $\partial_\mu A_\mu$ transforms like a Lorentz scalar under proper Lorentz transformations. However, this does *not* imply that $\langle \mathbf{n}|\partial_\mu A_\mu|\mathbf{p}\rangle$ is a Lorentz scalar. To see what is involved, consider as a specific example, say, the one-proton matrix element of the electric charge operator Q (which is also clearly a Lorentz scalar),

$$\langle \mathbf{p}|Q|\mathbf{p}\rangle = e\bar{u}(\mathbf{p})u(\mathbf{p}) = e\frac{M}{E(\mathbf{p})}, \quad (12.81)$$

where

$$\frac{M}{E} = \sqrt{1 - v^2}. \quad (12.82)$$

Thus the quantity $\sqrt{\frac{E}{M}}\langle \mathbf{p}|Q|\mathbf{p}\rangle\sqrt{\frac{E}{M}}$ *is* a Lorentz scalar, but $\langle \mathbf{p}|Q|\mathbf{p}\rangle$ is not. This means, more generally, that the quantity

$$\sqrt{\frac{E_n}{M_n}}\langle \mathbf{n}|\partial_\mu A_\mu|\mathbf{p}\rangle\sqrt{\frac{E(\mathbf{p})}{M}}, \quad (12.83)$$

where $\mathbf{p}_n{}^2 = -M_n{}^2$, transforms under proper Lorentz transformations like a Lorentz scalar, which tells us how $\langle \mathbf{n}|\partial_\mu A_\mu|\mathbf{p}\rangle$ transforms. In particular, in Eq. (12.80), we may make the substitution

$$\langle \mathbf{n}|\partial_\mu A_\mu|\mathbf{p}\rangle = \langle n|\partial_\mu A_\mu|\mathbf{0}\rangle\sqrt{\frac{M}{E(\mathbf{p})}} = \langle \mathbf{n}|\partial_\mu A_\mu|\mathbf{0}\rangle(1 - v^2)^{1/4}. \quad (12.84)$$

Moreover, in the frame in which the neutron is at rest,

$$q = \left(0, 0, \frac{-v\Delta E}{\sqrt{1 - v^2}}, i\frac{\Delta E}{\sqrt{1 - v^2}}\right). \quad (12.85)$$

If we call

$$q_0 = \frac{\Delta E}{\sqrt{1 - v^2}}, \quad (12.86)$$

then

$$\frac{dq_0}{q_0{}^2} = \frac{d\Delta E}{(\Delta E)^2}\sqrt{1 - v^2}. \tag{12.87}$$

Thus we can write the sum rule in the form

$$1 - v^2 g_A(0)^2 = \int \frac{dq_0}{2\pi q_0{}^2} \sum_n |\langle \mathbf{n}|\partial_\mu A_\mu|\mathbf{0}\rangle|^2 (2\pi)^4 \delta^4(p_n - p - q) \\ - (A_\mu \longleftrightarrow A_\mu{}^\dagger), \tag{12.88}$$

where the dependence of the matrix element on v has been removed and replaced by an implicit dependence on v in the δ^4 function.

It has been pointed out by Adler[20] that this sum rule can be checked, in principle, without making special assumptions about $\partial_\mu A_\mu$. This is in virtue of the fact that the matrix elements that enter the sum rule can be related directly to observable processes involving leptons. We follow here the discussion of Bell,[21] which illuminates the main ideas. The interested reader is referred to Adler's paper for the details.

Consider the reaction $\nu_\mu + N \longrightarrow \mu + n$, where n is any of the final states, $|n\rangle$, in the sum rule. The matrix element for this transition is given by

$$M = \frac{G}{\sqrt{2}} i\bar{u}_\mu \gamma_\mu (1 + \gamma_5) u_{\nu_\mu} \langle \mathbf{n}|J_\mu|\mathbf{0}\rangle, \tag{12.89}$$

where

$$J_\mu = V_\mu + A_\mu. \tag{12.90}$$

We consider the special kinematic situation in which the initial neutron is at rest, the initial neutrino is incident along the z-axis, and the final muon exits from the reaction also along the z-axis; that is, in the forward direction. We also suppose that the energies involved are so great that we can neglect, to a good approximation, the leptonic mass. Given the kinematics we can make an explicit calculation of $\bar{u}_\mu \gamma_\mu (1 + \gamma_5) u_{\nu_\mu}$ along the following lines. The incident neutrino is described by the spinor[22]

$$u_\nu = \begin{pmatrix} 0 \\ 1 \\ 0 \\ -1 \end{pmatrix} \frac{1}{\sqrt{2}}. \tag{12.91}$$

The final muon ($m_\mu = 0$) is also left-handed (a property of the $\gamma_\mu(1 + \gamma_5)$ interaction) so since the muon is also in the $+z$ direction,

$$\bar{u}_\mu = (0101)\frac{1}{\sqrt{2}}. \tag{12.92}$$

20. S. L. Adler, *Phys. Rev.* **140B**:736 (1965) and *Phys. Rev.* **143**:1144 (1965).
21. J. S. Bell, op. cit.
22. It is simple to verify that u_ν satisfies the Dirac equation for a mass-zero particle moving in the $+z$ direction and that

$$\gamma_5 u_\nu = u_\nu,$$

so that this function represents a left-handed neutrino.

We can now evaluate the four-vector $\bar{u}_\mu\gamma_\mu(1 + \gamma_5)u_{\nu_\mu}$ explicitly:[23]

$$i\bar{u}_\mu\gamma_\mu(1 + \gamma_5)u_{\nu_\mu} = (0101)i\gamma_\mu \begin{pmatrix} 0 \\ 1 \\ 0 \\ -1 \end{pmatrix} = (0, 0, 2, i2). \tag{12.93}$$

On the other hand, the four-vector $q_\mu = (\nu - \mu)_\mu = (p_n - p)_\mu$ is, in view of the assumed kinematics—that is,

$$\begin{aligned} \nu + M &= \mu + E_n \\ \nu - \mu &= p_n, \end{aligned} \tag{12.94}$$

simply

$$q = (0, 0, 2, i2)(E_n - M). \tag{12.95}$$

Hence, and this is the essential point,

$$M = iN(E_n)q_\mu\langle n|J_\mu|\mathbf{0}\rangle = N(E_n)\langle n|\partial_\mu J_\mu|\mathbf{0}\rangle, \tag{12.96}$$

where $N(E_n)$ is a kinematical factor whose exact form follows from the previous discussion and will not concern us here. Thus, making the conserved vector current hypothesis, $\partial_\mu V_\mu = 0$, we have

$$M_n = N(E_n)\langle \mathbf{n}|\partial_\mu A_\mu|\mathbf{0}\rangle. \tag{12.97}$$

M_n is precisely the quantity that enters the sum rule. Apart from kinematical factors, its square is the cross section for producing the state $|n\rangle$ along with a muon in the forward direction. We see from Eq. (12.85) that

$$q^2 = -(1 - v^2)q_0^2. \tag{12.98}$$

Thus if we fix q_0 and let $v \longrightarrow 1$, a trick that is employed, as we shall see shortly, in all of the evaluations of the sum rule, $q^2 \longrightarrow 0$, as it must in forward muon production when $m_\mu = 0$. Letting $v \longrightarrow 1$ and working out the connection between the square of the matrix element and the cross section, Adler finds leptonic sum rules of the form

$$1 - g_A{}^2(0) = \int \frac{dW f(W)}{G^2}\left[\underset{\nu\to\mu}{\overset{\text{forward}}{\sigma\,(W)}} - \underset{\bar{\nu}\to\bar{\mu}}{\overset{\text{forward}}{\sigma\,(W)}}\right], \tag{12.99}$$

where $f(W)$ is a simple function of the total center of mass energy W and G is the fundamental weak interaction constant.[24] An identical formula holds if we replace ν_μ by ν_e and μ by e. These results should provide important tests of the

23. $(1 + \gamma_5)u_\nu = 2u_\nu$.
24.

$$f(W) = \frac{MW8\pi^2}{[M^2 + 2ME - W^2]^2},$$

where E is the neutrino energy.

Gell-Mann commutation relations when high-energy neutrino scattering data become available.

We now turn to a detailed derivation of the Adler-Weisberger sum rule that involves pions, that is Eq. (12.50). All derivations of this sum rule involve making additional assumptions about $\partial_\mu A_\mu$ of the type used in the last chapter in the discussion of the *G-T* relation. We begin by making the specific assumption that

$$\partial_\mu A_\mu = C\pi^+, \tag{12.100}$$

where C is a constant, to be discussed, and π^+ is the pion field, defined in terms of the dynamics of any theory of pions and nucleons in which Eq. (12.100) is true. Later we shall discuss how this special assumption can be relaxed or modified. First, using the work of the last chapter, we fix the constant C in terms of experimental quantities. We can always choose the normalization of, say, $|\pi^-\rangle$, so that

$$\langle 0|\pi^+|\pi^-\rangle = \frac{1}{\sqrt{2E_\pi}}\frac{1}{(2\pi)^{3/2}}. \tag{12.101}$$

Thus

$$\frac{G}{\sqrt{2}}\langle 0|\partial_\mu A_\mu|\pi^-\rangle = \frac{C}{\sqrt{2E_\pi}}\frac{G}{\sqrt{2}(2\pi)^{3/2}} = \frac{m_\pi{}^2 g_\pi(-m_\pi{}^2)}{\sqrt{2E_\pi}}, \tag{12.102}$$

where, as in the last chapter, G is the fundamental weak constant and $g_\pi(-m_\pi{}^2)$ is related to the pion lifetime. Hence

$$C = \frac{m_\pi{}^2 g_\pi(-m_\pi{}^2)\sqrt{2}(2\pi)^{3/2}}{G}. \tag{12.103}$$

However, from the derivation of the *G-T* relation,

$$\frac{g_\pi(-m_\pi{}^2)}{G} = \frac{M_P g_A(0)}{f_\pi D(0)}, \tag{12.104}$$

where f_π is the pion-nucleon coupling constant and $D(q^2)$ is the pion-nucleon vertex form factor defined so that[25]

$$D(-m_\pi{}^2) = 1. \tag{12.105}$$

Thus

$$C = \frac{\sqrt{2}(2\pi)^{3/2} M_P g_A(0) m_\pi{}^2}{f_\pi D(0)}. \tag{12.106}$$

Hence, with these assumptions, the sum rule takes the form

---

25. The empirical success of the *G-T* relation shows that

$$D(0) \simeq D(-m_\pi{}^2) = 1.$$

$$1 - v^2 g_A^2(0) = \left(\frac{\sqrt{2} M_P g_A(0) m_\pi^2 (2\pi)^{3/2}}{f_\pi D(0)}\right)^2$$

$$\left\{\int \frac{dq_0}{2\pi q_0^2} \sum_n |\langle n|\pi^+|\mathbf{0}\rangle|^2 (2\pi)^4 \delta^4(p_n - p - q) - (\pi^+ \longleftrightarrow \pi^-)\right\}. \quad (12.107)$$

If we fix q_0 and let $v \longrightarrow 1$, we have

$$1 - g_A^2(0) = \frac{2(2\pi)^3 M_P^2 g_A^2(0) m_\pi^4}{f_\pi^2 D^2(0)}$$

$$\left\{\int \frac{dq_0}{2\pi q_0^2} \sum_n |\langle n|\pi^+|\mathbf{0}\rangle|^2 (2\pi)^4 \delta^4(p_n - p - q) - (\pi^+ \longleftrightarrow \pi^-)\right\}. \quad (12.108)$$

It is clear that the quantity $\langle n|\pi^+|\mathbf{0}\rangle$ has to do entirely with the strong pion-nucleon interaction. It is the sum and integral over the squares of these quantities that is related to the integral over pion-nucleon cross sections that appears in the Adler-Weisberger sum rule. To appreciate the connection we must return to the method discussed in the beginning of the book for creating and annihilating particles appearing in state vectors. In particular, the in (out) state with one nucleon of momentum **0** and a positive pion of momentum **q** can be written

$$|\mathbf{0q}\rangle_{\substack{\text{in} \\ \text{out}}} = \lim_{\substack{t \to -\infty \\ +\infty}} \int d\mathbf{r} \left(\pi^+(\mathbf{r}, t) \frac{\partial}{\partial t} f(q) - \frac{\partial}{\partial t} \pi^+(\mathbf{r}, t) f(q)\right)|\mathbf{0}\rangle, \quad (12.109)$$

where, as we have defined it,

$$f(q) = \frac{e^{i(qx)}}{(2\pi)^{3/2}\sqrt{2q_0}}, \quad (12.110)$$

so that

$$\left(-\nabla^2 + \frac{\partial^2}{\partial t^2} + m_\pi^2\right) f = (-\Box^2 + m_\pi^2) f = 0. \quad (12.111)$$

We may now use a string of identities employing Eq. (12.111). We assume throughout that when partial integrations are done over *spatial* variables, infinitely remote surface terms can be dropped. Thus

$$\begin{aligned} |\mathbf{0q}\rangle_{\text{out}} - |\mathbf{0q}\rangle_{\text{in}} &= i \int dt\, d\mathbf{r}\, \partial_t \left[\pi^+(\mathbf{r}, t) \dot{f} - \dot{\pi}^+(\mathbf{r}, t) f\right] |\mathbf{0}\rangle \\ &= \int d^4x \left[\pi^+ \frac{\partial^2}{\partial t^2} f - \frac{\partial^2}{\partial t^2} \pi^+ f\right] |\mathbf{0}\rangle \\ &= \int d^4x \left[\pi^+ \left(-\nabla^2 + \frac{\partial^2}{\partial t^2} + m_\pi^2\right) f \right. \\ &\qquad \left. - f \left(-\nabla^2 + \frac{\partial^2}{\partial t^2} + m_\pi^2\right) \pi^+ \right] |\mathbf{0}\rangle \\ &= \int d^4x\, (\Box^2 - m_\pi^2) \pi^+ f |\mathbf{0}\rangle = -\int d^4x\, j_{\pi^+} f |\mathbf{0}\rangle, \end{aligned} \quad (12.112)$$

where the pion "source" function j_{π^+} is defined by the equation

$$(-\Box^2 + m_\pi^2)\pi^+ = j_{\pi^+} \tag{12.113}$$

Thus we have the fundamental identity

$$\begin{aligned}\langle n|_{\text{out}}\mathbf{0q}\rangle_{\text{in}} &= \langle n|_{\text{out}}\mathbf{0q}\rangle_{\text{out}} + \int d^4x \frac{e^{iqx}}{(2\pi)^{3/2}\sqrt{2q_0}}(-\Box^2 + m_\pi^2)\langle n|_{\text{out}}\pi^+(x)|\mathbf{0}\rangle \\ &= \langle n|_{\text{out}}\mathbf{0q}\rangle_{\text{out}} + \int \frac{d^4x\, e^{i(q-(p_n-p))x}}{(2\pi)^{3/2}\sqrt{2q_0}}((p_n - p)^2 + m_\pi^2)\langle n|_{\text{out}}\pi^+(0)|\mathbf{0}\rangle \\ &= \langle n|_{\text{out}}\mathbf{0q}\rangle_{\text{out}} + i\frac{(2\pi)^{5/2}}{\sqrt{2q_0}}\delta^4(q-(p_n-p))(q^2+m_\pi^2)\langle n|_{\text{out}}\pi^+(0)|\mathbf{0}\rangle.\end{aligned} \tag{12.114}$$

The quantity[26]

$$\langle n|_{\text{out}}\mathbf{0q}\rangle_{\text{in}} - \langle n|_{\text{out}}\mathbf{0q}\rangle_{\text{out}} \cdot\!\equiv\!\cdot\, T(n;\mathbf{0},\mathbf{q}) \tag{12.115}$$

is the transition matrix for a transition from a state with one neutron, at rest, and a pion, of momentum $\mathbf{q}$, incident to the final state $\langle n|$. The cross section for this process is given by the relation[27]

$$\sigma_n{}^+(\mathbf{q})\times(\text{flux}) = |T(n;\mathbf{0},\mathbf{q})|^2 = \frac{\pi}{q_0}\delta^4(q-(p_n-p))|(q^2+m_\pi^2)\langle n|_{\text{out}}\pi^+(0)|\mathbf{0}\rangle|^2. \tag{12.116}$$

In Eq. (12.108) we have taken $v \longrightarrow 1$, which corresponds to $q^2 \longrightarrow 0$ for fixed q_0. Taking this limit in Eq. (12.116) we can now write the sum rule in the form

$$1 - g_A{}^2(0) = \frac{2M_P{}^2 g_A{}^2(0)}{f_\pi{}^2 D^2(0)}\frac{1}{\pi}\int \frac{dq_0}{q_0}[\sigma_+{}^N(q_0) - \sigma_-{}^N(q_0)], \tag{12.117}$$

where $\sigma_\pm{}^N(q_0)$ are now the *total* cross sections for pions of energy q_0 and *zero* rest mass ($q^2 = 0$) incident on *neutrons* at rest. (For a zero mass pion the flux, with our pion field normalization, is unity.) It is more useful, in the comparison with experiment, to replace neutron total cross sections by proton cross sections. The proton is stable and it is a much better target, so that these cross sections are rather well known. Making this transformation is trivial. In fact if we call $|N\rangle$ the neutron state and $|P\rangle$ the proton state, then

$$\langle n|\pi^+|N\rangle = -\langle n|e^{i\pi T_2}\pi^-|P\rangle, \tag{12.118}$$

where $\langle n|$ is an arbitrary eigenstate of P_μ and $e^{i\pi T_2}$ is the isotopic rotation operator discussed in previous chapters. Since $e^{i\pi T_2}$ commutes with the strong interaction

26. See, for example, S. Schweber, op. cit., for a discussion of the T matrix and how it is connected to the cross section.
27. As is usual, in this connection, the square of the factor $(2\pi)^4\,\delta^4(q-(p_n-p))$ becomes simply $(2\pi)^4\,\delta^4(q-(p_n-p))$ when the normalization volumes are taken properly into account.

Hamiltonian, and, more generally, with P_μ, the states $\langle n|e^{i\pi T_2}$ are also complete, so that we have the identity

$$\sigma_\pm{}^N(q_0) = \sigma_\pm{}^P(q_0). \tag{12.119}$$

Thus

$$1 - \frac{1}{g_A{}^2(0)} = \frac{2M_P{}^2}{f_\pi{}^2 D^2(0)} \frac{1}{\pi} \int \frac{dq_0}{q_0} [\sigma_+{}^P(q_0) - \sigma_-{}^P(q_0)]. \tag{12.120}$$

Weisberger[28] has evaluated the integral in Eq. (12.120) between the limits $m_\pi \leq q_0 \leq \infty$, using the measured cross sections and correcting for the finite pion mass by inserting the flux factor $v = |\mathbf{q}|/q_0$ so that

$$\int \frac{dq_0}{q_0} \longrightarrow \int dq_0 \frac{|\mathbf{q}|}{q_0{}^2}$$

in Eq. (12.120). He finds that this predicts a value of $|g_A(0)|$, which is

$$|g_A(0)| = 1.15.$$

(Clearly the sign of $g_A(0)$ cannot be determined from the sum rule.) This is to be compared with the experimental value of

$$|g_A(0)| = 1.18 \pm 0.02.$$

Apart from the excellent numerical agreement, one of the most interesting aspects of Weisberger's evaluation is that most of the contribution to the integral comes from the well-known N^* resonance, the broad resonance[29] in pion-nucleon scattering that occurs in the energy range (for the incident pion) lying between 100 and 300 MeV. The N^* resonance, which is centered at about 190 MeV, occurs in the $J = \frac{3}{2}$, $T = \frac{3}{2}$ channel. A pion, $T = 1$, and a nucleon, $T = \frac{1}{2}$, can combine (by the addition theorem for angular momenta) in a $T = \frac{3}{2}$ or a $T = \frac{1}{2}$ state. The elastic scattering amplitudes for $\pi^+ + p \longrightarrow \pi^+ + p$ and $\pi^- + p \longrightarrow \pi^- + p$ can be decomposed, by the usual Clebsch-Gordon technique, into two non-interfering amplitudes, f_T, with $T = \frac{3}{2}$ and $\frac{1}{2}$. In particular,[30]

$$\begin{aligned} &\pi^+ + p \longrightarrow \pi^+ + p \longleftrightarrow f_{3/2}, \\ &\pi^- + p \longrightarrow \pi^- + p \longleftrightarrow \tfrac{2}{3} f_{1/2} + \tfrac{1}{3} f_{3/2}. \end{aligned} \tag{12.121}$$

Thus, insofar as the integral is dominated by the $T = \frac{3}{2}$, N^* resonance in pion-nucleon elastic scattering we have, in the resonance region,

$$\sigma_+{}^P(q_0) \gg \sigma_-{}^P(q_0)$$

and

$$g_A{}^2 > 1.$$

28. W. I. Weisberger, op. cit., and *Phys. Rev.* **143**:1302 (1965).
29. See, for example, J. D. Jackson, *The Physics of Elementary Particles*, Princeton University Press, p. 11, for a discussion of this resonance and its theoretical treatment.
30. Remember that $\pi^+ p$ is only in the $T = \frac{3}{2}$ state.

In fact, if this were the *only* contribution to the sum rule integral, then, Weisberger shows, that it would predict

$$g_A(0) \simeq 1.3.$$

There are resonances at higher energies, which are suppressed in the sum rule because of the $1/q_0$ factor, but which occur in the $T = \frac{1}{2}$ channel and serve to reduce this number to Weisberger's final value of $g_A(0) = 1.15$. We may wonder why the integral over cross section differences is convergent at all, apart from the $1/q_0$ factor, which would make it converge as long as the cross section showed any decrease with energy. The convergence appears to be guaranteed by the validity of the so-called Pomeranchuk theorem, which implies, under rather general assumptions, that particle and antiparticle (in this case $\pi^{\pm}$) cross sections should approach each other at high energies. The pionic data from 5 to 20 BeV[31] can be fitted with the formula

$$\sigma_+{}^P(q_0) - \sigma_-{}^P(q_0) = 7.73 \text{ mb} \left[\frac{|\mathbf{q}|}{\text{BeV}}\right]^{-0.7},$$

so that the cross sections do appear, experimentally, to be approaching each other. Using this formula Adler and Weisberger find that the high-energy region of the integral gives only a small contribution. Adler's method of correcting for the off-mass shell character of the pion is somewhat different from Weisberger's and is sufficiently complicated that the interested reader is best referred to his paper. He finds

$$|g_A(0)| = 1.24.$$

The difference between this result and Weisberger's 1.15 constitutes a measure of the theoretical uncertainties involved in using the sum rule. The striking features of this work are the closeness of $g_A(0)$ to one and the fact that $g_A(0) > 1$ is intimately connected to the pion dynamics. There does not seem to be a general symmetry principle, or at least no one has found one, which would guarantee these results a priori. They are, from this point of view, connected to the isotopic character of the pion-nucleon resonances.

The derivation of the Adler-Weisberger sum rule that we have given, and which follows closely the original derivations, appears to depend heavily on the assumption that $\partial_\mu A_\mu = C\pi^+$. We now give a brief discussion of how we might try to relax this assumption, in the spirit of the work of the last chapter on the Goldberger-Tremain relation. Hence we consider a theory in which

$$\partial_\mu A_\mu = O_5{}^+(\mathbf{r}, t) \qquad (12.122)$$

and in which there exists a pion field satisfying

31. G. von Dardel et al., *Phys. Rev. Letters* **8**:173 (1962). See, for example, the Proceedings of the 1962 International Conference on High Energy Physics, p. 897, for a review of the Pomeranchuk theorems given by S. Drell:

$$1 \text{ mb} = 10^{-27} \text{ cm}^2.$$

$$(-\Box^2 + m_\pi^2)\pi^+(\mathbf{r}, t) = j_{\pi^+}(\mathbf{r}, t) \tag{12.123}$$

such that

$$O_5^+(\mathbf{r}, t) \neq C\pi_+(\mathbf{r}, t). \tag{12.124}$$

We may still normalize π^+ so that

$$\langle 0|\pi^+(\mathbf{r}, t)|\pi^-\rangle = \frac{1}{\sqrt{2E_\pi}(2\pi)^{3/2}}. \tag{12.125}$$

Now, by the familiar arguments,

$$\frac{G}{\sqrt{2}}\langle 0|O_5^+(\mathbf{r}, t)|\pi^-\rangle = \frac{m_\pi^2}{\sqrt{2E_\pi}} g_\pi(-m_\pi^2) = (2\pi)^{3/2} m_\pi^2 g_\pi(-m_\pi^2)\langle 0|\pi^+|\pi^-\rangle, \tag{12.126}$$

so that

$$\langle 0|\pi^+|\pi^-\rangle = \gamma\langle 0|O_5^+|\pi^-\rangle, \tag{12.127}$$

where

$$\gamma = \frac{G}{\sqrt{2}} \frac{1}{m_\pi^2 g_\pi(-m_\pi^2)(2\pi)^{3/2}}. \tag{12.128}$$

Equation (12.127) means that O_5^+ has nonvanishing components in its Fourier decomposition such that $q^2 = -m_\pi^2$. Thus the state

$$|\mathbf{0q}\rangle_{\substack{\text{in}\\ \text{out}}} = \lim_{t\to\pm\infty} i\left\{\int d\mathbf{r}\left(O_5^+(\mathbf{r}, t)\frac{\partial}{\partial t} f(q) - \frac{\partial}{\partial t} O_5^+(\mathbf{r}, t) f(q)\right)\right\}|\mathbf{0}\rangle \tag{12.129}$$

will also be an eigenstate of $\mathbf{P}_\mu$ with the same quantum numbers as the state

$$|\mathbf{0q}\rangle_{\substack{\text{in}\\ \text{out}}} = \lim_{t\to\pm\infty} i\left\{\int d\mathbf{r}\left(\pi^+(\mathbf{r}, t)\frac{\partial}{\partial t} f(q) - \frac{\partial}{\partial t}\pi^+(\mathbf{r}, t) f(q)\right)\right\}|\mathbf{0}\rangle. \tag{12.130}$$

Hence, these states can differ at most by their normalizations.[32] Calling the normalization factor $N(\mathbf{0}, \mathbf{q})$, we can write

$$|\mathbf{0q}\rangle' = N(\mathbf{0}, \mathbf{q})|\mathbf{0}, \mathbf{q}\rangle. \tag{12.131}$$

By the preceding work we can conclude that

$$i\int d^4x\,(-\Box^2 + m_\pi^2)O_5^+(\mathbf{r}, t)f(q)|\mathbf{0}\rangle$$
$$= N(\mathbf{0}, \mathbf{q})i\int d^4x\,(-\Box^2 + m_\pi^2)\pi^+(\mathbf{r}, t)f(q)|\mathbf{0}\rangle, \tag{12.132}$$

with f, as usual, given by

$$f(q) = \frac{e^{iqx}}{(2\pi)^{3/2}\sqrt{2q_0}}. \tag{12.133}$$

32. There are no degeneracies in the spectrum of $\mathbf{P}_\mu$ once all the strong interaction quantum numbers have been specified.

We can use this equation to determine $N(\mathbf{0}, \mathbf{q})$. Let us multiply both sides by a one-proton state $\langle \mathbf{p}|$. Thus

$$\langle \mathbf{p}|O_5{}^+(0)|\mathbf{0}\rangle = N(\mathbf{0}, \mathbf{q})\langle \mathbf{p}|\pi^+|\mathbf{0}\rangle. \tag{12.134}$$

The left side of this equation involves the weak form factors of the axial current and the right side involves the form factor of the strong pion-nucleon vertex. But for the sum rule, $q^2 = 0$. Thus

$$2M_P g_A(0) = N(\mathbf{0}, \mathbf{q})\frac{f_\pi D(0)\sqrt{2}}{(2\pi)^{3/2}m_\pi{}^2}, \tag{12.135}$$

so that

$$N(\mathbf{0}, \mathbf{q}) = \frac{(2\pi)^{3/2}\sqrt{2}M_P m_\pi{}^2 g_A(0)}{f_\pi D(0)}. \tag{12.136}$$

If, in addition, the theory gives the G-T relation, then we see at once that to the extent that the relation is satisfied,

$$N(\mathbf{0}, \mathbf{q}) = \frac{1}{\gamma}. \tag{12.137}$$

where γ was defined in Eq. (12.128). We now have (always with $q^2 = 0$) a connection between the matrix elements of $\partial_\mu A_\mu$ in the sum rule and the T matrix for pion scattering defined in terms of π^+ (Eq. (12.115)):

$$\langle n|_{\text{out}}\mathbf{0q}\rangle'_{\text{in}} - \langle n|_{\text{out}}\mathbf{0q}\rangle'_{\text{out}} = \frac{\sqrt{2}M_P g_A(0)m_\pi{}^2(2\pi)^{3/2}}{f_\pi D(0)}\, T(n; \mathbf{0}, \mathbf{q}). \tag{12.138}$$

The rest of the derivation of the sum rule now proceeds from Eq. (12.116) exactly as in the case in which $\partial_\mu A_\mu = C\pi^+$.

It is interesting to study what happens if, following Adler, we evaluate Eq. (12.56) between one-pion states. Since the pion has $T = 1$, for this case the sum rule reads

$$2 = \int \frac{dq_0}{2\pi q_0{}^2}\left\{\sum_n |\langle n|O_5{}^+|\mathbf{0}\rangle|^2(2\pi)^4\delta^4(p_n - p - q) - (O_5{}^+ \longleftrightarrow O_5{}^-)\right\}, \tag{12.139}$$

where $|\mathbf{0}\rangle$ now stands for a one-pion state, with the pion at rest. Assuming that A_μ is first-class Weinberg, $\langle n|$ must have[33] $G = +1$, so that $\langle n|$ cannot be a one-pion state. Carrying out the same set of manipulations that lead to Eq. (12.120), we find the sum rule

$$\frac{2}{g_A{}^2} = \frac{2M_P{}^2}{f_\pi{}^2 D^2(0)}\frac{1}{\pi}\int \frac{dq_0}{q_0{}^2}\,[\sigma_-{}^{\pi^+}(q_0) - \sigma_+{}^{\pi^+}(q_0)], \tag{12.140}$$

where $\sigma_\pm{}^{\pi^+}(q_0)$ are the total cross sections for zero-mass $\pi^\pm$ scattering from a $\pi^\pm$

33. $|\mathbf{0}\rangle$ and $O_5{}^+$ are odd under G.
34. S. L. Adler, *Phys. Rev.* **140**:736 (1965), op. cit.
35. See *Scalar Mesons* by L. M. Brown, Northwestern University.

at rest. These cross sections are not known experimentally. Adler has estimated the integral[34] by assuming that it is dominated by the known resonances such as the ρ meson. He finds that the sum rule *fails* badly and concludes that it can only be saved if there exists a low energy π, π resonance in the $T = 0$, $l = 0$ channel. There may be indications of such a resonance[35] so that the sum rule philosophy, which works so well elsewhere, may indeed be saved here.

Earlier in the chapter we remarked on the fact that it is not obvious how to make a Lagrangian that generates an axial current that is partially conserved, $\partial_\mu A_\mu = C\pi^+$, and which, at least in the limit $m_\pi = 0$, is invariant under the group $SU_2 \otimes SU_2$. The latter would imply that the weak charges obey the Gell-Mann commutator algebra. In particular, we have seen that it is possible to find a simple theory, the pseudovector theory, in which we have a partially conserved A_μ, but the associated charges do *not* obey the commutator algebra. It is easy to understand why this is true in this theory, and this understanding has led to some very interesting new developments, which we discuss, in more detail, in Chapter 15.[36] Here we simply sketch the essentials.

In the pseudovector theory, A_μ was generated by the transformations

$$\begin{aligned} \psi &\longrightarrow \psi, \\ \phi &\longrightarrow \phi + a\Lambda \cdot \phi. \end{aligned} \tag{12.141}$$

These are the transformations of an *Abelian* group while $SU_2 \otimes SU_2$ is a *non-Abelian* group. Hence the charges cannot possibly commute like the generators of $SU_2 \otimes SU_2$. Within the context of the Yukawa couplings contained in the pseudovector Lagrangian there is no way of fixing this situation. The only way is to enlarge the Lagrangian. The enlargement should retain the partial current conservation and it should also be invariant under $SU_2 \otimes SU_2$ when $m_\pi = 0$. In Chapter 15 we give the exact solution to this problem in the form suggested by the work of Schwinger,[37] and here we present a partial solution. We write for the pion-nucleon Lagrangian,

$$\begin{aligned} \mathfrak{L} = &-\tfrac{1}{2}[(\partial_\mu\phi)^2 + m_\pi^2\phi^2] - \bar{\psi}(\partial_\mu\gamma_\mu + \mu)\psi \\ &+ \frac{f}{m_\pi} i\bar{\psi}(\gamma_5\gamma_\mu\tau)\psi \cdot \partial_\mu\phi - \left(\frac{f_0}{m_\pi}\right)^2 \bar{\psi}\gamma_\mu\tau\psi \cdot \phi \times \partial_\mu\phi. \end{aligned} \tag{12.142}$$

The new feature here is the last term. The last term corresponds to pion-nucleon S-wave scattering. Indeed, if this term is used in the Born approximation, f_0 may be determined directly from the low-energy scattering data:

$$f_0 \simeq 0.8.$$

Such a Lagrangian is regarded as "effective," which means that its matrix elements are taken in first order and the constants are adjusted to fit the data. We

36. The reader will also find in Chapter 15 a detailed list of references to the literature on this subject.
37. J. Schwinger; see references in Chapter 15.

return in Chapter 15 to discussion of this philosophy. Clearly the Lagrangian is invariant under the ordinary isotopic spin group, and hence it leads to a conserved isotopic vector current whose exact functional form need not concern us. However, in the limit $m_\pi = 0$, it is also invariant under the combined transformations

$$\begin{aligned} \boldsymbol{\phi} &\longrightarrow \boldsymbol{\phi} + a\boldsymbol{\Lambda}, \\ \psi &\longrightarrow \left[1 + i\left(\frac{f_0}{m_\pi}\right)^2 \boldsymbol{\tau}\cdot(\boldsymbol{\phi} \times a\boldsymbol{\Lambda})\right]\psi, \end{aligned} \tag{12.143}$$

if one considers the change in $\mathcal{L}$ to first order in $\boldsymbol{\phi}$. Indeed these transformations are those of $SU_2 \otimes SU_2$ in the neighborhood of $\boldsymbol{\phi} = 0$. In Chapter 15 we exhibit the full nonlinear Lagrangian that is invariant under $SU_2 \otimes SU_2$ for finite $\boldsymbol{\phi}$. We can use this infinitesimal part of the $SU_2 \otimes SU_2$ group to generate an axial vector current that is partially conserved in the strong sense. This $\mathbf{A}\mu$ is

$$\mathbf{A}_\mu = \frac{f}{f_0}\,\bar{\psi} i\gamma_\mu\gamma_5 \frac{\boldsymbol{\tau}}{2}\psi - \frac{m_\pi}{2f_0}\,\partial_\mu\boldsymbol{\phi} + \cdots. \tag{12.144}$$

The dots refer to terms of order at least $\boldsymbol{\phi}^2$, which enter because of the non-Yukawa character of the Lagrangian. The interesting thing about this current is its *scale*. It is the combination f/f_0 that multiplies the usual nuclear part of A_μ, and it is easy to see that the corresponding part of $\mathbf{V}_\mu$ is simply $i\bar{\psi}\gamma_\mu\boldsymbol{\tau}\psi$, which has *unit* scale. Thus in extending the pseudovector Lagrangian so that it is $SU_2 \otimes SU_2$ invariant, we are forced to weak currents $\mathbf{V}_\mu$ and $\mathbf{A}_\mu$, which have the relative scale

$$\frac{f}{f_0} \simeq 1.2, \tag{12.145}$$

providing that we demand that the effective pion-nucleon Lagrangian has matrix elements that give the observed pion-nucleon scattering. This constitutes an alternative approach to an explanation of the size of the axial coupling constant in which the $SU_2 \otimes SU_2$ group is used, but not the algebra of currents. In Chapter 15 we discuss these matters further. For the moment we leave the sum rules and turn to strange particles and SU_3.

13

Strange Particles: An Introduction to SU_3

So much is now known about the strange particles that a detailed survey is well beyond the scope of this book.[1] For our purposes the point to be emphasized is that the strong interactions of the strange particles are characterized by a new conserved quantum number in addition to the isotopic spin. This quantum number is the "strangeness," S, or, equivalently, the "hypercharge" Y. Isotopic spin and strangeness (or hypercharge) are connected to the electric charge by the well-known formula

$$Q = \frac{B}{2} + T_3 + \frac{S}{2} = T_3 + \frac{Y}{2}, \tag{13.1}$$

where

$$Y = B + S. \tag{13.2}$$

In all strong interactions $\Delta Y = \Delta B = 0$. We list the more familiar strange particles, their masses, and their Y and T assignments:

Y	T	*Baryons*	*Mesons*
$+1$	$\frac{1}{2}$	p, n	K^+, K^0
0	1	$\Sigma^+, \Sigma^0, \Sigma^-$	π^+, π^0, π^-
0	0	Λ^0	η^0
-1	$\frac{1}{2}$	Ξ^0, Ξ^-	$\bar{K}^0, K^-$

The associated masses are (in MeV):[2]

	Baryons		*Mesons*
p, n	938.256 ± 0.005,	K^+, K^0	493.82 ± 0.11,
	939.550 ± 0.005		497.87 ± 0.16
$\Sigma^+, \Sigma^0, \Sigma^-$	1189.47 ± 0.08,	π^+, π^0, π^-	139.579 ± 0.014,
	$1192.56 \pm .11$,		134.975 ± 0.014,
	1197.44 ± 0.09		139.579 ± 0.014

1. For an excellent survey see, for example, *Elementary Particle Physics* by S. Gasiorowicz, Wiley, 1966.
2. These masses are taken from A. H. Rosenfeld et al., *Rev. Mod. Phys.* **39**:1 (1967).

Baryons		*Mesons*	
Λ^0	$1115.58 \pm .10$	η^0	548.6 ± 0.4
Ξ^0, Ξ^-	$1314.7 \pm 1.0,$	$\bar{K}^0, K^-$	$497.87 \pm 0.16,$
	1321.2 ± 0.2		493.82 ± 0.82

As we have seen in Chapter 3 the most general exact invariance[3] that the theory of pions and nucleons, in the absence of electromagnetism, exhibits is the invariance under SU_2, the isotopic spin group. As we have indicated, the simplest nontrivial unitary representation of this group is the set of all 2×2 matrices of the form

$$U(\boldsymbol{\Lambda}) = e^{i\boldsymbol{\Lambda}\cdot\boldsymbol{\tau}}, \tag{13.3}$$

where the $\boldsymbol{\tau}$ are the Pauli spin matrices with

$$\mathrm{Tr}\,(\boldsymbol{\tau}) = 0 \tag{13.4}$$

and $\boldsymbol{\Lambda}$ is a real 3×1 vector. This assures us that the $U(\boldsymbol{\Lambda})$ are unitary and that

$$\det\,(U(\boldsymbol{\Lambda})) = 1. \tag{13.5}$$

The theory is also invariant under the phase transformations $e^{i\Lambda}$ which, as we have argued, guarantees the conservation of baryon number. The quantum numbers correspond to the eigenvalues of mutually commuting "charges" and serve to characterize completely the various isotopic multiplets.

If the strange particles are added to the picture and if the Lagrangian that characterizes their interactions with each other, and with the non-strange particles, is to have a higher symmetry than SU_2, then this symmetry must correspond to a larger group, and, in fact, to a group that contains SU_2 as a subgroup. Isotopic spin is still a good quantum number for the strange particle interactions; witness, for example, the isotopic multiplet structure in the table of strange particle masses given above. After innumerable false starts and trials, it now appears as if the generalization of the SU_2 group that works for the strange particles is simply the group SU_3 that can be characterized as the abstract group having the same multiplication table as the set of all 3×3 unitary, unimodular matrices. These, in fact, form its simplest nontrivial representation.[4]

We begin our discussion by a characterization of the traceless 3×3 matrices (universally called in the literature λ_i) that replace the three τ_i of SU_2. Any 3×3 matrix can be written as a linear combination of nine fundamental matrices; for example, $\begin{pmatrix} 1 & 0 & 0 \\ 0 & 0 & 0 \\ 0 & 0 & 0 \end{pmatrix}, \begin{pmatrix} 0 & 1 & 0 \\ 0 & 0 & 0 \\ 0 & 0 & 0 \end{pmatrix}$, etc. The λ_i, however, must be both traceless and Hermitian if $e^{i\Sigma\Lambda_i\lambda_i}$ is to be unitary and unimodular. A trace-

3. By "invariance" we mean here invariance under groups of continuous transformations, Lie groups, exclusive of Lorentz invariance, and the invariance under such discrete transformations as P, C, and T.

less 3×3 matrix has only eight independent components, since the trace condition eliminates one of the diagonal matrix elements in terms of the other two. Hence there are only eight λ_i. Three of λ_i, say $\lambda_1, \lambda_2, \lambda_3$, must generate the isotopic spin. Thus

$$\lambda_1 = \begin{pmatrix} 0 & 1 & 0 \\ 1 & 0 & 0 \\ 0 & 0 & 0 \end{pmatrix}, \quad \lambda_2 = \begin{pmatrix} 0 & -i & 0 \\ i & 0 & 0 \\ 0 & 0 & 0 \end{pmatrix}, \quad \lambda_3 = \begin{pmatrix} 1 & 0 & 0 \\ 0 & -1 & 0 \\ 0 & 0 & 0 \end{pmatrix}.$$

Gell-Mann[5] has produced a standard choice for the other five λ's as follows:

$$\lambda_4 = \begin{pmatrix} 0 & 0 & 1 \\ 0 & 0 & 0 \\ 1 & 0 & 0 \end{pmatrix}, \quad \lambda_5 = \begin{pmatrix} 0 & 0 & -i \\ 0 & 0 & 0 \\ i & 0 & 0 \end{pmatrix}, \quad \lambda_6 = \begin{pmatrix} 0 & 0 & 0 \\ 0 & 0 & 1 \\ 0 & 1 & 0 \end{pmatrix},$$

$$\lambda_7 = \begin{pmatrix} 0 & 0 & 0 \\ 0 & 0 & -i \\ 0 & i & 0 \end{pmatrix}, \quad \lambda_8 = \frac{1}{\sqrt{3}}\begin{pmatrix} 1 & 0 & 0 \\ 0 & 1 & 0 \\ 0 & 0 & -2 \end{pmatrix}.$$

Clearly, by inspection, these λ's are traceless and Hermitian. They are also "orthogonal" in the sense that

$$\mathrm{Tr}\,[\lambda_i\lambda_j] = 2\delta_{ij}. \tag{13.6}$$

We also see that λ_3 and λ_8 are diagonal. Moreover,

$$[\lambda_3, \lambda_8]_- = 0,$$

which, as we shall soon discover, means that an SU_3 invariant theory admits two commuting charges T_3 and Y. As with SU_2, for the $e^{i\Sigma_i\Lambda_i\lambda_i}$ to form a group, the commutator algebra of the λ_i must close; that is, we must have

$$[\lambda_i, \lambda_j]_- = 2if_{ijk}\lambda_k. \tag{13.7}$$

Here the f_{ijk} are the "structure constants" of SU_3, which generalize the totally antisymmetric ϵ_{ijk} structure constants of SU_2. The simplest way to compute the f_{ijk} is to use the identity

$$\mathrm{Tr}\,[\lambda_k[\lambda_i, \lambda_j]_-] = 4if_{ijk}, \tag{13.8}$$

which is a consequence of the trace orthogonality of the λ_i (Eq. (13.6)). From the properties of the trace, it follows that f_{ijk} is totally antisymmetric in i, j, k. The nonvanishing elements of f_{ijk} are

ijk	f_{ijk}
123	1
147	$\frac{1}{2}$
156	$-\frac{1}{2}$
246	$\frac{1}{2}$

4. For a very complete collection of fundamental papers on SU_3, with commentary, see *The Eightfold Way*, by M. Gell-Mann and Y. Ne'eman, Benjamin, 1964.
5. M. Gell-Mann, *Phys. Rev.* **125**:1067 (1962).

ijk	f_{ijk}
257	$\frac{1}{2}$
345	$\frac{1}{2}$
367	$-\frac{1}{2}$
458	$\sqrt{3}/2$
678	$\sqrt{3}/2$

The 3×3 representation of SU_3 also has the special algebraic property[6]

$$[\lambda_i, \lambda_j]_+ = \tfrac{4}{3}\delta_{ij}I + 2d_{ijk}\lambda_k, \tag{13.9}$$

where d_{ijk} is a totally symmetric function of the indices *ijk*. It can be computed from the identity

$$\mathrm{Tr}\,[\lambda_k[\lambda_i, \lambda_j]_+] = 4d_{ijk}. \tag{13.10}$$

The nonvanishing components of d_{ijk} are

ijk	d_{ijk}	ijk	d_{ijk}
118	$1/\sqrt{3}$	355	$\frac{1}{2}$
146	$\frac{1}{2}$	366	$-\frac{1}{2}$
157	$\frac{1}{2}$	377	$-\frac{1}{2}$
228	$1/\sqrt{3}$	448	$-1/2\sqrt{3}$
247	$-\frac{1}{2}$	558	$-1/2\sqrt{3}$
256	$\frac{1}{2}$	668	$-1/2\sqrt{3}$
338	$1/\sqrt{3}$	778	$-1/2\sqrt{3}$
344	$\frac{1}{2}$	888	$-1/\sqrt{3}$

The simplest realization of an SU_3 invariant field theory is given by the Lagrangian, $\mathcal{L}$, for three degenerate (in mass) Dirac fields, which we call q, for "quark,"[7]

$$q = \begin{pmatrix} q_1 \\ q_2 \\ q_3 \end{pmatrix}. \tag{13.11}$$

That is,

$$\mathcal{L} = -\bar{q}(x)[\gamma_\mu \partial_\mu + m_q]q(x). \tag{13.12}$$

Clearly this $\mathcal{L}$ is invariant under all transformations of the form

$$q' = u(\mathbf{\Lambda})q, \tag{13.13}$$

with

$$u(\mathbf{\Lambda}) = u(\Lambda_1 \ldots \Lambda_8) = e^{i\sum_{i=1}^{8}\Lambda_i\lambda_i}, \tag{13.14}$$

where the Λ_i are real numbers and the λ_i are the eight 3×3 matrices given above. Hence the eight-vector currents

6. $I = \begin{pmatrix} 1 & 0 & 0 \\ 0 & 1 & 0 \\ 0 & 0 & 1 \end{pmatrix}.$

$$V_{\mu i} = i\bar{q}\gamma_\mu \frac{\lambda_i}{2} q \tag{13.15}$$

are conserved, by $\mathcal{L}$, along with the baryon number current B_μ,

$$B_\mu = i\bar{q}\gamma_\mu q, \tag{13.16}$$

whose conservation corresponds to the invariance of $\mathcal{L}$ under the simple phase transformations

$$q' = e^{i\Lambda}q. \tag{13.17}$$

Thus the eight charges

$$Q_i = \int d\mathbf{r} q^\dagger(\mathbf{r}, 0) \frac{\lambda_i}{2} q(\mathbf{r}, 0) \tag{13.18}$$

are constants of the motion, and generators of SU_3, in the sense that[8]

$$[Q_i, Q_j]_- = if_{ijk}Q_k. \tag{13.19}$$

From what has been said above (or from the values of the f's), Q_3 and Q_8 commute with each other and with B, the baryon number,

$$B = \int d\mathbf{r} q^\dagger q. \tag{13.20}$$

For reasons that will become clear, the hypercharge, Y, is defined to be (for quarks)

$$Y = \frac{2}{\sqrt{3}} Q_8, \tag{13.21}$$

and the electric charge, Q, is then

$$Q = Q_3 + \frac{1}{\sqrt{3}} Q_8. \tag{13.22}$$

Again, from the equal time relations for the q (distinct q anticommute at equal times), we have

$$[Q_i, q]_- = \left[\int d\mathbf{r}' q^\dagger(\mathbf{r}', t) \frac{\lambda_i}{2} q(\mathbf{r}', t), q(\mathbf{r}, t)\right]_- = -\frac{\lambda_i}{2} q(\mathbf{r}, t) \tag{13.23}$$

and

$$[Q_i, q^\dagger]_- = q^\dagger \frac{\lambda_i}{2}. \tag{13.24}$$

We can construct three quark states, from the vacuum, in the form of vectors in a three-dimensional space,

$$\begin{pmatrix} |q\rangle \\ 0 \\ 0 \end{pmatrix}, \begin{pmatrix} 0 \\ |q\rangle \\ 0 \end{pmatrix}, \begin{pmatrix} 0 \\ 0 \\ |q\rangle \end{pmatrix}. \tag{13.25}$$

---

7. See M. Gell-Mann, *Physics Letters* **8**:214 (1964), and G. Zweig (unpublished CERN report), where the quarks first were introduced.
8. Use the equal time anticommutation relations among the quarks and the algebra of the λ_i.

These states represent an isotopic doublet with $Q_3 = \pm\frac{1}{2}$ and an isosinglet with $Q_3 = 0$, and hypercharges $\frac{1}{3}$, $\frac{1}{3}$, $-\frac{2}{3}$ and hence electric charges, in units of e, with values $\frac{2}{3}$, $-\frac{1}{3}$, $-\frac{1}{3}$. These are the celebrated quark fractional charges. The quarks and antiquarks can be represented pictorially as follows:

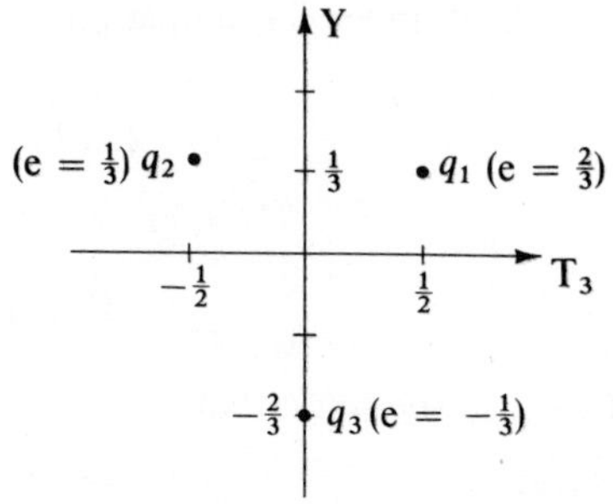

3

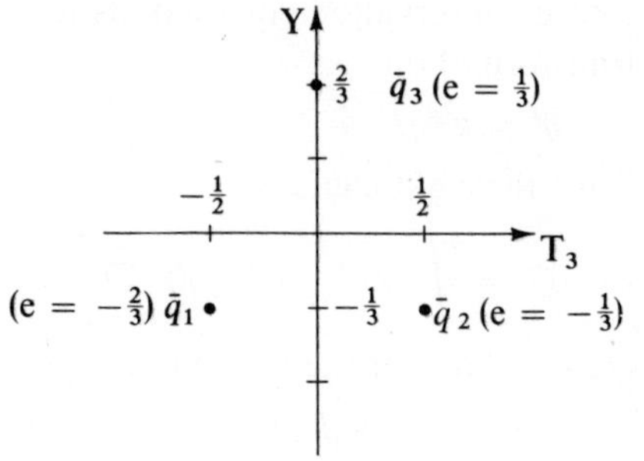

3*

It is an open question as to whether such particles exist.

We may also construct eight axial vector charges Q_{5i}, with

$$Q_{5i}(t) = \int q^\dagger(\mathbf{r}, t)\gamma_5 \frac{\lambda_i}{2} q(\mathbf{r}, t)\, d\mathbf{r}. \tag{13.26}$$

The Q_{5i} are not conserved, even if we consider free quark fields, because of the quark mass. However, using the equal time commutation relations, we have explicitly

$$[Q_{5i}(t), Q_{5j}(t)]_- = \int d\mathbf{r} q^\dagger(\mathbf{r}, t)\left[\frac{\lambda_i\lambda_j - \lambda_j\lambda_i}{4}\right] q(\mathbf{r}, t) = if_{ijk}Q_k. \tag{13.27}$$

That is, axial charge commuted with axial charge yields vector charge, a situation now familiar to us from our work, in the last chapter, with SU_2. Finally, we have, again from the equal time commutation relations,

$$[Q_{5i}(t), Q_{5j}(t)]_- = if_{ijk}Q_k. \tag{13.28}$$

In addition to these relations among the charges we can also derive for the eight vector quark currents and eight axial vector currents,

$$\begin{aligned} V_{\mu i} &= i\bar{q}\gamma_\mu \frac{\lambda_i}{2} q, \\ A_{\mu i} &= i\bar{q}\gamma_\mu\gamma_5 \frac{\lambda_i}{2} q, \end{aligned} \tag{13.29}$$

the commutation relations, at equal times,

$$\begin{aligned} [Q_i, V_{\mu j}]_- &= if_{ijk}V_{\mu k}, \\ [Q_i, A_{\mu j}]_- &= if_{ijk}A_{\mu k}, \\ [Q_{5i}, V_{\mu j}]_- &= if_{ijk}A_{\mu k}, \\ [Q_{5i}, A_{\mu j}]_- &= if_{ijk}V_{\mu k}. \end{aligned} \tag{13.30}$$

The first two of these relations carry the information that $V_{\mu j}$ and $A_{\mu j}$ transform like "vectors" under the SU_3 group, in the same sense that an isotopic vector such as $J_{\mu i}{}^V$ transforms under SU_2 according to the relation

$$[T_i, J_{\mu j}{}^V]_- = i\epsilon_{ijk}J_{\mu k}{}^V. \tag{13.31}$$

Often, in the literature on this subject, these transformation properties of $V_{\mu i}$ and $A_{\mu i}$ are summarized by saying that the currents belong to an "octet representation" of SU_3. We shall clarify this terminology later.

All of these transformation properties hold for interacting quarks and quark currents, providing that the interaction is an SU_3 scalar and does not have derivative couplings, which would modify the equal-time commutation relations. We might be tempted to consider the commutation relations among the current densities. Again, a naive application of the equal time commutation relations yields

$$[V_{0i}(0, 0), V_{\mu j}(\mathbf{r}, 0)]_- = [A_{0i}(0, 0), A_{\mu j}(\mathbf{r}, 0)]_- = if_{ijk}V_{\mu k}(\mathbf{r}, 0)\,\delta^3(\mathbf{r}) \tag{13.32}$$

and

$$[V_{0i}(0, 0), A_{\mu j}(\mathbf{r}, 0)]_- = if_{ijk}A_{\mu k}(\mathbf{r}, 0)\,\delta^3(\mathbf{r}). \tag{13.33}$$

Although these commutation relations look very plausible, it was pointed out, most recently by Schwinger,[9] that they lead to paradox. Consider the commutator $[V_{0i}(0, 0), V_{\mu j}(\mathbf{r}, 0)]_-$ for the special case $i = j$ and $\mu = 1, 2, 3$. From the antisymmetry of f_{ijk} we have

$$[V_{0i}(0, 0), \mathbf{V}_i(\mathbf{r}, 0)]_- = 0. \tag{13.34}$$

Since

$$\partial_\mu V_{\mu i} = 0, \tag{13.35}$$

we have

$$\nabla\cdot\mathbf{V}_i = -\frac{\partial}{\partial t}V_{0i}. \tag{13.36}$$

Thus Eq. (13.34) implies that

$$[V_{0i}(0, 0), \dot{V}_{0i}(\mathbf{r}, 0)]_- = 0. \tag{13.37}$$

We can take the vacuum expectation value of this equation:

$$\begin{aligned} 0 &= \langle 0|[V_{0i}(0, 0), \dot{V}_{0i}(\mathbf{r}, 0)]_-|0\rangle \\ &= \langle 0|[V_{0i}(0, 0), [H, V_{0i}(\mathbf{r}, 0)]_-]_-|0\rangle, \end{aligned} \tag{13.38}$$

where H is the strong interaction Hamiltonian. Thus inserting a complete set of states, $|n\rangle$,

$$\begin{aligned} 0 &= \sum_n \langle 0|V_{0i}(0, 0)|n\rangle\langle n|[H, V_{0i}(\mathbf{r}, 0)]_-|0\rangle - \sum_n \langle 0|[H, V_{0i}(\mathbf{r}, 0)]_-|n\rangle\langle n|V_{0i}(0, 0)|0\rangle \\ &= \sum_n |\langle 0|V_{0i}(0)|n\rangle|^2 E_n \cos(\mathbf{p}_n\cdot\mathbf{r}). \end{aligned} \tag{13.39}$$

---

9. J. Schwinger, *Phys. Rev. Letters* **3**:276 (1959). See also T. Imanura and T. Goto, *Prog. Theor. Physics* **14**:395 (1955).

According to the naive commutation relations this formula is meant to hold for all $\mathbf{r}$ and hence for $\mathbf{r} = 0$, so that

$$\sum_n |\langle 0|V_{0i}(0, 0)|n\rangle|^2 E_n = 0, \tag{13.40}$$

which is nonsense because $E_n \geq 0$. The obvious conclusion is that the commutator $[V_{0i}(\mathbf{r}, t), \mathbf{V}_j(\mathbf{r}', t)]_-$ is not simply proportional to $\delta^3(\mathbf{r}' - \mathbf{r})$. Put in another way, we must be exceedingly careful in treating the products of currents evaluated at distinct space points in the limit in which these points are allowed to coincide. No problems appear to arise for commutators of charges with charges, charges with currents, or even charge densities with charge densities. Schwinger[10] has suggested that the correct form of the commutation relations, which follow, if the limiting process is done carefully, might, for $V_{\mu i}$, look like

$$[V_{0i}(0, 0), \mathbf{V}_j(\mathbf{r}, 0)]_- = if_{ijk}\mathbf{V}_k(\mathbf{r}, 0)\, \delta^3(\mathbf{r}) + C_{ijk}\, \nabla\delta^3(\mathbf{r}) + \cdots, \tag{13.41}$$

where the dots stand for the possibility of having even higher derivatives of the δ^3 function. So long as the C_{ijk} are *numbers*, and not operators, these gradient terms will not effect the integrated commutation relations. The study of these "Schwinger" terms is one of the most active in the theory of current algebras and the interested reader is referred to the growing literature. For the rest of this book we will make the optimistic assumption that

$$[V_{0i}(0, 0), V_{0j}(\mathbf{r}, 0)]_- = if_{ijk}V_{0k}(\mathbf{r}, 0)\, \delta^3(\mathbf{r}), \tag{13.42}$$

with no Schwinger terms, and we will otherwise be careful to commute spatially integrated quantities with the hope that the Schwinger terms will disappear for them. We shall return later to some of the applications of these density commutation relations.

Since the quarks have not yet exhibited themselves experimentally, all of the interesting applications of the SU_3-extended Gell-Mann current algebra come about because of the connection (theoretical) between the quarks and the less peculiar particles. This connection arises out of two assumptions, both of which have a good deal of experimental and theoretical support.

1. The familiar particles can, in some sense, be regarded as bound states of quarks.

2. The dynamics of this binding and the effective dynamics of the resultant quark-constituted particles is, at least in an approximation that is physically interesting, invariant under SU_3.

The rest of this chapter is concerned with elucidating these two assumptions. To explore the first assumption, the simplest system to begin with is that of the ordinary pseudoscalar mesons, π^+, π^0, π^-, η^0, K^+, K^0, $\bar{K}^0$, K^-, eight objects in all, similar to each other to the extent that they all exhibit the same spin and parity. The object of the game is to start with the three quarks, q_1, q_2, q_3 dia-

10. Op. cit.
11. Taking products of q_i with q_j, quark with quark, leads to particles with quantum numbers, so far never seen, so this does not appear to be a physically interesting thing to do. As

grammed in Fig. 1 and the three antiquarks $\bar{q}_1, \bar{q}_2, \bar{q}_3$, put them together in pairs, and study the quantum numbers T, T_3, Y of the pairs. It is, of course, familiar from the usual addition of ordinary angular momentum, that arbitrary products of q's and $\bar{q}$'s will not necessarily have well-defined quantum numbers. We will have to choose suitable linear combinations of products, in some cases, to obtain states that are eigenfunctions of T, T_3 and Y. The nine products of q_i and $\bar{q}_i$ are given below,[11] along with their T_3 and Y, read directly from Table 1.

Table 1

Product	T_3	Y
$q_1\bar{q}_1$	0	0
$q_1\bar{q}_2$	1	0
$q_1\bar{q}_3$	$+\frac{1}{2}$	1
$q_2\bar{q}_1$	-1	0
$q_2\bar{q}_2$	0	0
$q_2\bar{q}_3$	$-\frac{1}{2}$	1
$q_3\bar{q}_1$	$-\frac{1}{2}$	-1
$q_3\bar{q}_2$	$\frac{1}{2}$	-1
$q_3\bar{q}_3$	0	0

These products are not eigenstates of T^2 nor do they have simple transformation properties under SU_3. Howcver, it is easy to manufacture linear combinations of the products that *do* transform simply under SU_3. We recall that for any of the three quarks we have the transformation ($\alpha \ll 1$)

$$e^{i\alpha Q_i} q e^{-i\alpha Q_i} = q + i\alpha[Q_i, q]_- = q - i\alpha \frac{\lambda_i}{2} q, \tag{13.43}$$

and for the antiquarks,

$$e^{iQ_i\alpha} \bar{q} e^{-i\alpha Q_i} = \bar{q} + i\alpha\bar{q}\lambda_i, \tag{13.44}$$

where

$$q = \begin{pmatrix} q_1 \\ q_2 \\ q_3 \end{pmatrix}. \tag{13.45}$$

Now, with $\alpha \ll 1$,

$$e^{iQ_i\alpha} \bar{q} q e^{-iQ_i\alpha} = \left(\bar{q} + i\alpha\bar{q}\frac{\lambda_i}{2}\right)\left(q - i\alpha\frac{\lambda_i}{2}q\right) = \bar{q}q. \tag{13.46}$$

Thus, the combination of field operators

$$\bar{q}q = \bar{q}_1 q_1 + \bar{q}_2 q_2 + \bar{q}_3 q_3 \tag{13.47}$$

is an SU_3 "scalar" or singlet. For want of anything else, and to conform to the

usual, in a direct product the corresponding quantum numbers are the algebraic sum of the quantum numbers belonging to the factors.

literature, we call the corresponding particle[12] the x^0. We can use these field operators, acting on the vacuum, to make the x^0 state and we can then use what we know about quarks and antiquarks to compute the quantum numbers of this state. There is one point in this evaluation to be careful about. If the q_1, q_2 isotopic doublet is represented by the isospinors

$$q_1 = \begin{pmatrix}1\\0\end{pmatrix}, \qquad q_2 = \begin{pmatrix}0\\1\end{pmatrix}, \tag{13.48}$$

then

$$\bar{q}_1 = \begin{pmatrix}0\\-1\end{pmatrix}, \qquad \bar{q}_2 = \begin{pmatrix}1\\0\end{pmatrix}. \tag{13.49}$$

This minus sign in $\bar{q}_1$ arises when we go from particle to antiparticle and is familiar from the treatment of the $n, p; \bar{n}, \bar{p}$ system.[13] Using Fig. 1 and the identity[14]

$$T^2 = T_3^2 + T_3 + T_- T_+, \tag{13.50}$$

we easily see that x^0 has the quantum numbers, $T^2 = T_3 = Y = Q = 0$. Having separated off the x^0, this still leaves eight independent combinations of q and $\bar{q}$ and these, as we will now show, can be constructed so as to transform like an SU_3 octet. This means that, if these combinations are called ϕ_j, and, if the eight generators of SU_3 are called Q_i, then ($\alpha \ll 1$)

$$e^{i\alpha Q_i}\phi_j e^{-i\alpha Q_i} = \phi_j + i\alpha[Q_i, \phi_j]_- = \phi_j - \alpha f_{ijk}\phi_k, \tag{13.51}$$

where the f_{ijk} are the SU_3 structure constants introduced earlier. The first step in Eq. (13.51) is an algebraic identity. The second step is a definition of what it means for the eight ϕ_i to transform like an octet; that is, $[Q_i, \phi_j]_- = if_{ijk}\phi_k$. We may now easily verify that, if we define

$$\phi_j = \bar{q}\lambda_j q, \tag{13.52}$$

and take for the Q_i the quark generators

$$Q_i = \int d\mathbf{r}\, q^\dagger \frac{\lambda_i}{2} q, \tag{13.53}$$

we have, from Eqs. (13.43) and (13.44), the equation ($\alpha \ll 1$)

$$\begin{aligned} e^{i\alpha Q_i}\phi_j e^{-i\alpha Q_i} &= e^{i\alpha Q_i}\bar{q}e^{-i\alpha Q_i}\lambda_j e^{iQ_i\alpha} q e^{-iQ_i\alpha} = \bar{q}\lambda_j q + i\alpha\bar{q}\tfrac{1}{2}[\lambda_i\lambda_j - \lambda_j\lambda_i]q \\ &= \phi_j - \alpha f_{ijk}\phi_k. \end{aligned} \tag{13.54}$$

12. Sometimes it is also called the $\eta^{0\prime}$.
13. See, e.g., S. Gasiorowicz, *Elementary Particle Physics*, Wiley (1966), p. 246.
14. Here $T_\pm$, etc., act in the product space of q and $\bar{q}$. Thus, for example

$$T^+(\bar{q}_1 q_1 + \bar{q}_2 q_2) = \left(\frac{\tau_+}{2} + \frac{\bar{\tau}_+}{2}\right)\left(\begin{pmatrix}0\\-1\end{pmatrix}\begin{pmatrix}1\\0\end{pmatrix} + \begin{pmatrix}1\\0\end{pmatrix}\begin{pmatrix}0\\1\end{pmatrix}\right) = 0.$$

Thus the ϕ_i transform like an octet. However, they do not have a definite T_3. (Remember, from SU_2, that the pion fields ϕ_1 and ϕ_2 did not have a definite T_3 either.) However, with the tools at hand it is easy to verify the validity of the following table, which gives the linear combinations of quarks and antiquarks that do have a definite T_3.

Table 2

Particle	*T*	T_3	*Y*	*Quark Combination*
π^+	1	1	0	$\bar{q}\left(\frac{\lambda_1 - i\lambda_2}{2}\right)q = \bar{q}_2q_1$
π^0	1	0	0	$\bar{q}\frac{\lambda_3}{\sqrt{2}}q = \frac{\bar{q}_1q_1 - \bar{q}_2q_2}{\sqrt{2}}$
π^-	1	-1	0	$\bar{q}\frac{(\lambda_1 + i\lambda_2)}{2}q = \bar{q}_1q_2$
K^+	$\frac{1}{2}$	$\frac{1}{2}$	$+1$	$\bar{q}\frac{(\lambda_4 - i\lambda_5)}{2}q = \bar{q}_3q_1$
K^0	$\frac{1}{2}$	$-\frac{1}{2}$	$+1$	$\bar{q}\frac{(\lambda_6 - i\lambda_7)}{2}q = \bar{q}_3q_2$
K^-	$\frac{1}{2}$	$-\frac{1}{2}$	-1	$\bar{q}\frac{(\lambda_4 + i\lambda_5)}{2}q = \bar{q}_1q_3$
$\overline{K^0}$	$\frac{1}{2}$	$\frac{1}{2}$	-1	$\bar{q}\frac{(\lambda_6 + i\lambda_7)}{2}q = \bar{q}_2q_3$
η^0	0	0	0	$\bar{q}\frac{\lambda_8}{\sqrt{2}}q = \frac{(\bar{q}_1q_1 + \bar{q}_2q_2 - 2\bar{q}_3q_3)}{\sqrt{2}}$

Note: These states must be multiplied by the appropriate ordinary Pauli spin function $\frac{(\uparrow\downarrow - \downarrow\uparrow)}{\sqrt{2}}$ to give them the correct spin.

The octet representation can be pictured in diagrammatic form:

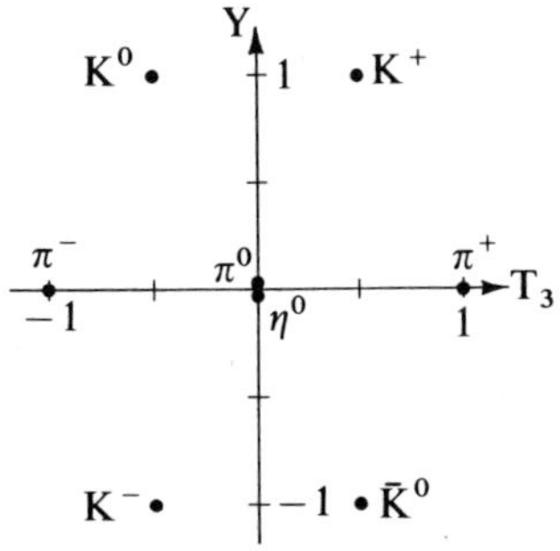

(The combinations of quark fields given in the table *annihilate* the particles indicated so that the Hermitian conjugate fields are the ones that create these particles.) In mathematical language the constructions of Table 2 and Eq. (13.47) show that

$$3 \times 3^* = 8 + 1,$$

which is to say that the product of the triplet and antitriplet representations yields an octet and a singlet. (Numbers refer to representations and $\times$ stands for the "direct product" of two representations.) The eight pseudoscalars in the octet are all well-known particles and the singlet, x^0, may be identifiable with an object observed at 958.3 ± 0.8 MeV with a narrow width, less than 4 MeV, whose principal decay mode is $x^0 \longrightarrow \eta^0 + 2\pi$.[15]

The assignment of baryon numbers to the quarks has been so chosen that products of three quarks qqq (no antiquarks) have the baryon numbers and hypercharges of the familiar baryons. By using somewhat more sophisticated arguments[16] we can show that

$$3 \times 3 \times 3 = 1 + 8 + 8 + 10,$$

which is to say that by tacking together three quarks it is possible to construct a singlet, an octet[17] (corresponding to the familiar eight baryons n, p, Λ, Σ^+, Σ^0, Σ^-, Ξ^+, Ξ^0), and a decimet (containing the $J = \frac{3}{2}$ resonances like the N^*).

We can exploit the work done so far to make predictions about experimentally accessible phenomena if we adopt the attitude, as a reasonable first approximation, that the symmetries of the quark model are actually shared by the interactions of the known particles. To see how this works we begin by

---

15. A. H. Rosenfeld et al., *Rev. Mod. Phys.* **39**:1 (1967).
16. See, for example, R. Dalitz, *Les Houches Lectures 1965*, p. 253, Gordon and Breach, and Chapter 14.
17. Octets have the property that $8 = 8^*$, which is, essentially, why two 8's appear in the reduction of $3 \times 3 \times 3$. All of these matters are fully treated in *The Eightfold Way* by M. Gell-Mann and Y. Ne'eman (op. cit.). See also Chapter 14.
18. The discussion given here follows that of R. Gatto, *Theoretical Physics*, International Atomic Energy Agency, Vienna (1963), p. 197 and S. Coleman and R. L. Glashow, *Phys. Rev. Letters* **21**:1423 (1961).
19. From the values of the f_{ijk} we also have

$$[Q_7, F(J_\mu{}^\gamma)]_- = 0.$$

Hence the combinations

$$U_\pm = Q_6 \pm iQ_7$$

also commute with $F(J_\mu{}^\gamma)$. Moreover, using the f_{ijk} we easily see that

$$[U_+, U_-]_- = \tfrac{3}{2}Y - T_3 \cdot\equiv\cdot 2U_3.$$

Thus $U_\pm$ and $\frac{1}{2}[U_+, U_-]_-$ have the commutation relations of a spin: U-spin. In this language the current,

$$J_\mu{}^\gamma = V_{\mu 3} + \frac{1}{\sqrt{3}} V_{\mu 8},$$

is a U-spin scalar. Thus the linear combination of SU_3 states

$$|\alpha\rangle = |A_3\rangle + \frac{1}{\sqrt{3}}|A_8\rangle$$

discussing the electromagnetic interactions.[18] The quark electromagnetic current, as in Eq. (13.22), is given by

$$J_\mu{}^\gamma = V_{\mu 3} + \frac{1}{\sqrt{3}} V_{\mu 8}. \tag{13.55}$$

We then assume that the *actual* electromagnetic current has the same transformation properties under SU_3 as this $J_\mu{}^\gamma$; that is, we suppose that there exists an octet of vector currents $V_{\mu i}$ that transform under the generators of SU_3, Q_i, according to the rule

$$[Q_i, V_{\mu j}]_- = if_{ijk} V_{\mu k}, \tag{13.56}$$

and that

$$J_\mu{}^\gamma = V_{\mu 3} + \frac{1}{\sqrt{3}} V_{\mu 8}, \tag{13.57}$$

as in the quark model. From the table of values of f_{ijk} we see at once that

$$[Q_3, J_\mu{}^\gamma]_- = [Q_8, J_\mu{}^\gamma]_- = [Q_6, J_\mu{}^\gamma]_- = 0. \tag{13.58}$$

This equation expresses the fact that $J_\mu{}^\gamma$ transfers neither charge nor hypercharge. In addition there is a new symmetry that we now exploit. In particular, if $|B\rangle$ and $|A\rangle$ are any two states and $F(J_\mu{}^\gamma)$ is any function of $J_\mu{}^\gamma$, we have[19]

---

must be a U-spin singlet; that is,

$$U^2|\alpha\rangle = 0.$$

(The reader who doubts this can apply the formulas to be given shortly to show it directly.) Therefore the orthogonal combination

$$|\beta\rangle = \frac{1}{\sqrt{3}}|A_3\rangle - |A_8\rangle$$

must be a part of a U^2 triplet (the $U_3 = 0$ part). Since the vacuum is a U-spin singlet we have

$$\langle 0|F(J^\gamma)|\beta\rangle = 0$$

or

$$\langle 0|F(J^\gamma)|A_3\rangle = \sqrt{3}\,\langle 0|F(J^\gamma)|A_8\rangle.$$

If we take for $F(J^\gamma)$, say, the product of two currents, $J_\mu{}^\gamma(x)J_\nu{}^\gamma(x')$, which is the operator that gives rise to the two-photon decay of A_3 or A_8, we have the theorem

$$\langle\gamma\gamma|A_3\rangle = \sqrt{3}\,\langle\gamma\gamma|A_8\rangle,$$

which is the celebrated SU_3 prediction relating the two-photon decay of η^0 to that of π^0:

$$\text{amp}\,(\eta^0 \longrightarrow \gamma\gamma) = \frac{1}{\sqrt{3}}\,\text{amp}\,(\pi^0 \longrightarrow \gamma\gamma).$$

In testing this prediction, like all SU_3 predictions, one puts in the *physical* η^0 and π^0 masses in the phase space. (In the good SU_3 limit these masses would be identical.) The η^0 lifetime is not yet known well enough to allow a test of this prediction.

$$\langle A|[Q_6, F(J_\mu{}^\gamma)]_-|B\rangle = 0. \tag{13.59}$$

We now suppose that the states $|A\rangle$ and $|B\rangle$ are members of octets, say A_i and B_i, with $i = 1, \ldots, 8$. Symbolically, we can create these states out of the vacuum, $|0\rangle$:

$$\begin{aligned} |A_i\rangle &= A_i|0\rangle, \\ |B_i\rangle &= B_i|0\rangle, \end{aligned} \tag{13.60}$$

where A_i and B_i are shorthand for the creation operators and the limiting process. In the good SU_3 limit

$$Q_i|0\rangle = 0, \tag{13.61}$$

since the vacuum must be SU_3-invariant. Thus

$$Q_i|A_j\rangle = [Q_i, A_j]_-|0\rangle = if_{ijk}|A_k\rangle. \tag{13.62}$$

Thus Eq. (13.59) becomes

$$\begin{aligned} \langle A_i|Q_6F(J^\gamma)|B_j\rangle = \langle A_i|F(J^\gamma)Q_6|B_j\rangle &= \langle A_i|F(J^\gamma)|B_k\rangle if_{6jk} \\ &= -if_{6il}\langle A_l|F(J^\gamma)|B_j\rangle. \end{aligned} \tag{13.63}$$

This equation, along with the values of the f_{ijk}, leads to several identities among the matrix elements. Some of the most interesting ones are

$$\begin{aligned} \langle A_1|F(J^\gamma)|B_1\rangle &= \langle A_5|F(J^\gamma)|B_5\rangle, \\ \langle A_2|F(J^\gamma)|B_2\rangle &= \langle A_4|F(J^\gamma)|B_4\rangle, \\ \langle A_2|F(J^\gamma)|B_1\rangle &= -\langle A_4|F(J^\gamma)|B_5\rangle, \\ \langle A_1|F(J^\gamma)|B_2\rangle &= -\langle A_5|F(J^\gamma)|B_4\rangle, \end{aligned} \tag{13.64}$$

or

$$\langle A_1 + iA_2|F(J^\gamma)|A_1 + iA_2\rangle = \langle A_4 + iA_5|F(J^\gamma)|A_4 + iA_5\rangle. \tag{13.65}$$

If we take $F(J) = J_\mu{}^\gamma$ and apply Eq. (13.65) to the pseudoscalar octet, we have the relation

$$\langle \pi^+|J_\mu{}^\gamma|\pi^+\rangle = \langle K^+|J_\mu{}^\gamma|K^+\rangle, \tag{13.66}$$

which would say, for example, that the π^+ charge radius is identical to the K^+ charge radius in the good SU_3 limit. Neither radius has been measured.

Of more practical interest are the SU_3 predictions for the electromagnetic properties of the baryon octet, which is represented by the diagram below.

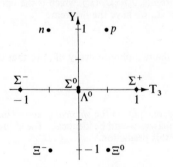

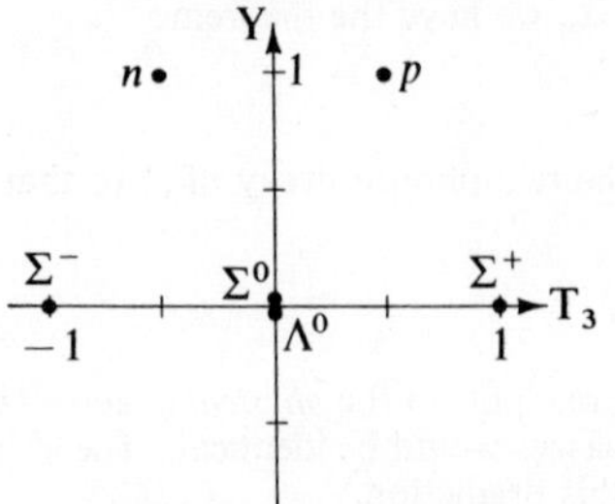

We can take over the octet labeling from the pseudoscalar case and hence have the immediate result

$$\langle\Sigma^+|J_\mu{}^\gamma|\Sigma^+\rangle = \langle P|J_\mu{}^\gamma|P\rangle, \tag{13.67}$$

which says, in particular, that, for magnetic moments,

$$\mu_P = \mu_{\Sigma^+}. \tag{13.68}$$

Recent determinations of the Σ^+ moment[20] give (in proton magnetons $e/2M_P$)

$$\mu_{\Sigma^+} = 2.3 \pm .6 \tag{13.69}$$

and

$$\mu_P = 2.793,$$

which constitutes good agreement. More generally we have

$$\langle\Sigma^+|F(J^\gamma)|\Sigma^+\rangle = \langle P|F(J^\gamma)|P\rangle, \tag{13.70}$$

and by passing from the states $A_1 + iA_2$ to $A_1 - iA_2$ we find

$$\langle\Sigma^-|F(J^\gamma)|\Sigma^-\rangle = \langle\Xi^-|F(J^\gamma)|\Xi^-\rangle. \tag{13.71}$$

We easily show that

$$\langle A_6|F(J^\gamma)|B_7\rangle = \langle A_7|F(J^\gamma)|B_6\rangle = 0, \tag{13.72}$$

so that

$$\langle K^0|F(J^\gamma)|K^0\rangle = \langle\bar{K}^0|F(J^\gamma)|\bar{K}^0\rangle \tag{13.73}$$

and

$$\langle\Xi^0|F(J^\gamma)|\Xi^0\rangle = \langle N|F(J^\gamma)|N\rangle. \tag{13.74}$$

These equations also lead to interesting consequences. If we set $F(J^\gamma) = J_\mu{}^\gamma$, we learn that

$$\begin{aligned} \mu_{\Sigma^-} &= \mu_{\Xi^-} \\ \mu_N &= \mu_{\Xi^0}. \end{aligned} \tag{13.75}$$

Moreover,

$$\langle K^0|J_\mu{}^\gamma|K^0\rangle = \langle\bar{K}^0|J_\mu{}^\gamma|\bar{K}^0\rangle = -\langle K^0|J_\mu{}^\gamma|K^0\rangle. \tag{13.76}$$

The last step is a consequence of charge conjugation and shows that in the good SU_3 limit the K^0 charge radius is zero.[21] Any finite charge radius would arise from SU_3-breaking interactions.[22] We can also take

$$F(J^\gamma) = J_\mu{}^\gamma(x)J_\nu{}^\gamma(0) \tag{13.77}$$

---

20. A. H. Rosenfeld et al., *Rev. Mod. Phys.* **39**:1 (1967).
21. Like the neutron, the K^0, though neutral, can have electromagnetic properties, since $K^0 \neq \bar{K}^0$.
22. A recent non-SU_3 estimate by N. M. Kroll, T. D. Lee, and B. Zumino (to be published) gives

$$\langle \mathbf{r}^2\rangle_{K^0} = -\langle \mathbf{r}^2\rangle_{\bar{K}^0} \simeq 7.0 \times 10^{-28}\ \text{cm}^2.$$

and use Eqs. (13.70), (13.71), and (13.74) to derive a set of SU_3 predictions for second-order electromagnetic processes.[23] If real photons are attached to the $J_\mu{}^\gamma$, then these matrix elements correspond to Compton scattering:

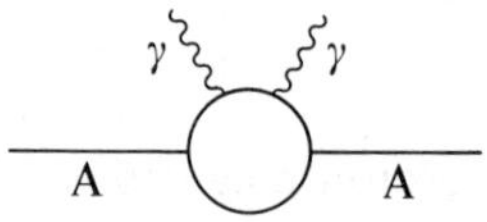

Such matrix elements do not lead to predictions of much practical interest since these phenomena are all but unobservable (except for proton Compton scattering), as all of the particles involved are highly unstable. However, if we couple the $J_\mu{}^\gamma$'s to *virtual* quanta, then we are led to a set of predictions that *are* physically interesting. The graph

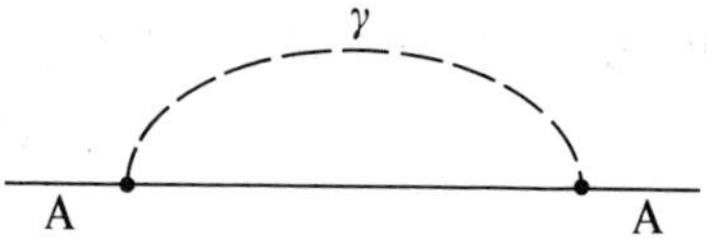

represents the leading field theoretic contribution to the electromagnetic self energy or self-mass. In perturbation theory these self-mass terms are generally infinite and, at best, involve arbitrary constants representing cutoff parameters fed into the theory to suppress the infinities. However, SU_3 makes the following predictions as to how these quantities are related to each other. For the pseudo-scalar mesons[24] we have

$$\begin{aligned} \delta m_{\pi^+} &= \delta m_{K^+}, \\ \delta m_{\pi^-} &= \delta m_{K^-}, \\ \delta m_{K^0} &= \delta m_{\overline{K^0}}. \end{aligned} \tag{13.78}$$

The corresponding baryon formulas state that

$$\begin{aligned} \delta m_P &= \delta m_{\Sigma^+}, \\ \delta m_{\Sigma^-} &= \delta m_{\Xi^-}, \\ \delta m_N &= \delta m_{\Xi^0}. \end{aligned} \tag{13.79}$$

Thus we have the Coleman-Glashow relation[25]

$$\delta m_{\Xi^-} - \delta m_{\Xi^0} = \delta m_P - \delta m_N + \delta m_{\Sigma^-} - \delta m_{\Sigma^+}. \tag{13.80}$$

Referring to the table of masses we find

---

23. These would, strictly speaking, involve $(J_\mu{}^\gamma(x)J_\nu{}^\gamma(0))_+$, the time-ordered product, but this does not affect the argument.
24. For bosons, since the mass enters quadratically in the Klein-Gordon equation and Green's functions, the mass formulas are taken to refer to squares of the masses.

$$\begin{aligned} \delta m_{\Xi^-} - \delta m_{\Xi^0} &= 6.5 \pm 1.5 \text{ MeV}, \\ \delta m_P - \delta m_N &= -1.3 \text{ MeV}, \\ \delta m_{\Sigma^-} - \delta m_{\Sigma^+} &= 7.97 \pm 0.01 \text{ MeV}, \end{aligned} \tag{13.81}$$

which gives excellent agreement with Eq. (13.80). There is another set of relations given by SU_3. The simplest way to get at these is by extending the discussion (footnote on p. 222) in which U-spin was defined. As shown there, the three operators,

$$U_\pm = F_6 \pm iF_7 \tag{13.82}$$

and

$$\tfrac{1}{2}[U_+, U_-]_- = U_3, \tag{13.83}$$

commute among each other like the SU_2 generators of an isospin group. We also argued that $J_\mu{}^\gamma$ was a U-spin scalar; that is,

$$[J_\mu{}^\gamma, \mathbf{U}]_- = 0, \tag{13.84}$$

since $J_\mu{}^\gamma$ commutes with F_6 and F_7. We furthermore indicated that the SU_3 octet could be decomposed into eigenfunctions of U^2 and, below, we fill out this list. (The states have all been normalized to one.)

Baryons		
Singlet	*Doublets*	*Triplet*
$\frac{1}{2}(\sqrt{3}\Sigma^0 + \Lambda^0)$	(P, Σ^+) (Σ^-, Ξ^-)	$(N, \frac{1}{2}(\sqrt{3}\Lambda^0 - \Sigma^0), \Xi^0)$
U_3 0	$\frac{1}{2}\ -\frac{1}{2}$ $\frac{1}{2}\ -\frac{1}{2}$	1, 0, -1
	Mesons	
$\frac{1}{2}(\sqrt{3}\pi^0 + \eta^0)$	(K^+, π^+) (π^-, K^-)	$(K^0, \frac{1}{2}(\sqrt{3}\eta^0 - \pi^0), \bar{K}^0)$
U_3 0	$\frac{1}{2}, -\frac{1}{2}$ $\frac{1}{2}, -\frac{1}{2}$	1, 0, -1

This table and the matrix relation (see Eq. (10.19)),

$$U_\pm |UU_3\rangle = \sqrt{(U \mp U_3)(U \pm U_3 + 1)}\, |UU_3 \pm 1\rangle, \tag{13.85}$$

give us enough information to derive all the relations so far given and in addition we get some new ones (which, of course, could have been obtained with the previous technique).

In particular, since $F(J^\gamma)$ cannot transfer U-spin,[26] we have

$$\langle \sqrt{3}\Sigma^0 + \Lambda^0 | F(J^\gamma) | \sqrt{3}\Lambda^0 - \Sigma^0 \rangle = 0 \tag{13.86}$$

or

25. S. Coleman and S. L. Glashow, *Phys. Rev. Letters* **6**:1423 (1961).
26. By writing this equation in the opposite order we see that

$$\langle \Lambda^0 | F(J^\gamma) | \Sigma^0 \rangle = \langle \Sigma^0 | F(J^\gamma) | \Lambda^0 \rangle.$$

$$-\sqrt{3}\,\langle\Sigma^0|F(J^\gamma)|\Sigma^0\rangle + \sqrt{3}\,\langle\Lambda^0|F(J^\gamma)|\Lambda^0\rangle + 2\langle\Sigma^0|F(J^\gamma)|\Lambda^0\rangle = 0. \tag{13.87}$$

Moreover, since $F(J^\gamma)$ is a U-spin scalar we have

$$\begin{aligned}\langle N|F(J^\gamma)|N\rangle &= \tfrac{1}{4}\langle\sqrt{3}\Lambda^0 - \Sigma^0|F(J^\gamma)|\sqrt{3}\Lambda^0 - \Sigma^0\rangle \\ &= \tfrac{3}{4}\langle\Lambda^0|F(J^\gamma)|\Lambda^0\rangle + \tfrac{1}{4}\langle\Sigma^0|F(J^\gamma)|\Sigma^0\rangle - \frac{\sqrt{3}}{2}\langle\Sigma^0|F(J^\gamma)|\Lambda^0\rangle.\end{aligned} \tag{13.88}$$

This is as far as one can go with U-spin conservation alone. If we count up the number of relations that we have found, involving diagonal matrix elements, we see that there are five. There are, however, eight magnetic moments in the octet. Of these, assuming only SU_2 invariance, only seven are independent, for the electric current has the form

$$J_\mu{}^\gamma = V_{\mu 3} + \frac{1}{\sqrt{3}} V_{\mu 8}, \tag{13.89}$$

which means that we can write, for the isotopic triplet $\boldsymbol{\Sigma}$,

$$\langle\Sigma^+|J_\mu{}^\gamma|\Sigma^+\rangle = \langle\boldsymbol{\Sigma}|J_{\mu 3}{}^V|\boldsymbol{\Sigma}\rangle + \langle\boldsymbol{\Sigma}|J_\mu{}^S|\boldsymbol{\Sigma}\rangle, \tag{13.90}$$

where the superscripts V and S refer to the isovector and isoscalar parts of $J_\mu{}^\gamma$:

$$\langle\Sigma^0|J_\mu{}^\gamma|\Sigma^0\rangle = \langle\boldsymbol{\Sigma}|J_\mu{}^S|\boldsymbol{\Sigma}\rangle \tag{13.91}$$

and

$$\langle\Sigma^-|J_\mu{}^\gamma|\Sigma^-\rangle = \langle\boldsymbol{\Sigma}|J_\mu{}^S|\boldsymbol{\Sigma}\rangle - \langle\boldsymbol{\Sigma}|J_{\mu 3}{}^V|\boldsymbol{\Sigma}\rangle. \tag{13.92}$$

Thus

$$\frac{\langle\Sigma^+|J_\mu{}^\gamma|\Sigma^+\rangle + \langle\Sigma^-|J_\mu{}^\gamma|\Sigma^-\rangle}{2} = \langle\Sigma^0|J_\mu{}^\gamma|\Sigma^0\rangle, \tag{13.93}$$

so that, in particular,

$$\frac{\mu_{\Sigma^+} + \mu_{\Sigma^-}}{2} = \mu_{\Sigma^0}. \tag{13.94}$$

Thus, only two of the three Σ baryons have independent electromagnetic properties.

There is yet another relation among the magnetic moments or, more generally, among the form factors, that follows from the explicit form of our current. The essential point is that $J_\mu{}^\gamma$ is the sum of two currents, $V_{\mu 3}$ and $(1/\sqrt{3})V_{\mu 8}$, that transform like members of an octet. In principle we could have added to $J_\mu{}^\gamma$ a current that transforms like an SU_3 scalar and, hence like a U-spin scalar, without changing any of the conclusions, such as the Coleman-Glashow relations, that we have so far drawn. However, the next relation depends explicitly on the absence of such a scalar term. We argue as follows; consider

$$\sum_{i=1}^{8} \langle A_i|J_\mu{}^\gamma|A_i\rangle,$$

where $|A_i\rangle$ is a particle in an octet. Clearly, since we are summing over *all* i,

$$\sum_{i=1}^{8} \langle A_i|J_\mu{}^\gamma|A_i\rangle = \sum_{i=1}^{8} \langle A_i'|J_\mu{}^\gamma|A_i'\rangle = \sum \langle A_i|U^{-1}J_\mu{}^\gamma U|A_i\rangle, \tag{13.95}$$

where

$$|A_i'\rangle = U|A_i\rangle \tag{13.96}$$

is any SU_3 transformation of the states A_i. But $U^{-1}J_\mu{}^\gamma U$ is an SU_3 transformation of $J_\mu{}^\gamma$ and under such a transformation $J_\mu{}^\gamma$ transforms like an octet. Thus

$$\sum_{i=1}^{8} \langle A_i|J_\mu{}^\gamma|A_i\rangle = 0, \tag{13.97}$$

since it is required to transform both like a scalar and a vector under SU_3. In fact, both

$$\sum_{i=1}^{8} \langle A_i|V_{\mu 3}|A_i\rangle = 0 \tag{13.98}$$

and

$$\sum_{i=1}^{8} \left\langle A_i \left| \frac{1}{\sqrt{3}} V_{\mu 8} \right| A_i \right\rangle = 0. \tag{13.99}$$

In particular, taking A_i to be the eigenstates of U^2 and U_3, which are, clearly, complete, we have for the baryon octet[27]

$$\begin{aligned} &\langle P|J_\mu{}^\gamma|P\rangle + \langle \Sigma^+|J_\mu{}^\gamma|\Sigma^+\rangle + \langle \Sigma^-|J_\mu{}^\gamma|\Sigma^-\rangle + \langle \Xi^-|J_\mu{}^\gamma|\Xi^-\rangle \\ &\quad + \tfrac{1}{4}\langle \sqrt{3}\Sigma^0 + \Lambda^0|J_\mu{}^\gamma|\sqrt{3}\Sigma^0 + \Lambda^0\rangle + \langle N|J_\mu{}^\gamma|N\rangle + \langle \Xi^0|J_\mu{}^\gamma|\Xi^0\rangle \\ &\quad + \tfrac{1}{4}\langle \sqrt{3}\Lambda^0 - \Sigma^0|J_\mu{}^\gamma|\sqrt{3}\Lambda^0 - \Sigma^0\rangle = 0. \end{aligned} \tag{13.100}$$

We can make use of Eq. (13.100), along with what we have learned so far, in various ways. In particular, using Eq. (13.93), we have a relation involving N, Σ^0 and Λ^0 alone:

$$\begin{aligned} &4\langle \Sigma^0|J_\mu{}^\gamma|\Sigma^0\rangle + 2\langle N|J_\mu{}^\gamma|N\rangle + \tfrac{1}{4}\langle \sqrt{3}\Sigma^0 + \Lambda^0|J_\mu{}^\gamma|\sqrt{3}\Sigma^0 + \Lambda^0\rangle \\ &\quad + \tfrac{1}{4}\langle \sqrt{3}\Lambda^0 - \Sigma^0|J_\mu{}^\gamma|\sqrt{3}\Lambda^0 - \Sigma^0\rangle = 0 \end{aligned} \tag{13.101}$$

or

$$2\langle N|J_\mu{}^\gamma|N\rangle + \langle \Lambda^0|J_\mu{}^\gamma|\Lambda^0\rangle + 5\langle \Sigma^0|J_\mu{}^\gamma|\Sigma^0\rangle = 0. \tag{13.102}$$

Combining Eqs. (13.102), (13.88), and (13.87) we have the well-known SU_3 prediction that

$$\langle \Lambda^0|J_\mu{}^\gamma|\Lambda^0\rangle = \tfrac{1}{2}\langle N|J_\mu{}^\gamma|N\rangle, \tag{13.103}$$

so that

$$\mu_{\Lambda^0} = \tfrac{1}{2}\mu_N. \tag{13.104}$$

---

27. The pseudoscalar octet contains both the particles and the antiparticles. Thus, using the oddness of $J_\mu{}^\gamma$ under charge conjugation Eq. (13.97) leads simply to 0 = 0 for the pseudoscalars.

Experimentally,[28]

$$\mu_{\Lambda^0} = -0.73 \pm 0.16 \tag{13.105}$$

and

$$\mu_N = -1.91. \tag{13.106}$$

(These numbers are in terms of proton magnetons, $e/2M_P$.) We may also conclude from these equations that

$$\langle \Sigma^0 | J_\mu{}^\gamma | \Sigma^0 \rangle = -\langle N | J_\mu{}^\gamma | N \rangle \tag{13.107}$$

and that

$$\langle \Sigma^0 | J_\mu{}^\gamma | \Lambda^0 \rangle = -\frac{\sqrt{3}}{2} \langle N | J_\mu{}^\gamma | N \rangle. \tag{13.108}$$

The latter matrix element determines the decay $\Sigma^0 \longrightarrow \Lambda^0 + \gamma$, which is known, experimentally, to occur at a rate

$$\tau_{\Sigma^0} < 1.0 \times 10^{-14} \text{ sec.} \tag{13.109}$$

Indeed, in general ($m_{\Sigma^0} \neq m_{\Lambda^0}$),

$$\langle \Lambda^0 | J_\mu{}^\gamma | \Sigma^0 \rangle = i\bar{u}(\Lambda^0)[\gamma_\mu F_1{}^{\Sigma\Lambda}(q^2) + \sigma_{\mu\nu} q_\nu F_2{}^{\Sigma\Lambda}(q^2) + q_\mu F_3{}^{\Sigma\Lambda}(q^2)]u(\Sigma^0). \tag{13.110}$$

For real photon emission,

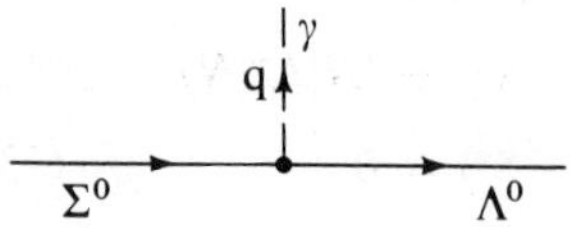

we have $q^2 = 0$, and current conservation then implies that[29]

$$F_1{}^{\Sigma\Lambda}(0) = 0. \tag{13.111}$$

Since the photon is real, the term proportional to $F_3{}^{\Sigma\Lambda}$ cannot contribute to $\Sigma^0 \longrightarrow \Lambda^0 + \gamma$ because of the transversality of the photon field to the photon momentum q_μ. Thus

$$\langle \Lambda^0 | J_\mu{}^\gamma | \Sigma^0 \rangle = i\bar{u}(\Lambda^0)\sigma_{\mu\nu} q_\nu F_2{}^{\Sigma\Lambda}(0) u(\Sigma^0). \tag{13.112}$$

In the good SU_3 approximation

$$F_2{}^{\Sigma\Lambda}(0) = -\frac{\sqrt{3}}{2} F_2{}^N(0), \tag{13.113}$$

---

28. A. H. Rosenfeld et al., *Rev. Mod. Phys.* **39**:1 (1967).
29. Multiply Eq. (13.110) by q_μ and set the result equal to zero. In writing Eq. (13.110) we assume Σ and Λ have even relative parity, which is known, experimentally, to be the case and must be the case if they are members of the same octet.
30. R. van Royen and V. F. Weisskopf (to be published). In this evaluation one uses the physical masses in the phase space.
31. A. J. MacFarlane and E. G. C. Sudarshan, *Nuovo Cimento* **31**:1176 (1964), and R. H.

and the rate is fixed to be[30]

$$\tau_{\Sigma^0} = 0.7 \times 10^{-17} \text{ sec.} \tag{13.114}$$

We can extract another interesting application of Eqs. (13.87) and (13.88) alone (the conservation of U-spin). Combining the two equations we learn that

$$\sqrt{3}\,\langle\Sigma^0|F(J^\gamma)|\Lambda^0\rangle = \langle\Sigma^0|F(J^\gamma)|\Sigma^0\rangle - \langle N|F(J^\gamma)|N\rangle, \tag{13.115}$$

so that (using Eq. (13.67)) and taking $F(J^\gamma) = J_\mu{}^\gamma J_\nu{}^\gamma$ we have

$$\sqrt{3}\delta(\Lambda^0\Sigma^0) = (\delta m_{\Sigma^0} - \delta m_{\Sigma^+}) + (\delta m_P - \delta m_N), \tag{13.116}$$

which gives[31]

$$\delta(\Lambda^0\Sigma^0) \simeq 1 \text{ MeV}. \tag{13.117}$$

The quantity $\delta(\Lambda^0\Sigma^0)$ has the physical interpretation of a "mixing parameter" measuring the magnitude of the transition

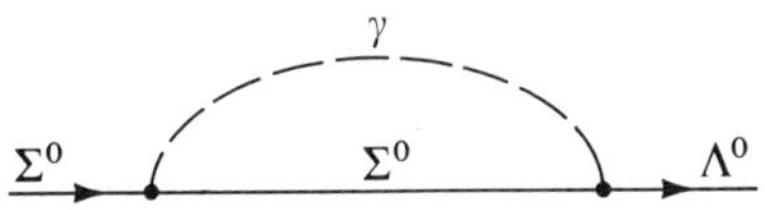

which can occur virtually to order α. Because of it the physical Σ^0 and Λ^0 will be mixtures of the "bare" Σ^0 and Λ^0; that is, mixtures of the Λ^0 and Σ^0 eigenfunctions of the strong interaction Hamiltonian, with the electromagnetic field turned off. Indeed, to order α,[32]

$$\begin{aligned} \Lambda^0{}_{\text{phys}} &= \Lambda^0 + \frac{\delta(\Sigma^0\Lambda^0)\Sigma^0}{(E_{\Lambda^0} - E_{\Sigma^0})}, \\ \Sigma^0{}_{\text{phys}} &= \Sigma^0 - \frac{\delta(\Sigma^0\Lambda^0)}{(E_{\Lambda^0} - E_{\Sigma^0})}\Lambda^0. \end{aligned} \tag{13.118}$$

This mixing would show up as a breakdown of isotopic spin symmetry in strong interactions involving Λ^0's and Σ^0's. An effect of this kind is reported[32] in the nonequality of the binding energies of the mirror hypernuclei $_\Lambda He^4$ and $_\Lambda H^4$.

We may summarize what we have learned about the electromagnetic properties of the baryons in the good SU_3 limit by the statement that, in this limit, *all* of the one-baryon matrix elements of $J_\mu{}^\gamma$ are determined by *four* real form factors $F_1{}^N$, $F_1{}^P$, $F_2{}^N$, $F_2{}^P$, or, equivalently, $F_1{}^S$, $F_1{}^V$, $F_2{}^S$, $F_2{}^V$, providing that

$$J_\mu{}^\gamma = V_{\mu 3} + \frac{1}{\sqrt{3}} V_{\mu 8}, \tag{13.119}$$

Dalitz and F. Von Hippel, *Phys. Letters* **10**:153 (1964). We can also derive, for the pseudoscalar octet,

$$\frac{1}{\sqrt{3}}\delta(\pi^0\eta^0) = (\delta m_{\pi^0} - \delta m_{\pi^+}) + (\delta m_{K^+} - \delta m_{K^0}).$$

32. See R. H. Dalitz, *Les Houches Lectures* 1965, p. 203 (op. cit.) for a more complete discussion with references to the experimental literature.

so that J_μ^γ transforms like an octet. We would now like to give an alternative, although equivalent, discussion of this observation, based on a generalization of the Wigner-Eckart theorem, which fixes the number of independent matrix elements in situations where the ordinary rotation group and the conservation of angular momentum express the symmetries of the system. The simplest way to deal with this question in SU_3 is to introduce the 3×3 matrices,

$$B = \frac{1}{\sqrt{2}} \sum_{i=1}^{8} \lambda_i \psi_i \tag{13.120}$$

and

$$M = \frac{1}{\sqrt{2}} \sum_{i=1}^{8} \lambda_i \phi_i, \tag{13.121}$$

where the ϕ_i are the meson fields and the ψ_i are the eight baryon fields in the octet. B and M are 3×3 traceless matrices and are given explicitly in terms of the particles in the multiplet by

$$M = \begin{pmatrix} \frac{1}{\sqrt{2}}\pi^0 + \frac{1}{\sqrt{6}}\eta^0 & \pi^+ & K^+ \\ \pi^- & -\frac{1}{\sqrt{2}}\pi^0 + \frac{1}{\sqrt{6}}\eta^0 & K^0 \\ K^- & \bar{K}^0 & -\frac{2}{\sqrt{6}}\eta^0 \end{pmatrix},$$

$$B = \begin{pmatrix} \frac{1}{\sqrt{2}}\Sigma^0 + \frac{1}{\sqrt{6}}\Lambda^0 & \Sigma^+ & P \\ \Sigma^- & -\frac{1}{\sqrt{2}}\Sigma^0 + \frac{1}{\sqrt{6}}\Lambda^0 & N \\ -\Xi^- & \Xi^0 & -\frac{2}{\sqrt{6}}\Lambda^0 \end{pmatrix}. \tag{13.122}$$

The conjugate matrix $\bar{B}$ represents the antibaryons

$$\bar{B} = \begin{pmatrix} \frac{1}{\sqrt{2}}\bar{\Sigma}^0 + \frac{1}{\sqrt{6}}\bar{\Lambda}^0 & \bar{\Sigma}^- & -\bar{\Xi}^- \\ \bar{\Sigma}^+ & -\frac{1}{\sqrt{2}}\bar{\Sigma}^0 + \frac{1}{\sqrt{6}}\bar{\Lambda}^0 & \bar{\Xi}^0 \\ \bar{P} & \bar{N} & -\frac{2}{\sqrt{6}}\bar{\Lambda}^0 \end{pmatrix}. \tag{13.123}$$

To every unitary SU_3 transformation, on fields in the octet, there corresponds a 3×3 unitary, unimodular transformation u on the matrices B and $\bar{B}$, such that

$$\begin{aligned} B &\longrightarrow uBu^{-1}, \\ \bar{B} &\longrightarrow u\bar{B}u^{-1}. \end{aligned} \tag{13.124}$$

We can use this correspondence to find combinations of the fields that are invariant under SU_3, and thus to construct SU_3-invariant Lagrangians. We exploit some elementary properties of the traces of finite matrices—in particular, the fact that for any finite matrices A and B,

$$\mathrm{Tr}\,(AB) = \mathrm{Tr}\,(BA). \tag{13.125}$$

Thus, from Eqs. (13.122) and (13.123),

$$\mathrm{Tr}\,(\bar{B}B) = \mathrm{Tr}\,(u\bar{B}u^{-1}uBu^{-1}) \tag{13.126}$$

is an SU_3-invariant. Therefore, with m the octet baryon mass and μ the octet meson mass,

$$\mathfrak{L}^0 = -\mathrm{Tr}\,[\bar{B}(\partial_\mu\gamma_\mu + m)B] - \tfrac{1}{2}\,\mathrm{Tr}\,[(\partial_\mu M)^2 + \mu^2 M^2] \tag{13.127}$$

is the SU_3-invariant free particle Lagrangian expressed in terms of the observed baryons and mesons. If the quark picture turns out to be correct, then this Lagrangian would be a phenomenological description of the symmetries of the bound quark states rather than the "fundamental" Lagrangian of the theory. We can also write down SU_3-invariant Yukawa couplings in this notation. These couplings will involve M, linearly, and, B and $\bar{B}$, bilinearly. There are in fact *two* such SU_3-invariant Yukawa forms which can be written either as[33]

$$\mathrm{Tr}\,[\bar{B}\gamma_5 BM] \qquad \text{and} \qquad \mathrm{Tr}\,[\bar{B}\gamma_5 MB]$$

or, using the fact that

$$\mathrm{Tr}\,[ABC] = \mathrm{Tr}\,[CAB], \tag{13.128}$$

as

$$\mathrm{Tr}\,[[\overline{B}, \gamma_5 B]_- M] \qquad \text{and} \qquad \mathrm{Tr}\,[[\overline{B}, \gamma_5 B]_+ M].$$

The commutator coupling is known as F-type coupling and the anticommutator type is known as D-type coupling. It is now straightforward to introduce the electromagnetic field in an SU_3- and gauge-invariant way. We call the traceless 3×3 charge matrix Q,

$$Q = \lambda_3 + \frac{1}{\sqrt{3}}\lambda_8 = \frac{1}{3}\begin{pmatrix} 2 & 0 & 0 \\ 0 & -1 & 0 \\ 0 & 0 & -1 \end{pmatrix}, \tag{13.129}$$

and make the transformation

$$\partial_\mu \longrightarrow \partial_\mu - ieQA_\mu \tag{13.130}$$

in $\mathfrak{L}^0$. (For simplicity we will assume that $\mathfrak{L}'$, the coupling term, contains no derivatives.) Since Q does not commute with B, $\overline{B}$, or M, we should, strictly speaking, symmetrize $\mathfrak{L}^0$ with respect to B and $\overline{B}$ and then make the transformation above. A computation shows[34] that, applying Eq. (13.130) to $\mathfrak{L}^0$,

---

33. Considered as 3×3 matrices, M and B do not commute.
34. S. Coleman and S. L. Glashow, op. cit.

$$\mathcal{L}_{\rm em} = ieA_\mu \, \mathrm{Tr}\, [\overline{B}\gamma_\mu[B, Q]_- - \partial_\mu M[M, Q]_-] \\ + \tfrac{1}{2}e^2A^2 \, \mathrm{Tr}\, [[M, Q]_-[M, Q]_-]. \quad (13.131)$$

We are interested in the one-particle matrix elements of the electric current $J_\mu{}^\gamma$; that is, $\delta\mathcal{L}_{\rm em}/\delta A_\mu$. We note that $\langle \overline{B}|J_\mu{}^\gamma|B\rangle$ is a Lorentz four-vector, SU_3-invariant function that is linear and homogeneous in Q. Suppressing the Lorentz index, for a moment, there are two and only two such possibilities: Tr $[\overline{B}BQ]$ and Tr $[\overline{B}QB]$. Thus

$$\langle \overline{B}|J_\mu{}^\gamma|B\rangle = i[F_1(q^2)\, \mathrm{Tr}\, [\overline{B}\gamma_\mu QB] + F_1'(q^2)\, \mathrm{Tr}\, [\overline{B}\gamma_\mu BQ] \\ + F_2(q^2)q_\nu \, \mathrm{Tr}\, [\overline{B}\sigma_{\mu\nu}QB] + F_2'(q^2)q_\nu \, \mathrm{Tr}\, [\overline{B}\sigma_{\mu\nu}BQ]], \quad (13.132)$$

which again shows that four form factors are sufficient to describe the mass-shell electromagnetic properties of the baryon octet. Carrying out the details of the trace calculation leads, again, to the formulas we have given above connecting the various electromagnetic phenomena.

It is natural to wonder whether there is any way of taking the calculation one step further so as to express all of the baryonic quantities in terms of *two* form factors, say those of the proton. This subject has an extensive and vexed history which the interested reader may find reviewed in the article of Dalitz[35] already cited. What appears to have survived this discussion is the idea that under certain assumptions about quark binding one may relate the electromagnetic properties of neutron and proton. In particular, if one assumes that the quarks are Fermions (one might, more generally, assume that the quarks, although spin $\frac{1}{2}$, did not obey simple Fermi statistics but rather "parastatistics" [36]), then the "ground state" qqq system corresponding to the baryons would be totally antisymmetric in all the coordinates. The simplest such state corresponding to $l = 0$ has the curious form (1, 2, 3 are the labels for the three quarks)

$$\psi_{\rm space} = (\mathbf{r}_1{}^2 - \mathbf{r}_2{}^2)(\mathbf{r}_2{}^2 - \mathbf{r}_3{}^2)(\mathbf{r}_3{}^2 - \mathbf{r}_1{}^2)\phi_{\rm sym}(\mathbf{r}_1, \mathbf{r}_2, \mathbf{r}_3), \quad (13.133)$$

where $\phi_{\rm sym}$ is a symmetric function of 1, 2, 3. This is to be multiplied by a function of the spins and quark labels that is symmetric in 1, 2, 3. The explicit form of this function is given by Dalitz[37] and will not concern us. By attributing to each quark a total magnetic moment equal to the protonic moment, $\mu_P = 2.79(e/2M_P)$ and supposing that it is legitimate to evaluate the neutron and proton moments by simply attributing them, additively, to the moments of the underlying quarks, one may derive the statement that

35. R. H. Dalitz, *Les Houches* Summer School, op. cit. and also *Symmetries and the Strong Interactions*, International Conference on High Energy Physics, at Berkeley, September 1966.
36. O. W. Greenberg and A. Messiah, *Phys. Rev.* **138B**:1155 (1965).
37. Op. cit.

$$\frac{\mu_P}{\mu_N} = -\frac{3}{2}, \tag{13.134}$$

which is in astonishingly good agreement with the empirical ratio of -1.46. This along with all of the other successes of the naive quark model, some of which we will return to in the next chapter, leads one to believe that there must be something right about it. But, at this writing, it is difficult to give a fundamental explanation of what this is.

We turn, in the next chapter, to a discussion of "broken SU_3" and the vector mesons.

14
More on SU_3

We can summarize the gist of what we have learned in the last chapter as follows: if we label the three quarks q_α and the three antiquarks $\bar{q}_\alpha \cdot\equiv\cdot q^\alpha$, then we can manufacture, at least symbolically, all of the pseudoscalar mesons and baryons by putting q_α and q^α together to form "irreducible" tensors—tensors of lowest rank that transform according to a definite representation of SU_3. Two irreducible tensors can be formed from q_α and q^β alone:

$$S = q^\alpha q_\alpha \tag{14.1}$$

and

$$T_\beta{}^\alpha = q^\alpha q_\beta - \tfrac{1}{3}\delta_\beta{}^\alpha S. \tag{14.2}$$

Clearly, S transforms like an SU_3 singlet, and represents the x^0 meson, and $T_\beta{}^\alpha$ transforms like an octet. $T_\beta{}^\alpha$ is manifestly traceless and this representation is essentially equivalent to the introduction of the matrix M in the last chapter; Eq. (13.122). The object $\delta_\beta{}^\alpha$ is an SU_3-invariant tensor since under all elements of the group

$$\delta_\beta{}^\alpha \longrightarrow \delta_\beta{}^\alpha. \tag{14.3}$$

Both $\epsilon_{\alpha\beta\gamma}$ and $\epsilon^{\alpha\beta\gamma} = \delta_{\alpha'}{}^\alpha\delta_{\beta'}{}^\beta\delta_{\gamma'}{}^\gamma\epsilon_{\alpha'\beta'\gamma'}$ are also SU_3-invariant tensors. A simple calculation shows that under any u in SU_3

$$\epsilon_{\alpha\beta\gamma} \longrightarrow \det u\,\epsilon_{\alpha\beta\gamma}. \tag{14.4}$$

But, by definition $\det u = 1$. The quantity

$$T_1{}^1 = \tfrac{2}{3}q^1q_1 - \tfrac{1}{3}(q^2q_2 + q^3q_3) \tag{14.5}$$

transforms under SU_3 like the matrix

$$Q = \frac{1}{3}\begin{pmatrix} 2 & 0 & 0 \\ 0 & -1 & 0 \\ 0 & 0 & -1 \end{pmatrix}, \tag{14.6}$$

like the electric charge or the electric current. The baryon octet is obtained by taking $T_{\alpha\beta\gamma} = q_\alpha q_\beta q_\gamma$ and reducing it by contracting indices,

$$T_\beta{}^\alpha = \epsilon^{\alpha\lambda\mu}T_{\beta\lambda\mu}, \tag{14.7}$$

and then letting $B_\beta{}^\alpha$, the baryon octet, be defined as the traceless tensor

$$B_\beta{}^\alpha = \epsilon^{\alpha\lambda\mu}T_{\beta\lambda\mu} - \tfrac{1}{3}\delta_\beta{}^\alpha\epsilon^{\mu\nu\lambda}T_{\mu\nu\lambda}. \tag{14.8}$$

The trace, $\epsilon^{\mu\nu\lambda}T_{\mu\nu\lambda}$, would correspond to a baryon singlet, if and when such a particle is found. In addition to the singlet and octet we can construct the ten-component, irreducible, totally symmetric tensor

$$B_{\alpha\beta\gamma} = \tfrac{1}{6}[T_{\alpha\beta\gamma} + T_{\beta\alpha\gamma} + T_{\beta\gamma\alpha} + T_{\gamma\beta\alpha} + T_{\gamma\alpha\beta} + T_{\alpha\gamma\beta}], \tag{14.9}$$

which corresponds to the famous SU_3 decimet (decouplet). It can be diagrammed as follows:

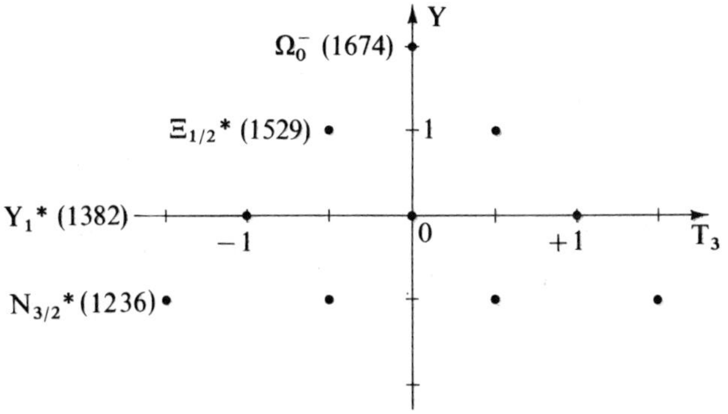

In the diagram the subscripts on the particle labels stand for the total isospin and the numbers in parentheses refer to the central masses in the mass spectrum corresponding to the fact that these "particles" are, in some cases, fairly broad resonances. Those particles in the decimet, whose spins have been measured, have $J = \frac{3}{2}$; and for the SU_3 interpretation to hold it is crucial that they all have $J = \frac{3}{2}$ and the same relative parity. The table below lists their important decays.

The Decimet Table

Particle	*T*	*Y*	*M (MeV)*	*Width (MeV)*	*Decay Mode*
N^*	$\frac{3}{2}$	−1	1236.0 ± 0.6	120 ± 2	πN
Y_1^*	1	0	1382.2 ± 0.9	37 ± 3 3.1 ± 1.5	$\pi\Lambda$ $\pi\Sigma$
Ξ^*	$\frac{1}{2}$	1	1528.9 ± 1.1	7.3 ± 1.7	$\pi\Xi$
Ω^-	0	2	1674	Long-lived	$\Xi\pi$ $\Lambda\bar{K}$

Source: Data taken from A. H. Rosenfeld et al., *Rev. Mod. Phys.* **39**:1 (1967).

There is an inequivalent 10* representation that can be manufactured from three antiquarks and is diagrammed by an inverted triangle, but, at this writing none of its members has been definitely identified. We can summarize the predictions as to the sorts of baryons that can be manufactured from three quarks in the group theoretic statement (a multiplication table for irreducible representations of SU_3):

$$3 \times 3 \times 3 = 1 + 8 + 8 + 10.$$

The philosophy of the quark model differs, in this respect, from that of the "eight-fold way" according to which the basic representation of the group, realized in nature, is the 8 and the pion-baryon resonances are then to be constructed by multiplying the 8 baryons and 8 mesons according to the formula

$$8 \times 8 = 1 + 8 + 8 + 10 + 10^* + 27,$$

which would predict the existence of a 27-fold representation, so far not seen. It is a nice feature of the quark scheme, that all, and only, those SU_3 multiplets that have actually been identified arise from multiplying quark-antiquark or three quarks in the simplest way.

The experimental study of the electromagnetic properties of the decimet is still in its infancy so that we shall only discuss one interesting prediction of SU_3 involving decays of the form $B_{10} \longrightarrow B_8 + \gamma$—that is, radiative decays, from the decimet to the baryon octet. In SU_3 these are given by the matrix element $\langle B_8|J_\mu{}^\gamma|B_{10}\rangle$, where $J_\mu{}^\gamma$ is assumed to transform like $T_1{}^1$. For the matrix element to be an SU_3-invariant we must combine B_8 and B_{10} to form an object that transforms like $T_1{}^1$. It turns out that this can only be done in one way since

$$8 \times 10 = 35 + 27 + 10 + 8.$$

The 8 occurs only once in the direct product of 8 with 10. In fact this combination is given by

$$T_1{}^1 = \epsilon_{1\lambda\mu} T^{1\lambda\sigma} B_\sigma{}^\mu. \tag{14.10}$$

This means that *all* of these decays are determined by the parameters measured in any one of them. A computation shows[1] that

$$\begin{aligned}\langle P|J^\gamma|N^{*+}\rangle &= -\langle\Sigma^+|J^\gamma|Y^{*+}\rangle = \langle N|J^\gamma|N^{*0}\rangle \\ &= 2\langle\Sigma^0|J^\gamma|Y^{*0}\rangle = \frac{2}{\sqrt{3}}\langle\Lambda^0|J^\gamma|Y^{*0}\rangle = \langle\Xi^0|J^\gamma|\Xi^{*0}\rangle. \end{aligned} \tag{14.11}$$

It also follows from Eq. (14.10) that all of the static electromagnetic properties within the decimet are fixed by any given matrix element. No experimental evidence is available on either of these matters. With special assumptions about the symmetry of the quark wave functions[2] and magnetic moments, one can show that for magnetic dipole transitions,

$$\langle P|\mu|N^{*+}\rangle = \frac{2\sqrt{2}}{3}\mu_P, \tag{14.12}$$

where μ_P is the proton moment in nucleon magnetons, which would fix the magnetic dipole transition rates for the whole decimet. We shall return later, briefly, to a few additional comments on the decimet.

The reader familiar with the recent developments in elementary particle

---

1. M. A. B. Beg, B. W. Lee, and A. Pais, *Phys. Rev. Letters* **13**:514 (1964).
2. Again, see R. H. Dalitz, *Les Houches Lectures*, op. cit.

physics may well wonder what has happened to the vector mesons ϕ, ω, ρ, etc., in our discussion of SU_3. There are nine such mesons at comparable masses whose properties we list below in a table.

Vector Meson Table (all of these objects are 1^-.)

Particle	*T*	T_3	*Y*	*M (MeV)*	Γ *(MeV)*	*Principal Decays*
ρ^+	1	+1	0	778	160	$\pi\pi \sim 100\%$
						$4\pi < 1\%$
ρ^0	1	0	0	770	140	$\pi\gamma < .4$
ρ^-	1	−1	0	778	160	$e^+e^- \sim .0065$
						$\mu^+\mu^- \sim .0033$
K^{*+}	$\frac{1}{2}$	$\frac{1}{2}$	1	892.4 ± 0.8	49.8 ± 1.7	$K\pi \sim 100\%$
K^{*0}	$\frac{1}{2}$	$-\frac{1}{2}$	1	892.4 ± 0.8	49.8 ± 1.7	$K\pi\pi < 0.2$
$\overline{K^{*0}}$	$\frac{1}{2}$	$\frac{1}{2}$	−1	$892.4 + 0.8$	49.8 ± 1.7	
K^*	$\frac{1}{2}$	$-\frac{1}{2}$	−1	892.4 ± 0.8	49.8 ± 1.7	
						$K_1K_2 \sim 40\%$
						$K^+K^- \sim 48\%$
ϕ	0	0	0	1018.6 ± 0.5	4.0 ± 1.0	$\pi\rho + 3\pi \sim 12\%$
						$\eta + \text{neut} < 12\%$
						$\mu^+ + \mu^- < 0.5\%$
						$e^+ + e^- < 0.4\%$
						$\pi^+\pi^-\pi^0 \sim 90\%$
						$\pi^+\pi^-$?
						$\pi^0\gamma$ $9.7 \pm .8$
ω^0	0	0	0	783.4 ± 0.1	11.9 ± 1.5	$\eta + \text{neut} < 1.5$
						$\pi^+\pi^-\gamma < 5$
						$e^+ + e^- = 0.012 \pm .003$
						$\mu^+ + \mu^- < 0.10$

Source: The data in this table are taken from A. H. Rosenfeld et al., *Rev. Mod. Phys.* **39**:1 (1967), where references to the original experimental papers may be found. The fact that the ρ meson masses and widths are given without experimental errors reflects the fact that these quantities are not precisely known. The numbers given are averaged over a variety of different experiments.

These particles can be arranged, in the now familiar way, in an octet and singlet. This is done conventionally as follows:

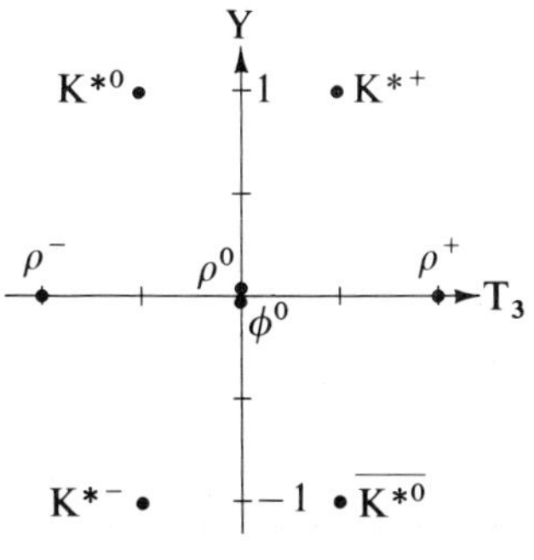

vector meson octet

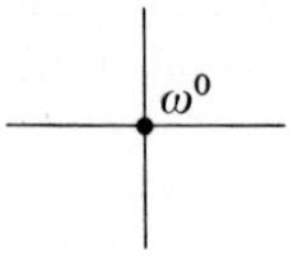

vector meson singlet

We can now also proceed to treat the vector mesons and their decays in terms of the quark model and SU_3, just as we treated the baryons and pseudoscalar mesons. (We can clearly define the vector mesons V_i to be the quark combinations $\bar{q}\lambda_i\gamma_\mu q$ for the octet and $\bar{q}\gamma_\mu q$ for the singlet.) We begin with the decays. It is energetically possible to have the decay $V_i \longrightarrow P_j + \gamma$; a one-photon decay of a vector meson into a pseudoscalar meson. Here V_i and P_j are members of their respective octets. There are also octet-singlet transitions, which we shall discuss shortly. The tensor notation is the simplest to use in making this study and we begin by proving two simple theorems.

1. If $A_\beta{}^\alpha$ is coupled to $\Gamma_\alpha{}^\beta$ and if $\mathrm{Tr}\,(\Gamma) = 0$, then, effectively, only the traceless part of A is coupled to Γ.

Proof.[3]

$$A_\beta{}^\alpha = (A_\beta{}^\alpha - \tfrac{1}{3}A_\lambda{}^\lambda\delta_\beta{}^\alpha) + \tfrac{1}{3}\delta_\beta{}^\alpha A_\lambda{}^\lambda. \tag{14.13}$$

Thus only $A_\beta{}^\alpha - \frac{1}{3}A_\lambda{}^\lambda\delta_\beta{}^\alpha$ has a nonvanishing coupling to $\Gamma_\alpha{}^\beta$.

2. If, as in the last chapter, we call

$$M = \begin{pmatrix} \dfrac{\pi^0}{\sqrt{2}} + \dfrac{\eta^0}{\sqrt{6}} & \pi^+ & K^+ \\ \pi^- & -\dfrac{\pi^0}{\sqrt{2}} + \dfrac{\eta^0}{\sqrt{6}} & K^0 \\ K^- & \overline{K^0} & -\dfrac{2\eta^0}{6} \end{pmatrix}, \tag{14.14}$$

then the 3×3 charge conjugation matrix $\mathscr{C}$ must (by inspection of M) have the property that

$$\mathscr{C}M\mathscr{C}^{-1} = M^T \tag{14.15}$$

(T is transpose.) Moreover, if we arrange the vector meson octet into the traceless matrix,

$$V = \begin{pmatrix} \dfrac{\rho^0}{\sqrt{2}} + \dfrac{\phi^0}{\sqrt{6}} & \rho^+ & K^{*+} \\ \rho^- & -\dfrac{\rho^0}{\sqrt{2}} + \dfrac{\phi^0}{\sqrt{6}} & K^{*0} \\ K^{*-} & \overline{K^{*0}} & -\dfrac{2\phi^0}{\sqrt{6}} \end{pmatrix}, \tag{14.16}$$

3. Here all indices run from 1 to 3 so that $\delta_\alpha{}^\alpha = 3$.

we define $\mathscr{C}$ such that

$$\mathscr{C}V\mathscr{C}^{-1} = -V^T. \tag{14.17}$$

This reflects the fact that ρ^0 is even under $G = e^{i\pi T_2}C$ ($\rho^0 \longrightarrow \pi^+ + \pi^-$ is allowed), and hence odd under C, while ϕ^0 is odd under G ($\phi \longrightarrow 3\pi$), and thus also odd under C. The charged particles go into their antiparticles under C and Eq. (14.17) defines the phase of this transformation. Thus under $\mathscr{C}$ the product[4] VM^T transforms as

$$\mathscr{C}VM^T\mathscr{C}^{-1} = -V^TM. \tag{14.18}$$

The photon field or, equivalently, the current to which it is coupled, transforms under SU_3 like a tensor $T_\beta{}^\alpha$, so that under $\mathscr{C}$ we must have

$$\mathscr{C}T\mathscr{C}^{-1} = -T. \tag{14.19}$$

Thus the product VM^TT becomes, under $\mathscr{C}$,

$$\mathscr{C}VM^TT\mathscr{C}^{-1} = V^TMT. \tag{14.20}$$

Hence

$$H_{\text{em}} = (VM^T + V^TM)T \tag{14.21}$$

is a $\mathscr{C}$-invariant vector meson-pseudoscalar photon coupling. Since Tr $(T) = 0$ and since Tr $(A) =$ Tr (A^T), the effective coupling (suppressing the Lorentz four-vector character) is of the form

$$H_{\text{em}} = (VM^T + V^TM - \tfrac{2}{3}I\,\text{Tr}\,(V^TM))T, \tag{14.22}$$

where I is the 3×3 unit matrix with Tr $(I) = 3$. Working out this expression and normalizing the $\rho^+ \longrightarrow \pi^+ + \gamma$ coefficient to unity we find that the vector mesons octet decays into the pseudoscalar octet (plus a photon) with the following relative coupling strengths:

$$\rho^+K^+ + K^{*+}K^+ + \rho^-\pi^- + K^{*-}K^- + \rho^0\pi^0 - \phi\eta^0 - 2K^{*0}K^0 - 2\overline{K^{*0}K^0}$$
$$+ \sqrt{3}\,\phi\pi^0 + \sqrt{3}\,\rho^0\eta^0. \tag{14.23}$$

Since the ω^0 is a U-spin singlet and since $J_\mu{}^\gamma$ cannot transfer U-spin, we must have

$$\langle\omega^0|J_\mu{}^\gamma|\sqrt{3}\,\eta^0 - \pi^0\rangle = 0 \tag{14.24}$$

or

$$\sqrt{3}\,\langle\omega^0|J_\mu{}^\gamma|\eta^0\rangle = \langle\omega^0|J_\mu{}^\gamma|\pi^0\rangle. \tag{14.25}$$

Energetically, the x^0 at 959 MeV can decay into the ρ^0 or the ω^0 plus a gamma. However, both the x^0 and ω^0 are unitary singlets and, if $J_\mu{}^\gamma$ transforms like an octet, as in the quark model, there can be no $x^0 \longrightarrow \omega^0 + \gamma$ transition in the good SU_3 limit. We have, on the other hand,

$$\langle x^0|J_\mu{}^\gamma|\sqrt{3}\,\phi - \rho^0\rangle = 0, \tag{14.26}$$

4. The coupling MV does not conserve charge.

so that

$$\sqrt{3}\,\langle x^0|J_\mu{}^\gamma|\phi\rangle = \langle x^0|J_\mu{}^\gamma|\rho^0\rangle. \tag{14.27}$$

The left side is the matrix element for $\phi \longrightarrow x^0 + \gamma$, which is also energetically possible.

In the naive quark model these SU_3 predictions can be supplemented by a specific calculation of the transition rate in terms of the quark electromagnetic properties.[5] In order to fit the baryon magnetic moments it is necessary, in the naive quark model, to give each quark a magnetic moment μ equal to the *total* proton moment $\mu_P = 2.79(e/2M_P)$. This means, by the way, if the usual electromagnetic ideas apply to quarks, that

$$\mu = g_q \frac{e}{2M_q} = \mu_P \frac{e}{2M_P}, \tag{14.28}$$

so that

$$g_q = 2.79 \frac{M_q}{M_P} \gtrsim 15, \tag{14.29}$$

since the lower experimental limit on quark masses appears to be in excess of 5 BeV. Thus, the quarks would have a very large g-value.

The vector and pseudoscalar states can be represented by quark wave functions.[6] For concreteness we consider the decay $\omega^0 \longrightarrow \pi^0 + \gamma$. The π^0 has the wave function (the second factor is ordinary spin):

$$\pi^0 = \frac{(q^1q_1 - q^2q_2)}{\sqrt{2}} \frac{(\uparrow\downarrow - \downarrow\uparrow)}{\sqrt{2}}, \tag{14.30}$$

and the $m = 0$ component of the ω^0 wave function has the form

$$\omega^0 = \frac{(q^1q_1 + q^2q_2 + q^3q_3)}{\sqrt{3}} \frac{(\uparrow\downarrow + \downarrow\uparrow)}{\sqrt{2}}. \tag{14.31}$$

This transition has the selection rule of an ordinary magnetic dipole transition induced by the spin flip of one of the quark spins. We can write (nonrelativistically) the magnetic dipole operator as

$$\frac{\mu_P}{e} \sum_{i=1}^{3} e_i \sigma_{zi} = M_z, \tag{14.32}$$

where the sum is over the three quarks weighted by their respective charges: $\frac{2}{3}, -\frac{1}{3}, -\frac{1}{3}$. Thus, putting in the factors, we have

---

5. See, for example, W. Thirring, *Phys. Letters* **16**:335 (1965); R. Van Royen and V. F. Weisskopf, op. cit.; and C. Becchi and R. Morpurgo, *Phys. Rev.* **140B**:687 (1965).
6. As is customary in this game, we do not exhibit the radial wave functions explicitly, and in taking the matrix elements we assume that the radial overlap integral is unity.

$$\langle \pi^0 | M_z | \omega^0 \rangle = \frac{2\mu_P}{\sqrt{6}}. \tag{14.33}$$

There is some ambiguity[7] on how this matrix element is to be used in view of the relativistic character of the emitted π^0. If the pion kinetic energy is neglected (the photon momentum k is taken to be small relative to the pion rest mass) then one finds that

$$\Gamma(\omega^0 \longrightarrow \pi^0 + \gamma) = \frac{2k^3\mu_P{}^2}{9\pi} = .78 \text{ MeV}. \tag{14.34}$$

The relativistic corrections tend to reduce this answer somewhat. The latest total width for the ω^0 is given as[8] 11.9 ± 1 MeV, while the branching ratio into $\pi^0 + \gamma$ is given experimentally as $9.7 \pm 0.890\%$, which perhaps can be regarded as a bit too high for complete agreement with this quark model calculation. (Shortly, we shall modify the calculation to include the effects of SU_3 symmetry breaking which increase the theoretical rate; Eq. (14.82).)

Up to this point we have systematically avoided the question of how accurate the predictions of SU_3 are meant to be. (This is, of course, a question that is distinct from estimating the probable reliability of the quark model, which could be zero if, in fact, there are no quarks.) With perfect SU_3 symmetry all of the particles within a given multiplet would have the same mass.[9] In fact a glance at the mass tables shows that for the pseudoscalar octet the masses range from 134.97 MeV (π^0) to 548.5 MeV (η^0), while for the vector mesons they range from 770 MeV (ρ^0) to 1019.5 MeV (ϕ). For the baryons the range for the octet is from 938.2 MeV (p) to 1329 ± 19 MeV (Ξ^0), while for the decimet the range is from 1236 MeV (N^*) to 1674 MeV (Ω^-). These mass splittings, taken at face value, show that SU_3 is indeed badly broken and cast doubt on the reliability of any of the predictions given by the theory. Fortunately it has turned out to be possible to give a partial theory of the symmetry breakings and, in some cases, to estimate their predicted magnitude. It is still something of a miracle that SU_3 holds as well as it appears to do.

The simplest way to begin the discussion of the symmetry breaking is, again, in terms of the quark model. In good SU_3 the three quarks belong to the threefold representation of SU_3 and have the same mass. The most straightforward way to break the symmetry is to suppose that the quarks have, in fact, distinct masses. Since q_1 and q_2 are members of an isodoublet it is natural to assume that

$$m_1 = m_2 = m, \tag{14.35}$$

and to break the symmetry by supposing that q_3, the "strange" quark, has a distinct mass $m_3 \neq m$. If we define an "average" quark mass $\overline{m}$ to be

---

7. For a careful discussion, see R. H. Dalitz, *Les Houches Lectures*, op. cit.
8. A. H. Rosenfeld et al., *Rev. Mod. Phys.* **39**:1 (1967).
9. The argument for this follows exactly the lines of the corresponding argument already given for the SU_2 multiplets.

$$\bar{m} = \frac{2m + m_3}{3}, \tag{14.36}$$

we can write the mass term in the quark Lagrangian in the form

$$\begin{aligned}\mathcal{L}_m &= m(q^1q_1 + q^2q_2) + m_3q^3q_3 = mq^\alpha q_\alpha + (m_3 - m)q^3q_3\\ &= \bar{m}q^\alpha q_\alpha + \tfrac{3}{2}(m_3 - \bar{m})(q^3q_3 - \tfrac{1}{3}q^\alpha q_\alpha).\end{aligned} \tag{14.37}$$

If the rest of the Lagrangian is SU_3 symmetric—that is, the part that accounts for quark binding and the strong quark couplings—then the symmetry breaking term transforms like the $T_3{}^3$ component of the traceless tensor $T_\beta{}^\alpha$ (Eq. (14.2)). It is this feature of the symmetry breaking coupling that we assume to hold in the general case, quarks or no. It is easy to see that the tensor $T_3{}^3$ transforms under SU_3 like λ_8

$$\lambda_8 = \frac{1}{\sqrt{3}}\begin{pmatrix}1 & 0 & 0\\ 0 & 1 & 0\\ 0 & 0 & -2\end{pmatrix}, \tag{14.38}$$

which means it is invariant under all SU_3 transformations generated by Q_3 and Q_8, the third component of isospin and the hypercharge, respectively.[10] This means that $T_3{}^3$ cannot transfer T, T_3, or Y when its matrix elements are taken. Moreover, if we recall from the last chapter that the third component of U-spin, U_3, is given as

$$U_3 = \tfrac{3}{4}Y - \frac{T_3}{2} = \frac{\sqrt{3}}{2}\,Q_8 - Q_3, \tag{14.39}$$

and the combination (electric charge)

$$U_S = Q_3 + \frac{1}{\sqrt{3}}\,Q_8 \tag{14.40}$$

transforms like a U-spin scalar, we see that

$$Q_8 = \frac{\sqrt{3}}{2}\,(U_S + U_3) \tag{14.41}$$

transforms like a linear combination of a U-spin scalar and the third component of a U-spin vector, and, hence so does $T_3{}^3$. This is all we need to know to derive the celebrated Gell-Mann-Okubo mass-splitting formulas.[11] We consider the U-spin baryon triplet $(N, \tfrac{1}{2}(\sqrt{3}\,\Lambda^0 - \Sigma^0), \Xi^0)$. From the transformation properties of U_S and U_3 it follows that

$$\begin{aligned}&\langle N|H_8|N\rangle = m_S + m_V,\\ &\tfrac{1}{4}\langle\sqrt{3}\,\Lambda^0 - \Sigma^0|H_8|\sqrt{3}\,\Lambda^0 - \Sigma^0\rangle = \tfrac{3}{4}\langle\Lambda^0|H_8|\Lambda^0\rangle + \tfrac{1}{4}\langle\Sigma^0|H_8|\Sigma^0\rangle = m_S,\end{aligned} \tag{14.42}$$

10. The total isospin, which transforms like $\lambda_1{}^2 + \lambda_2{}^2 + \lambda_3{}^2$, also commutes with λ_8.
11. M. Gell-Mann, Cal. Tech. Report CTSL-20 (1961) (unpublished), and S. Okubo, *Prog.*

and

$$\langle \Xi^0 | H_8 | \Xi^0 \rangle = m_S - m_V. \tag{14.43}$$

(We call the symmetry breaking Hamiltonian H_8.) We have called the contributions to the mass from U_S and U_3 respectively, m_S and m_V and, in the middle equation, we have used the fact that H_8 does not transfer total isotopic spin in order to eliminate the matrix elements $\langle \Lambda^0 | H_8 | \Sigma^0 \rangle$. Thus

$$\frac{\langle N | H_8 | N \rangle}{2} + \frac{\langle \Xi^0 | H_8 | \Xi^0 \rangle}{2} = \tfrac{3}{4} \langle \Lambda^0 | H_8 | \Lambda^0 \rangle + \tfrac{1}{4} \langle \Sigma^0 | H_8 | \Sigma^0 \rangle. \tag{14.44}$$

To interpret this result we must add to each matrix element of H_8 the mass M_0 which would characterize the entire unsplit octet in the limit of perfect SU_3 symmetry. We can add M_0 to each side of Eq. (14.44) in the form

$$\frac{M_0}{2} + \frac{M_0}{2} = \tfrac{3}{4} M_0 + \tfrac{1}{4} M_0 \tag{14.45}$$

and then equate

$$m_i = M_0 + \langle A_i | H_8 | A_i \rangle \tag{14.46}$$

to the experimental mass. It is interesting to remark that not all of the $\langle A_i | H_8 | A_i \rangle$ can be positive. This is a consequence of the trace condition

$$\sum_{i=1}^{8} \langle A_i | H_8 | A_i \rangle = 0, \tag{14.47}$$

which, in view of the $\Delta T = 0$ selection rule obeyed by H_8, reduces to

$$\langle \Xi^0 | H_8 | \Xi^0 \rangle + \langle \Xi^- | H_8 | \Xi^- \rangle + \langle P | H_8 | P \rangle + \langle N | H_8 | N \rangle$$
$$+ \langle \Sigma^+ | H_8 | \Sigma^+ \rangle + \langle \Lambda^0 | H_8 | \Lambda^0 \rangle + \langle \Sigma^0 | H_8 | \Sigma^0 \rangle + \langle \Sigma^- | H_8 | \Sigma^- \rangle = 0. \tag{14.48}$$

The quantity

$$\delta = (2m_N + 2m_{\Xi^0} - 3m_{\Lambda^0} - m_{\Sigma^0}) \simeq -26 \text{ MeV} \tag{14.49}$$

which would vanish if the mass splittings were given correctly by lowest order perturbation theory, gives some sort of measure of the size of H_8. In fact, it is possible to compute the mass splitting to the next order in H_8.[12] This is given in terms of the following parametric relation

$$M_8(T, Y) = M_0 + a_1 Y + b_1 \left(T(T+1) - \frac{Y^2}{4} \right) + Y^2 \delta \tag{14.50}$$

where the three real parameters M_0, a_1, b_1, are determined empirically to be

---

Theor. Phys. **27**:949 (1962). The derivation given here follows that of H. J. Lipkin, Argonne National Laboratory report (unpublished) (1963).

12. S. Okubo, op. cit., and M. A. Rashid and I. I. Yamanaka, *Phys. Rev.* **131**:2797 (1963).

$$M_0 = 1115 \text{ MeV},$$

$$a_1 \simeq -190 \text{ MeV},$$

$$b_1 \simeq 30 \text{ MeV},$$

and δ is defined as above. The lowest-order mass formula (14.44) is equivalent to simply taking $\delta = 0$ in Eq. (14.50). Thus $\frac{1}{4}\delta \sim -6.5$ MeV, as compared to a_1 or b_1, is a measure of $H_8{}^2$. Since, empirically, $\frac{1}{4}\delta$ is about the size of a typical electromagnetic mass splitting (characterized by $\alpha \sim \frac{1}{137}$), we can argue that H_8 involves a coupling of order $\frac{1}{10}$. Thus, despite the large mass splittings within the multiplets, the symmetry breaking term may be only 10%, or less, of the strong SU_3 symmetric coupling.

Using identical techniques a boson mass formula can also be derived. Noting that the K mesons are degenerate in mass (neglecting electromagnetic mass splittings) and using, as is customary, the (mass)^2 for bosons, one has

$$m_{\eta^0}{}^2 = \frac{(4m_K{}^2 - m_\pi{}^2)}{3} \tag{14.51}$$

which predicts $m_{\eta^0} = 566.5$ MeV as compared to the experimental value of 548.5 MeV, again showing that the mass-splitting is well accounted for by a "semi-strong" coupling. (We return later to a discussion of why the formula is not expected to give better agreement.) For the vector mesons one also has

$$m_{\phi^0}{}^2 = \frac{(4m_{K^*}{}^2 - m_\rho{}^2)}{3} = m_{\phi^0}{}^2 \simeq (928.1 \text{ MeV})^2, \tag{14.52}$$

which disagrees rather strongly with the experimental value $m_\phi = 1018.6$ MeV.

The way out of this *cul de sac* was actually suggested in Gell-Mann's first published paper on SU_3.[13] The ϕ^0 meson at 1018.6 MeV and the ω^0 meson at 783.4 Mev have the same quantum numbers $Y = T = 0$. Thus H_8, which cannot transfer these quantum numbers, *can* cause transitions between these states. In fact, in the absence of H_8 we can imagine that these states are degenerate in mass. From this point of view it is essentially arbitrary which of the two, ϕ^0 or ω^0, we call the SU_3 singlet and which we identify with the $Y = T = 0$ member of the octet. (In fact, in Gell-Mann's paper it was the ω^0 that was taken to be in the octet while the ϕ^0, which Gell-Mann called B^0, and had not yet been seen, was made the singlet.) The "physical particles" are neither ϕ^0 nor ω^0 but a linear combination ϕ^P, ω^P, which, if the states are normalized, we can write in the form[14]

---

13. M. Gell-Mann, *Phys. Rev.* **125**:1067 (1962). For a detailed working out of this idea, see J. J. Sakurai, *Phys. Rev. Letters* **9**:472 (1962), S. L. Glashaw, *Phys. Rev. Letters* **11**:48 (1963), and R. F. Dashen and D. H. Sharp, *Phys. Rev.* **133**:1585 (1964).
14. It is not completely obvious that the mixing parameters must be real. As will emerge shortly, this is a consequence of the *CP* invariance of H_8.

$$\begin{aligned}\omega^P &= \omega^0 \cos\theta_V + \phi^0 \sin\theta_V,\\ \phi^P &= \phi^0 \cos\theta_V - \omega^0 \sin\theta_V.\end{aligned} \tag{14.53}$$

Thus the physical particles ϕ^P and ω^P do not belong to a given representation of SU_3 and we must modify some of the analysis given above.

In the first place we can determine, at least up to a sign, the mixing angle, θ_V, in terms of the known masses as follows. The mixed states, and their energies, are found by solving the 2×2 "mass-matrix" problem (for the squared masses):

$$m_\pm{}^2 \begin{pmatrix} a_\pm \\ b_\pm \end{pmatrix} = \begin{pmatrix} M_8{}^2 + H_{88} & H_{81} \\ H_{81} & M_1{}^2 + H_{11} \end{pmatrix} \begin{pmatrix} a_\pm \\ b_\pm \end{pmatrix}. \tag{14.54}$$

Here $m_\pm$ are to be identified with the observed masses of ϕ^P and ω^P; the $a^\pm$, $b^\pm$ we will shortly identify with the $\cos\theta_V$ and $\sin\theta_V$, while H_{ij} are the symmetry-breaking matrix elements taken between ω^0 and ϕ^0, both diagonal and non-diagonal as indicated by the subscript indices. We have, implicitly, assumed the time-reversal invariance of the symmetry-breaking interaction in equating the two off-diagonal matrix elements.[15] The solution to this equation is

$$m_\pm{}^2 = \frac{(M_8{}^2 + H_{88})}{2} + \frac{(M_1{}^2 + H_{11})}{2} \pm \sqrt{\left[\frac{(M_8{}^2 + H_{88}) - (M_1{}^2 + H_{11})}{2}\right]^2 + H_{18}{}^2}. \tag{14.55}$$

It is natural to make the identifications

$$\begin{aligned} m_{\omega^0}{}^2 &= M_1{}^2 + H_{11},\\ m_{\phi^0}{}^2 &= M_8{}^2 + H_{88},\end{aligned} \tag{14.56}$$

and, since $m_+ > m_-$,

$$\begin{aligned} m_+{}^2 &= m_{\phi^P}{}^2,\\ m_-{}^2 &= m_{\omega^P}{}^2.\end{aligned} \tag{14.57}$$

We take m^{ϕ^0} from Eq. (14.52) so that

$$M_8{}^2 + H_{88} = (928.1 \text{ MeV})^2. \tag{14.58}$$

Clearly

$$m_+{}^2 + m_-{}^2 = m_{\omega^P}{}^2 + m_{\phi^P}{}^2 = m_{\phi^0}{}^2 + m_{\omega^0}{}^2. \tag{14.59}$$

This equation determines

$$M_1{}^2 + H_{11} = m_{\omega^0}{}^2 = (889.2 \text{ MeV})^2. \tag{14.60}$$

---

15. One of the recent attempts to account for the CP violation in K^0 decays involved assuming that the couplings that break SU_3 (that is, H_8) also violate CP or T. For a discussion, see J. Prentki, Oxford Conference Proceedings, op. cit., p. 47.

Now we can use Eq. (14.55) and the known parameters to compute $|H_{81}|$. (We cannot determine the sign of H_{81} this way, so that θ_V is not completely fixed.) Thus

$$\frac{(m_+{}^2 - m_-{}^2)^2}{4} = \frac{(m_{\phi^0}{}^2 - m_{\omega^0}{}^2)^2}{4} + H_{81}{}^2 \tag{14.61}$$

or

$$\begin{aligned} H_{81} &= \pm\tfrac{1}{2}\sqrt{(m_{\phi^P}{}^2 - m_{\omega^P}{}^2)^2 - (m_{\phi^0}{}^2 - m_{\omega^0}{}^2)^2} \\ &= \pm\frac{(m_{\omega^P} + m_{\phi^P})}{2}\sqrt{(m_{\phi^P} - m_{\omega^P})^2 - \frac{(m_{\phi^0}{}^2 - m_{\omega^0}{}^2)^2}{(m_{\phi^P} + m_{\omega^P})^2}} \\ &= \pm(m_{\omega^P} + m_{\phi^P})(108.5 \text{ MeV}). \end{aligned} \tag{14.62}$$

The states ϕ^P and ω^P are also readily constructed from Eq. (14.54). Indeed,

$$\frac{a_\pm}{b_\pm} = \frac{H_{81}}{m_\pm{}^2 - m_{\phi^0}{}^2} = \frac{m_\pm{}^2 - m_{\omega^0}{}^2}{H_{81}}, \tag{14.63}$$

which is to say, for example, that (remember the + components go with ϕ^0) using Eq. (14.53)

$$\frac{a_+}{b_+} = -\cot\theta_V = \frac{m_{\phi^P}{}^2 - m_{\omega^0}{}^2}{H_{81}}, \tag{14.64}$$

which gives, from Eq. (14.62) and Eq. (14.60),

$$\theta_V = \pm 40°. \tag{14.65}$$

The naive quark model gives us an insight into this result as follows.[16] The wave functions associated with ϕ^0 and ω^0 are

$$\begin{aligned} \phi^0 &= \frac{q^1q_1 + q^2q_2 - 2q^3q_3}{\sqrt{6}}, \\ \omega^0 &= \frac{(q^1q_1 + q^2q_2 + q^3q_3)}{\sqrt{3}}. \end{aligned} \tag{14.66}$$

We will suppose, for simplicity, that the potential, U, that binds the quarks together is independent of which quark it is that is being bound. Thus

$$\begin{aligned} m_{\phi^0} &= \tfrac{1}{6}(12m + 8\Delta) = \mu + \tfrac{4}{3}\Delta, \\ m_{\omega^0} &= \tfrac{1}{3}(6m + 2\Delta) = \mu + \tfrac{2}{3}\Delta, \end{aligned} \tag{14.67}$$

where Δ is the mass difference between q_1 or q_2 and q_3, and μ is the common mass that ϕ^0 and ω^0 would have if q_1, q_2, and q_3 were not split. The transition element can be computed by recalling that, in the quark model, Eq. (14.37) reads

$$H_8 = \Delta q^3 q_3, \tag{14.68}$$

16. Cf. R. H. Dalitz, *Les Houches Lectures*, op. cit.

so that H_{81} is proportional to[17] Δ. If we assume perfect overlap for the q_3 quark wave functions in ϕ^0 and ω^0, we then find that

$$H_{81} = -\frac{2\sqrt{2}}{3}\Delta, \tag{14.69}$$

which means that, using Eqs. (14.55) and (14.67) and the linear mass version of Eq. (14.55),[18]

$$m_\pm = \mu + \Delta \pm \sqrt{\Delta^2}, \tag{14.70}$$

so that in this model

$$\cot\theta_V = \frac{(\mu+\Delta) - (\mu + \frac{2}{3}\Delta)}{\frac{2}{3}\sqrt{2}\,\Delta} = \sqrt{2}, \tag{14.71}$$

and, picking the first quadrant for the angle,

$$\sin\theta_V = \frac{1}{\sqrt{3}} = \frac{1}{\sqrt{2}}\cos\theta_V. \tag{14.72}$$

We can compare θ_V, determined empirically, with this theoretical quark model angle, namely

$$(\tan\theta_V)_{\text{obs}} = \pm .83, \tag{14.73}$$

while

$$(\tan\theta_V)_{\text{quark}} = \frac{1}{\sqrt{2}} = .71, \qquad \theta_V \simeq 35°. \tag{14.74}$$

In this picture, the physical states with masses $m_\pm$ are given in terms of quarks by

$$\begin{aligned}\phi^P &= \sqrt{\frac{2}{3}}\left\{\frac{q^1q_1 + q^2q_2 - 2q^3q_3}{\sqrt{6}}\right\} - \sqrt{\frac{1}{3}}\left\{\frac{q^1q_1 + q^2q_2 + q^3q_3}{\sqrt{3}}\right\}\\ &= -q^3q_3,\end{aligned} \tag{14.75}$$

while

$$\begin{aligned}\omega^P &= \sqrt{\frac{2}{3}}\left\{\frac{q^1q_1 + q^2q_2 + q^3q_3}{\sqrt{3}}\right\} + \sqrt{\frac{1}{3}}\left\{\frac{q^1q_1 + q^2q_2 - 2q^3q_3}{\sqrt{6}}\right\}\\ &= \frac{1}{\sqrt{2}}[q^1q_1 + q^2q_2].\end{aligned} \tag{14.76}$$

Equation (14.69) also leads to a simple estimate of Δ, the q_1, q_3 mass difference:

$$m_{\phi^P} - m_{\omega^P} = 2\Delta \tag{14.77}$$

or

17. We use the *linear* vector meson mass formula, which for the vector mesons differs very little from the quadratic one so far as numerical results are concerned.
18. See R. H. Dalitz, op. cit.

$$\Delta \simeq 118 \text{ MeV}. \tag{14.78}$$

It is clear that the mixing phenomenon will modify the predictions for $\omega \longrightarrow \pi^0 + \gamma$ and $\phi \longrightarrow \pi^0 + \gamma$ given above. Indeed the decay of ω^P is given by the matrix element of M_z (Eq. (14.32)):

$$\langle \omega^P | M_z | \pi^0 \rangle = \langle \omega^0 | M_z | \pi^0 \rangle \cos \theta_V + \langle \phi^0 | M_z | \pi^0 \rangle \sin \theta_V. \tag{14.79}$$

Since ϕ^0 and ω^0 belong to different representations of SU_3, this is as far as one can get without invoking a model that relates the representations. In the naive quark model, using the ideas that lead to Eq. (14.33), we find

$$\langle \omega^P | M_z | \pi^0 \rangle = \mu_P \sqrt{\frac{2}{3}} \cos \theta_V + \frac{1}{\sqrt{3}} \mu_P \sin \theta_V. \tag{14.80}$$

Thus in the "ideal" quark model, $\tan \theta_V = 1/\sqrt{2}$,

$$\langle \omega^P | M_z | \pi^0 \rangle = \mu_P \tag{14.81}$$

and[19]

$$\Gamma(\omega^P \longrightarrow \pi^0 + \gamma) = 1.18 \text{ MeV}. \tag{14.82}$$

It is also easy to show in the quark model that

$$\langle \phi^P | M_z | \pi^0 \rangle = \sqrt{\tfrac{1}{3}} \cos \theta_V - \sqrt{\tfrac{2}{3}} \sin \theta_V, \tag{14.83}$$

so that in the "ideal" quark model

$$\langle \phi^P | M_z | \pi^0 \rangle = 0, \tag{14.84}$$

which is clear from Eq. (14.75) since, for this choice of angle, ϕ^P is made up only of strange quarks while π^0 has only nonstrange quarks.

An interesting test of many of these ideas and, in particular, of the ϕ, ω mixing will come when the ratio

$$\frac{\Gamma(\omega \longrightarrow e^+ + e^-)}{\Gamma(\phi \longrightarrow e^+ + e^-)}$$

becomes known. In lowest order in α, either of these processes is given by the diagram[20]

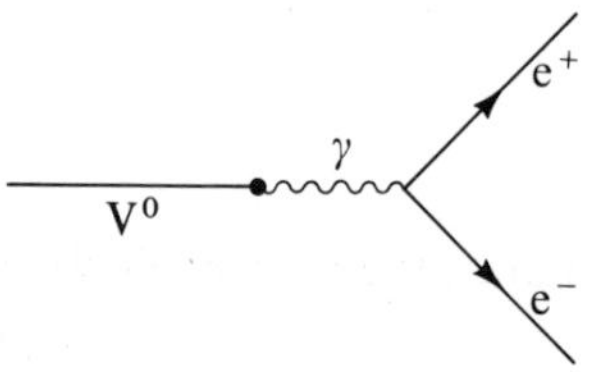

19. Compare with Eq. (14.34).
20. For spin-zero particles the corresponding diagram vanishes because it is proportional to

$$q_\mu(\bar{u}_{e^-}\gamma_\mu u_{e^+}) = 0.$$

21. For an unconventional but extremely interesting alternative treatment, see N. Kroll, T. D. Lee, and B. Zumino, *Phys. Rev.*, **157**:1376 (1967). We describe it briefly.

As we have argued in Chapter 6, the coupling between the photon and V^0, if treated in the conventional perturbation theory,[21] must be written in the form

$$e\lambda_V q^2 V_\mu A_\mu \tag{14.85}$$

with λ_V a dimensional constant (or, more generally, a form factor). The factor of q^2 is inserted to maintain the gauge invariance. In an arbitrary theory both ϕ^0 and ω^0 could couple directly to the photon. However, in SU_3 electrodynamics, assuming that $J_\mu{}^\gamma$ is composed of octet currents (no SU_3 scalar current), the matrix element

$$\langle 0|J_\mu{}^\gamma|\omega^0\rangle = 0, \tag{14.86}$$

since both $|\omega^0\rangle$ and $|0\rangle$ are, in good SU_3, invariant under the group, while $J_\mu{}^\gamma$ transforms like a "vector." In the mixing theory, Eq. (14.53), both ϕ^P and ω^P can couple to the photon with effective constants:

$$\begin{aligned} \lambda_{\omega^P} &= \lambda \sin \theta_V, \\ \lambda_{\phi^P} &= \lambda \cos \theta_V. \end{aligned} \tag{14.87}$$

The matrix element corresponding to Figure 4 is[22]

$$M = \frac{1}{\sqrt{2m_V}} e\lambda_V V_\mu \bar{u}_{e^-}\gamma_\mu u_{e^+}. \tag{14.88}$$

Working out the rates and taking the ratio,[23] one finds, theoretically,

$$\frac{\Gamma(\omega^P \longrightarrow e^+ + e^-)}{\Gamma(\phi^P \longrightarrow e^+ + e^-)} = \tan^2(\theta_V)\frac{m_{\omega^P}}{m_{\phi^P}}. \tag{14.89}$$

In the "ideal" quark model with $\tan^2\theta_V = \frac{1}{2}$ this ratio turns out to be .38. No leptonic ϕ^P decays have been seen at this writing, but a few leptonic ω^P decays have been reported and a branching ratio of $0.012 \pm .003$ is quoted.[24] Perhaps the most interesting feature of this result is that the leptonic ω^P mode is seen at all. This is, from the present point of view, a manifestation of SU_3 breaking via the mixing with the ϕ^0. However, one should keep in mind that we have assumed that $J_\mu{}^\gamma$ is purely an octet. It might be that $J_\mu{}^\gamma$ also has an SU_3 singlet component. This would allow the $\omega^0 - \gamma$ coupling. It would also modify the SU_3 prediction $\mu_\Lambda = \frac{1}{2}\mu_N$ as discussed in the last chapter. The experimental situation is still too uncertain to allow one to rule out such a possibility.

In Chapter 4 the prospect was raised that the vector mesons ρ, ω, and ϕ might be coupled to conserved currents.[25] We are now in a position to discuss this idea in more detail. The strong interactions admit five absolutely conserved

22. The photon propagator has cancelled because of the q^2 factor in the coupling.
23. R. H. Dalitz, *Proc. Sienna Intl. Conf.* (Italian Phys. Soc. Bologna 1963, p. 171).
24. A. H. Rosenfeld, *Rev. Mod. Phys.*, **39**:1 (1967).
25. The first suggestion for doing this can be found in J. J. Sakurai, *Ann. Phys.* **11**:1 (1960).

"charges": **T**, B, Y. It is very attractive to couple the five vector mesons, $\boldsymbol{\rho}$, ω, and ϕ to the corresponding currents. The natural coupling is[26] $\boldsymbol{\rho}$ with $J_\mu{}^V$, ω with $J_\mu{}^B$, and[27] ϕ with $J_\mu{}^Y$. The fact that these vector mesons are coupled to *conserved* currents implies a "universality" principle for the coupling constants as measured in various processes involving the vector mesons, just as the conservation of the electric current implies that the same electric charge governs all of the electromagnetic processes. In a recent paper[28] Sakurai has given a detailed study of how this works for the ρ meson. The ρ coupling constant, f_ρ, can be defined by the Lagrangian interaction

$$\mathcal{L}_\rho = f_\rho \boldsymbol{\rho}_\mu \left[i\bar{N}\frac{\boldsymbol{\tau}}{2}\gamma_\mu N - \boldsymbol{\pi} \times \partial_\mu \boldsymbol{\pi} + \cdots \right] \tag{14.90}$$

where the dots stand for the remaining fields, strange hyperons and mesons, that make up the rest of the isotopic spin current. The constant $f_\rho{}^2/4\pi$ can be measured directly from the rate of the fundamental ρ decay $\rho \longrightarrow 2\pi$. Indeed, using Eq. (14.90)

$$\Gamma(\rho \longrightarrow 2\pi) = \frac{2}{3}\frac{f_\rho{}^2}{4\pi}\frac{p_\pi{}^3}{m_\rho{}^2}. \tag{14.91}$$

Since, experimentally[29] $\Gamma(\rho^+ \longrightarrow \pi^+ + \pi^0) = 160$ MeV we have $f_\rho{}^2/4\pi \simeq 2.9$. In his paper Sakurai discusses several processes involving real and virtual ρ's and concludes that, despite the different approximations involved, all of them yield essentially the same coupling constant. One of the processes is $\rho^0 \longrightarrow \mu^+ + \mu^-$, which takes place via the graph

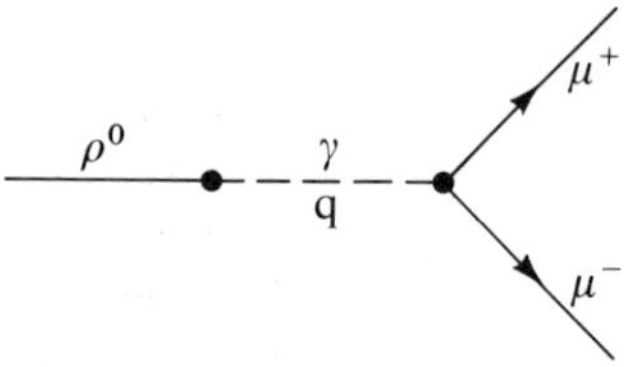

Here again we encounter the infamous $\rho^0 - \gamma$ coupling which, assuming ρ^0 is stable and is treated conventionally, must be written in the form

$$q^2 e \lambda_\rho \rho_\mu A_\mu.$$

The ρ^0 is, of course, far from being stable. It has been argued[30] that treating

26. N. Kroll, T. D. Lee, and B. Zumino, op. cit., have shown that if $\boldsymbol{\rho}$ is coupled to a conserved current **V**, then from the **T**, C, and Baryon number invariance of the theory one *must* have

$$\int V_{03}(\mathbf{r}, 0)\, d\mathbf{r} = \lambda T_3,$$

where λ is a constant.
27. Remember that $\phi \longrightarrow K\bar{K}$ is the main decay mode of the ϕ.
28. J. J. Sakurai, *Phys. Rev. Letters* **17**:1021 (1966).

the ρ^0 as a resonance allows one, effectively, to replace the factor $q^2\lambda_\rho$ by m_ρ^2/f_ρ, where m_ρ is the mass of the ρ^0 and f_ρ is the constant introduced above. Many physicists have been mystified by this result. Recently[31] a simple gauge invariant model has been found where it can be justified straightforwardly. In this model the ρ^0 is coupled to the isotopic spin current as in Eq. (14.90). However, and this is the novel idea, the *entire* isotopic vector electromagnetic current $J_\mu{}^V$ is given by

$$J_\mu{}^{V\gamma} = \frac{-m_\rho{}^2}{f_\rho}\,\rho_\mu{}^0, \tag{14.92}$$

where f_ρ is as above. This coupling factor is forced, in the model, by the consistency between the ρ^0 field equation

$$(-\Box^2 + m_\rho{}^2)\rho_\mu{}^0 = f_\rho J_{\mu 3}{}^V \tag{14.93}$$

and the photon equation which, in this model, where $J_{\mu 3}{}^V \neq J_\mu{}^{V\gamma}$, is

$$-\Box^2 A_\mu = J_\mu{}^{V\gamma} = \lambda\rho_\mu{}^0. \tag{14.94}$$

We can evaluate λ by remembering that for nonstrange mesons, in this model,

$$\begin{aligned} Q = \lambda \int \rho_0{}^0(\mathbf{r}, 0)\, d\mathbf{r} &= T_3 = \int J_{03}(\mathbf{r}, 0)\, d\mathbf{r} \\ = \int (-\Box^2 + m_\rho{}^2)\frac{\rho_0{}^0}{f_\rho}\, d\mathbf{r} &= \frac{-m_\rho{}^2}{f_\rho}\int \rho_0{}^0(\mathbf{r}, 0)\, d\mathbf{r}. \end{aligned} \tag{14.95}$$

Thus, in general, in this model, if $|A\rangle$ and $|B\rangle$ are any two states with the *same* charge

$$\langle A|J_\mu{}^{V\gamma}(0)|B\rangle = \frac{-m_\rho{}^2}{f_\rho}\langle A|\rho_\mu{}^0|B\rangle \tag{14.96}$$

The matrix element which is relevant to the $\rho^0 - \gamma$ transition is

$$\begin{aligned} \langle 0|J_\mu{}^{V\gamma}(0)|\rho_\lambda\rangle &= \frac{-m_\rho{}^2}{f_\rho}\langle 0|\rho_\mu{}^0(0)|\rho_\lambda\rangle \\ &= \frac{-m_\rho{}^2}{f_\rho}\frac{\epsilon_\mu\delta_{\lambda\mu}}{\sqrt{2m_\rho}} \qquad \text{(no sum on } \mu\text{)}, \end{aligned} \tag{14.97}$$

where ϵ_μ is the ρ^0 polarization vector. If this matrix element is used with Figure 5, one has for the rate for $\rho^0 \longrightarrow l^+ + l^-$, where l is a lepton,

---

29. In Sakurai's paper he used the value $\Gamma(\rho^+) = 128.7 \pm 7.7$ MeV and found $f_\rho{}^2/4\pi = 2.4 \pm 0.2$. We use the new value given in Rosenfeld et al., which raises f_ρ perhaps in better agreement with the values given by the alternate methods discussed by Sakurai. Rosenfeld et al. indicate that there is considerable experimental uncertainty about the ρ width as measured in different experiments. Hence these numbers are tentative. In fact, if we use the ρ^0 width of 140 Mev, we get $f_\rho{}^2/4\pi \simeq 2.55$.
30. M. Gell-Mann, D. Sharp, and W. G. Wagner, *Phys. Rev. Letters* **8**:261 (1962).
31. N. Kroll, T. D. Lee, and B. Zumino, op. cit.

$$\Gamma(\rho^0 \longrightarrow l^+ + l^-) = \left(\frac{e^2}{4\pi}\right)^2 \left(\frac{4\pi}{f_\rho^2}\right) \frac{m_\rho}{3} \left[1 + \frac{2m_l}{m_\rho}\right] \left[1 - 4\left(\frac{m_l}{m_\rho}\right)^2\right]^{1/2}. \tag{14.98}$$

The experiment[32] gives a branching ratio for $\rho^0 \longrightarrow \mu^+ + \mu^-$ equal to $0.33^{+0.16}_{-0.07} \times 10^{-2}$, which leads to

$$\frac{f_\rho^2}{4\pi} \simeq 2.7, \tag{14.99}$$

in reasonable agreement with the value of f_ρ obtained from $\rho^0 \longrightarrow \pi^+ + \pi^-$. It also follows from this model that

$$\langle A|J_\mu^{V\gamma}(0)|B\rangle = \frac{m_\rho^2}{q^2 + m_\rho^2} \langle A|J_\mu^\rho(0)|B\rangle, \tag{14.100}$$

where J_μ^ρ is the source current of the ρ field—that is, the isotopic spin. This means that in electron scattering the graph

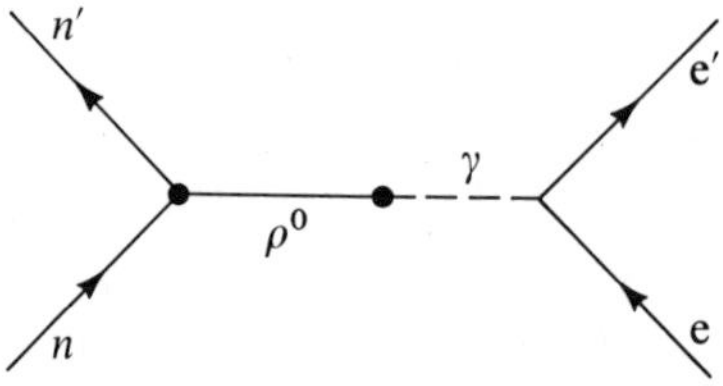

leads to isovector form factors of the form

$$F_{1,2}^V(q^2) = \frac{m_\rho^2}{q^2 + m_\rho^2} F_{1,2}^\rho(q^2), \tag{14.101}$$

so that, if the ρ-nucleon vertex form factors fall off as $1/q^2$, the electromagnetic form factors will fall off as $1/q^4$, in agreement with experiment.

A similar analysis can be applied to the isotopic scalar vector mesons, ω and ϕ. Only in the limit of good SU_3 can these particles be separated so that ω^0 couples to the baryon current J_μ^B alone while ϕ^0 couples to the hypercharge current J_μ^Y. In general, the mixing angles enter and the analysis is correspondingly complicated. All of these ideas have been combined in a recent analysis of electron-proton scattering,[33] which we now present. The basic quantities that enter are the vector meson-γ couplings and the vector meson-nucleon couplings. We first show, assuming the SU_3 properties of the currents and the mixing angle theory, that the photon-vector meson coupling is given, for the three mesons, in terms of one constant (really a form factor) $C_{V\gamma}$ and the ω^0, ϕ^0 mixing angle; θ_V. For the vector mesons the combination, $\sqrt{3}\,\phi^0 - \rho^0$, belongs to a U-spin triplet. Thus using the U-spin scalar character of J_μ^γ,

32. J. K. de Pagter et al., *Phys. Rev. Letters* **16**:35 (1966). In Sakurai, op. cit., a branching ratio of $0.44^{+0.21}_{-0.09} \times 10^{-2}$ is used leading, with the old value of $\Gamma(\rho^+ \longrightarrow \pi^+ + \pi^0) = 128.7 \pm 7.7$ to $f_\rho^2/4\pi = 2.5^{+0.6}_{-0.8}$.

$$\langle \sqrt{3}\,\phi^0 - \rho^0 | J_\mu^\gamma | 0 \rangle = 0, \tag{14.102}$$

so that

$$\sqrt{3}\, C_{\phi^0\gamma} = C_{\rho^0\gamma}. \tag{14.103}$$

Moreover, as we noted earlier,

$$C_{\omega^0\gamma} = 0. \tag{14.104}$$

Thus, if we identify the basic constant with $C_{\rho^0\gamma}$, we have for the physical ϕ and ω (mixed)[34]

$$\begin{aligned} C_{\phi^P\gamma} &= \frac{1}{\sqrt{3}}\, C_{\rho^0\gamma} \cos\theta_V, \\ C_{\omega^P\gamma} &= \frac{1}{\sqrt{3}}\, C_{\rho^0\gamma} \sin\theta_V. \end{aligned} \tag{14.105}$$

In terms of the work above,

$$C_{\rho\gamma} = \frac{m_\rho^2}{f_\rho}. \tag{14.106}$$

As we noted in the last chapter the most general form (suppressing Lorentz indices and Dirac matrices) of SU_3-invariant nucleon-meson Yukawa coupling can be written in the form

$$\mathcal{L}_{NNV} = f\,\mathrm{Tr}\,[[\bar{B}, B]_- V] + d\,\mathrm{Tr}\,[[\bar{B}, B]_+ V], \tag{14.107}$$

where $\bar{B}$, B, and V are the 3×3 particle matrices now familiar. Thus, working out the traces, one finds that the couplings are

$$\begin{aligned} \rho^0 &\leftrightarrow d + f, \\ \phi^0 &\leftrightarrow \sqrt{3}\, f - \frac{1}{\sqrt{3}}\, d. \end{aligned} \tag{14.108}$$

In fact, as discussed in Chapter 7, the vector meson, given time reversal invariance, has two Lorentz covariant couplings with the nucleons γ_μ and $\sigma_{\mu\nu}q_\nu$, so that there are four parameters (form factors) f_1, f_2 and d_1, d_2, corresponding to these possibilities. The ω^0, since it is an SU_3 singlet, is coupled to the nucleons with different coupling constants, d_{s_1} and d_{s_2} (there is only one SU_3 coupling type: $\mathrm{Tr}\,[[\bar{B}, B]_+]\omega^0$). Without additional theoretical assumptions there is no a priori connection between d_s and d. We can now compute the form factors in terms of the various quantities at hand. For simplicity we make $F_2(q^2)$ dimensionless by factoring out the nucleon magneton $e/2M$, so that the F_2 that appears here is proportional to the one used in our earlier work.

---

33. T. Massam and A. Zichichi, *Nuovo Cimento* **43**:1137 (1966), and T. Massam and A. Zichichi, *Nuovo Cimento* **44**:309 (1966).
34. With our choice of mixing this $C_{\phi^P\gamma}$ differs, by a sign, from that of Massam and Zichichi, op. cit.

Thus from the typical graph

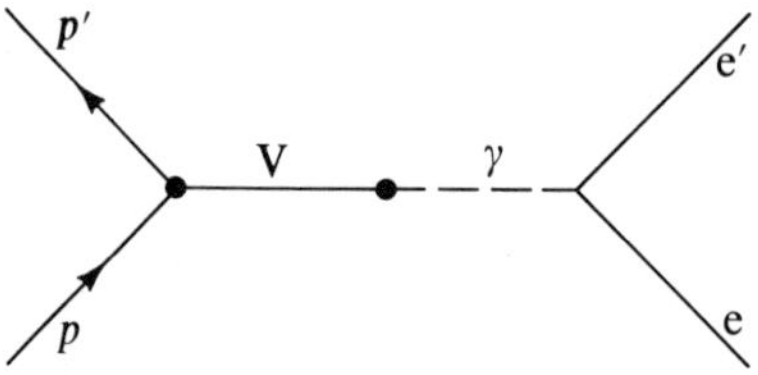

$$F_1^V(q^2) = \frac{(f_1 + d_1)}{f_\rho}\frac{1}{(1 + q^2/m_\rho^2)} \tag{14.102}$$

and

$$F_2^V(q^2) = \frac{(f_2 + d_2)}{f_\rho}\frac{1}{(1 + q^2/m_\rho^2)}. \tag{14.103}$$

The scalar form factors are complicated by the mixing angles. We assume that the nucleons emit ϕ^P and ω^P. Thus the form factors are given by

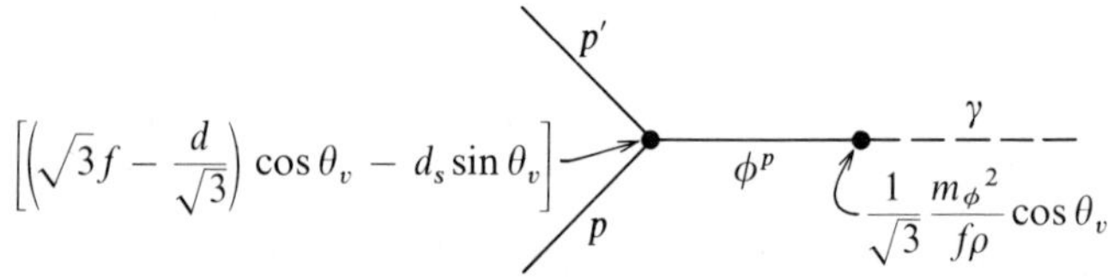

+

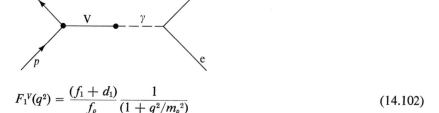

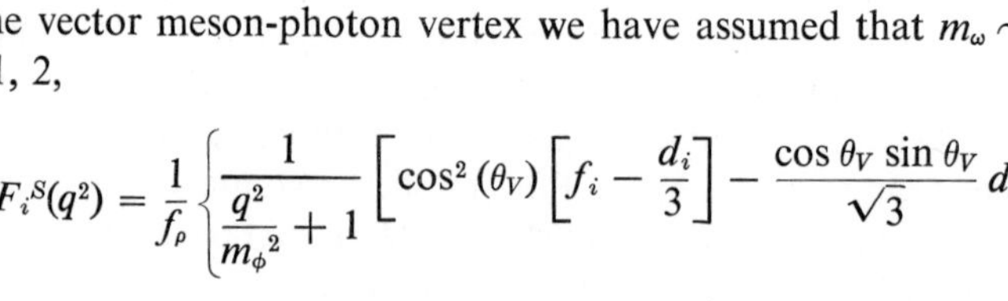

In the vector meson-photon vertex we have assumed that $m_\omega \sim m_\phi$. Thus, for $i = 1, 2$,

$$F_i^S(q^2) = \frac{1}{f_\rho}\left\{\frac{1}{\frac{q^2}{m_\phi{}^2} + 1}\left[\cos^2(\theta_V)\left[f_i - \frac{d_i}{3}\right] - \frac{\cos\theta_V \sin\theta_V}{\sqrt{3}} d_{S_i}\right]\right.$$
$$\left. + \frac{1}{\frac{q^2}{m_\omega{}^2} + 1}\left[\sin^2(\theta_V)\left[f_i - \frac{d_i}{3}\right] + \frac{\cos\theta_V \sin\theta_V}{\sqrt{3}} d_{S_i}\right]\right\}. \tag{14.104}$$

We know, from our previous work in electron scattering, that no fit of the large q^2 scattering data with form factors of this type is possible if the f's and d's are taken to be constants. Massam and Zichichi[35] make the plausible conjecture that the vertex form factors, the f's and d's, vary with q^2 as

$$\frac{1}{\left(\frac{q}{\Lambda}\right)^2 + 1},$$

where Λ is a "mass" which, in the absence of an alternate method, must be determined by fitting the electron scattering data. For simplicity, it is supposed that the same Λ obtains for the f_i's and d_i's. One could also imagine that the vector meson-photon vertex varies with q^2. In the model of Kroll et al.[36] this variation is essentially contained in the factors

$$\frac{1}{\left(\frac{q}{m_V}\right)^2 + 1}$$

with f_ρ a constant and, in any case, such additional q^2 dependence does not seem to be required by the data. There are now eight parameters in the fit. (We do not count f_ρ as a free parameter since we can either take it from the ρ meson data, or since it is common to all of the form factors, simply consider as the unknowns the ratios of the d's and f's to f_ρ.) However, there are four constraints among them from the charge and magnetic moments of N and P:

$$\frac{F_1^S(0) + F_1^V(0)}{2} = 1, \qquad F_1^S(0) - F_2^V(0) = 0, \tag{14.105}$$

and

$$\frac{F_2^S(0) + F_2^V(0)}{2} = 1.79, \qquad \frac{F_2^S(0) - F_2^V(0)}{2} = -1.91. \tag{14.106}$$

Equation (14.105) leads to the condition that

$$\frac{f_1}{f_\rho} = \frac{1}{2}, \tag{14.107}$$

and

$$d_1 = 0,$$

so that the "charge" coupling of the vector mesons to the nucleons is of pure F type. This condition arises naturally in a theory in which ρ and ϕ are coupled to the conserved currents that carry T and Y. The "fundamental" ρ coupling (see Eq. (13.131) of the last chapter) would be of the form

---

35. Op. cit.
36. Op. cit.

$$\rho_\mu{}^0 \operatorname{Tr}\,[\bar{B}\gamma_\mu[B, \lambda_3]_- - \partial_\mu M[M, \lambda_3]_-], \tag{14.108}$$

where M stands for the pseudoscalar octet matrix. From this point of view Eq. (14.107) would represent a "phenomenological" description of the ρ^0 baryon vertex, $\langle B|\rho^0|B\rangle$. The unmixed ϕ^0 would couple as

$$\phi_\mu{}^0 \operatorname{Tr}\left[\bar{B}\gamma_\mu\left[B, \frac{\lambda_8}{\sqrt{3}}\right]_- - \partial_\mu M\left[M, \frac{\lambda_8}{\sqrt{3}}\right]_-\right] \tag{14.109}$$

and the condition that

$$Q = \lambda_3 + \frac{1}{\sqrt{3}}\lambda_8 \tag{14.110}$$

then determines d_1 and f_1, as above. The f_2 and d_2 are, from this point of view, "induced" quantities since the basic coupling involves γ_μ alone. We can compute these parameters from μ_P and μ_N. Thus

$$\begin{aligned} \frac{f_2}{f_\rho} &= \mu_P + \frac{\mu_N}{2}, \\ \frac{d_2}{f_\rho} &= -\tfrac{3}{2}\mu_N. \end{aligned} \tag{14.111}$$

This still leaves the d_{S_i} undetermined. As Massam and Zichichi[37] note, higher symmetry schemes, such as SU_6 in which the nine vector mesons are treated symmetrically, make a definite connection between d_{S_i} and d_i:

$$d_{S_i} = 2\sqrt{\tfrac{2}{3}}\, d_i \tag{14.112}$$

and

$$\tan\theta_V = \frac{1}{\sqrt{2}}. \tag{14.113}$$

Making this assumption and analyzing the scattering data they find that

$$\Lambda = 980 \pm 18 \text{ MeV}.$$

This is not an implausible result since the vertex $\langle N|V|N\rangle$, whose form factors are given by this Λ, are expected to have a pole at the vector meson mass. Which mass this is depends on whether one assumes that the mixed, or unmixed, vector mesons are being emitted. In this analysis, these distinctions are averaged over by assuming that Λ is common to all the vertices, and hence it is not surprising that Λ lies somewhere between the ρ mass and the ϕ mass. Massam and Zichichi[38] have also made analysis of the data allowing the other parameters, θ_V and so on,

37. Op. cit. See A. Pais, *Rev. Mod. Phys.* **38**:215 (1966) for a detailed review of these higher symmetry predictions.
38. Op. cit.
39. A. H. Rosenfeld et al., *Rev. Mod. Phys.*, **39**:1 (1967).

to vary and find that the conclusions are consistent with what would be expected on the ideal quark model.

There is one process that occurs, experimentally, in violation of a selection rule given by the ideal quark model. With $\tan\theta_V = 1/\sqrt{2}$, as we have seen,

$$\phi^P = -q^3 q_3. \tag{14.114}$$

Thus all decays of the form $\phi^P \longrightarrow \rho\pi$ or $n\pi$ are strictly forbidden since ϕ^P is, in this limit, made up of "strange" quarks alone while ρ and π are made up of nonstrange quarks alone. But the decays $\phi \longrightarrow \rho\pi$, $\phi \longrightarrow 3\pi$ are known to occur with a branching ratio[39] of $12 \pm 4\%$. The other modes of the ϕ are $\phi \longrightarrow K^+ + K^-$, $48 \pm 3\%$, and[40] $\phi \longrightarrow K_1{}^0 + K_2{}^0$, $40 \pm 3\%$. This violation of the ideal quark selection rule has been analyzed by Glashow and Socolow,[41] who conclude that it is consistent with having $\theta_V = 39° \pm 1$. The mass formula gave $\theta_V = 40°$ and the ideal quark model gives $\theta_V \simeq 35°$. Several theoretical assumptions go into their analysis and the interested reader is advised to consult their paper.

The K meson modes of the ϕ are subject to an SU_3 analysis. The state $K\bar{K}$ resulting from the ϕ decay is a P state and since C, the charge conjugation operator, has an eigenvalue of $(-1)^l$, with l the orbital angular momentum, for a two-meson state of the form $A\bar{A}$, this state must be odd under C or antisymmetric under exchange of particle and antiparticle. This means that the ω^0 component of ϕ^P cannot couple to $K\bar{K}$ since, as we have seen, the only such SU_3 invariant coupling is[42] $\omega_\mu{}^0 \,\mathrm{Tr}\,[M\,\partial_\mu M]$. In this coupling K and $\bar{K}$ enter symmetrically. But the ϕ^0 can couple to $K\bar{K}$ by the antisymmetric octet coupling $\mathrm{Tr}\,[V_\mu[M, \partial_\mu M]_-]$. Using it we can relate $\phi^P \longrightarrow K\bar{K}$ to $\rho \longrightarrow \pi\pi$ in the ratio

$$\frac{\Gamma(\phi \longrightarrow K\bar{K})}{\Gamma(\rho^0 \longrightarrow \pi\pi)} = \tfrac{3}{4}\cos^2\theta_V. \tag{14.115}$$

This formula, which as it stands, gives very poor agreement with experiment, and must be corrected for phase space since, for example, the two charged K's have a combined mass of 987.4 MeV, which is close to the ϕ^P mass.[43] Using Eq. (14.115), with $\theta_V = 40°$, we find, correcting for phase space,

$$\Gamma(\phi \longrightarrow K^+ + K^-) = \tfrac{2}{3}(\tfrac{3}{4}\cos^2\theta_V)\frac{f_\rho{}^2}{4\pi}\left(\frac{p_K{}^3}{m_\phi{}^2}\right) \simeq 1.7 \text{ MeV}, \tag{14.116}$$

where we have used $f_\rho{}^2/4\pi \simeq 2.9$. This number is in reasonable agreement with the experimental number $\sim$1.9 MeV.

We close this chapter with a few additional notes on SU_3 breaking. From what has been said about the vector meson nonet it is clear that there will be a

---

40. $K_1{}^0$ and $K_2{}^0$ are the short- and long-lived neutral K mesons respectively.
41. S. L. Glashow and R. H. Socolow, *Phys. Rev. Letters* **15**:329 (1965).
42. Tr $(M^T M)$, for example, is neither SU_3 invariant nor charge conserving.
43. We recall that the *total* ϕ^P width is $\Gamma_{\phi^P} = 4.0 \pm 1.0$ MeV.

mixing phenomenon for the η^0, x^0 system among the pseudoscalars. Since the x^0 is at 958.3 ± 0.8 MeV while the η^0 is at 548.6 ± 0.4 MeV the octet and singlet states are rather well-separated so that we would expect θ_P, the pseudoscalar mixing angle, to be small compared to θ_V. If we write

$$\begin{aligned} x^P &= x^0 \cos\theta_P + \eta^0 \sin\theta_P, \\ \eta^P &= \eta^0 \cos\theta_P - x^0 \sin\theta_P, \end{aligned} \tag{14.117}$$

and use the formalism, as above, for the square masses, we learn that $\theta_P = \pm 10.8°$. The mixing will modify several of the predictions of SU_3 given above. As an example[44] we consider the three decays $x^P \longrightarrow \gamma + \gamma$, $\eta^P \longrightarrow \gamma + \gamma$, and $\pi^0 \longrightarrow \gamma + \gamma$. We call

$$\begin{aligned} \langle x^0|\gamma\gamma\rangle &= A, \\ \langle \eta^0|\gamma\gamma\rangle &= B, \\ \langle \pi^0|\gamma\gamma\rangle &= \sqrt{3}\,B. \end{aligned} \tag{14.118}$$

The last relation is a consequence of SU_3, as shown in the last chapter. Thus

$$\begin{aligned} \langle x^P|\gamma\gamma\rangle &= \cos\theta_P A + \sin\theta_P B, \\ \langle \eta^P|\gamma\gamma\rangle &= \cos\theta_P B - \sin\theta_P A, \\ \langle \pi^0|\gamma\gamma\rangle &= B\sqrt{3}. \end{aligned} \tag{14.119}$$

This is far as one can go without a model. A natural model to make[45] is given by graphs of the form

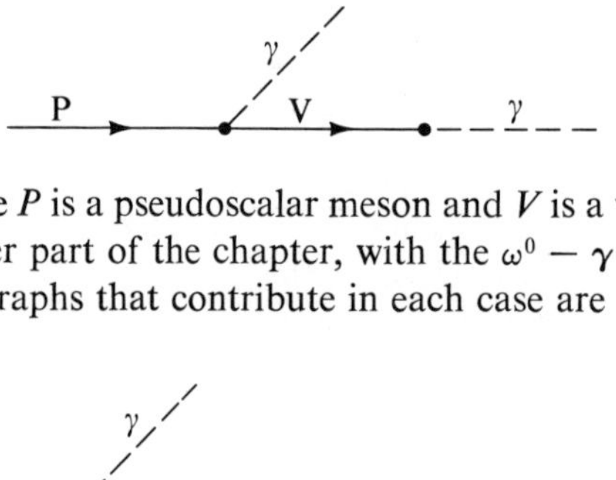

where P is a pseudoscalar meson and V is a vector meson. From the work of the earlier part of the chapter, with the $\omega^0 - \gamma$ coupling is forbidden in good SU_3, the graphs that contribute in each case are seen to be

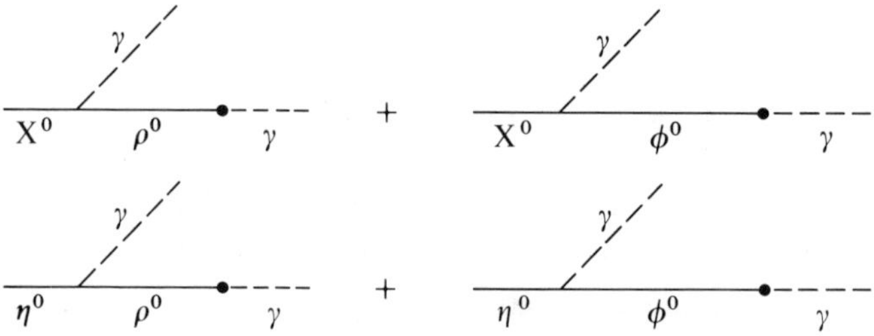

44. R. H. Dalitz, *Les Houches Lectures*, op. cit., R. H. Dalitz and D. G. Sutherland, *Nuovo Cimento* **37**:1777 (1965), and R. H. Dalitz (private communication).

assuming that x^0 and η^0 go into ϕ^P. In good SU_3 we have, by Eqs. (14.23) and (14.27),

$$\langle x^0|\rho^0\gamma\rangle = \sqrt{3}\,\langle x^0|\phi^0\gamma\rangle,$$
$$\langle \eta^0|\rho^0\gamma\rangle = -\sqrt{3}\,\langle \eta^0|\phi^0\gamma\rangle, \tag{14.120}$$

and

$$\sqrt{3}\,C_{\phi^0\gamma} = C_{\rho^0\gamma}, \tag{14.121}$$

so that

$$\frac{A}{B} = \frac{\sqrt{3} + 1/\sqrt{3}}{\sqrt{3} - 1/\sqrt{3}}\frac{\langle x^0|\rho^0\gamma\rangle}{\langle \eta^0|\rho^0\gamma\rangle} = 2\frac{\langle x^0|\rho^0\gamma\rangle}{\langle \eta^0|\rho^0\gamma\rangle}. \tag{14.122}$$

In the quark model

$$\rho^0 = \frac{1}{\sqrt{2}}(q^1q_1 - q^2q_2)(\uparrow\,\uparrow), \tag{14.123}$$

while

$$\eta^0 = \frac{1}{\sqrt{6}}(q^1q_1 + q^2q_2 - 2q^3q_3)\frac{(\uparrow\,\downarrow - \downarrow\,\uparrow)}{\sqrt{2}} \tag{14.124}$$

and

$$x^0 = \frac{1}{\sqrt{3}}(q^1q_1 + q^2q_2 + q^3q_3)\frac{(\uparrow\,\downarrow - \downarrow\,\uparrow)}{\sqrt{2}}. \tag{14.125}$$

Taking the matrix element to be the magnetic dipole operator (Eq. (14.32)), it follows at once (note that the q^3q_3 parts of x^0 and η^0 do not contribute) that

$$\frac{\langle x^0|\rho^0\gamma\rangle}{\langle \eta^0|\rho^0\gamma\rangle} = \sqrt{2}. \tag{14.126}$$

Thus,

$$R = \frac{\langle \eta^P|\gamma\gamma\rangle}{\langle \pi^0|\gamma\gamma\rangle} = \cos\theta_P\frac{\langle \eta^0|\gamma\gamma\rangle}{\langle \pi^0|\gamma\gamma\rangle} - \sin\theta_P\frac{\langle x^0|\gamma\gamma\rangle}{\langle \pi^0|\gamma\gamma\rangle}$$
$$= \frac{\cos\theta_P}{\sqrt{3}}[1 - \sqrt{2}\,2\tan\theta_P]. \tag{14.127}$$

The mass formula does not fix the sign of θ_P.[46] Thus with $\theta_P = \pm 10.8°$,

$$R \simeq \frac{1}{\sqrt{3}}[.98 \mp .51], \tag{14.128}$$

which means that the mixing makes a sizable change in the SU_3 prediction for R. The mixing will effect the rates of processes of the form $V^P \longrightarrow M^P + \gamma$ and the reader may amuse himself by working them out in the naive quark model.

---

45. M. Gell-Mann, D. Sharp, and W. Wagner, op. cit.
46. In the naive quark model; see R. H. Dalitz, *Les Houches Lectures*, op. cit.: $\theta_P = -10.8°$.

Finally we give a brief discussion of the effect of the symmetry breaking on the decimet. The simplest way to derive the mass formula for the decimet is again in terms of the quark model. Symbolically we can write the decimet in terms of its quark content as

$$\begin{aligned} N^* &\sim (q_1q_1q_1) \qquad && (\text{No } q_3), \\ Y^* &\sim (q_1q_1q_3) && (1\ q_3), \\ \Xi^* &\sim (q_1q_3q_3) && (2\ q_3), \\ \Omega^- &\sim (q_3q_3q_3) && (3\ q_3). \end{aligned} \tag{14.129}$$

Thus, with M_0 the "unsplit" mass and Δ the q_1, q_3 mass splitting, we have the well-known result[47]

$$\begin{aligned} M_{N^*} &= M_0, \\ M_{Y^*} &= M_0 + \Delta, \\ M_{\Xi^*} &= M_0 + 2\Delta, \\ M_{\Omega^-} &= M_0 + 3\Delta, \end{aligned} \tag{14.130}$$

which is to say that the decimet masses increase in steps of Δ with decreasing strangeness. From the known decimet masses one obtains

$$\Delta \simeq 147 \text{ MeV},$$

which is at least in the same ball park with the 118 MeV estimate of Δ from the vector mesons.[48]

The baryon and meson states treated in this chapter are, in some sense, the simplest states allowed in the quark model. More specifically, they are states whose total spin is given by combining the quark spins and supposing that there is no orbital angular momentum shared among the quarks. It is then not surprising that there exist "excited" states with higher spin and higher mass corresponding, perhaps, to quarks, more loosely bound, in states of nonzero l. The most clearly established set of such states is a family of spin 2 mesons, nine of them, at masses in the neighborhood of 1300 MeV, which fit nicely into the broken SU_3 octet-singlet scheme,[49] and no doubt other families will show up as time goes on.

In the next, and last, chapter we turn to a discussion of the weak decays of the strange particles.

---

47. S. Okubo, op. cit.
48. The pseudoscalar mesons when analyzed in terms of (mass)^2 lead to a Δ of comparable magnitude while the baryon octet mass splittings lead to $\Delta = 189$ MeV. For a discussion of these matters, see again R. H. Dalitz, *Les Houches Lectures*.
49. See, for example, S. L. Glashow and R. H. Socolow, *Phys. Rev. Letters* **15**:329 (1965) for a discussion.

15
Decays that Change Strangeness

The subject of the weak decays of the strange particles has grown so enormously in the past decade that, in this chapter, we cannot do much more than touch the surface. Weak strange particle decays[1] can be divided into two classes: leptonic as in $\Sigma^- \longrightarrow n + e^- + \bar{\nu}_e$ and $K^+ \longrightarrow \mu^+ + \nu_\mu$, and nonleptonic as in $K^+ \longrightarrow \pi^+ + \pi^0$ and $\Sigma^- \longrightarrow n + \pi^-$. It is clear, from the examples given, that the leptonic decays of strange particles involve modes that are strikingly similar to the decays such as $\pi^+ \longrightarrow \mu^+ + \nu_\mu$ and $n \longrightarrow p + e^- + \bar{\nu}_e$ that we have already studied. Moreover, in light of the connections between strange and nonstrange particles of the type given in SU_3 and the quark model, it is a reasonable first guess that the theory governing these decays might be similar. The nonleptonic decays are a new phenomenon and since all of the initial and final particles are strongly interacting one would imagine that a quantitative theory would be more difficult to invent. We will return to these decays later.

1. *Leptonic decays.* The simplest assumption that one can make, in light of what we have learned about ordinary weak decays, is that the leptonic decays of the strange particles are generated by currents, V_μ', A_μ', vector and axial vector, in the same sense that the corresponding decays of the ordinary particles are generated by V_μ and A_μ. It is clear that V_μ' and A_μ' must be different functions of the fields than V_μ and A_μ, since V_μ' and A_μ' change strangeness and V_μ and A_μ do not. We may begin the analysis without making any specific assumptions about the functional forms of V_μ' and A_μ'. However, we shall assume that S_μ, the total strange current is given by

$$S_\mu = V_\mu' + A_\mu', \tag{15.1}$$

an analogy to the nonstrange current. This is to say that we attribute any differences in the constants associated with V_μ' and A_μ' to renormalization effects in the matrix elements. Since the K meson is a pseudoscalar[2] we can write, as

1. Recall that strangeness and hypercharge are related by the formula $Y = B + S$. Since $\Delta B = 0$ in all reactions, we have $\Delta Y = \Delta S$ in strange particle decays. We prefer, in conformity with the literature, to use the "strangeness" rather than the "hypercharge" language in this discussion.
2. The experimental evidence for this involves a detailed analysis of the reaction $K^- + He^4 \longrightarrow {}_\Lambda H^4 + \pi^0$ (see for example, S. Gasiorowicz, *Elementary Particle Physics*, Wiley, 1966, p. 239) into which we will not go here. We only remind the reader that the SU_3 representation of K's and π's as members of the same octet requires that they have the same parity.

in the decay $\pi \longrightarrow \mu + \nu_\mu$,

$$\frac{G}{\sqrt{2}}\langle 0|S_\mu(0)|\mathbf{p}_K\rangle = \frac{G}{\sqrt{2}}\langle 0|A_\mu'(0)|\mathbf{p}_K\rangle = \frac{ip_{K\mu}}{\sqrt{2E_K}}g_K(-p_K{}^2). \tag{15.2}$$

Assuming T invariance for the strange particle decays,[3] the kaon form factor, as written, is real. We have assumed here that the G that governs the strange particle decays is identical to the one that governs the nonstrange particle decays. As we shall see, there is good reason to doubt this, and part of the concern of this chapter is to see why and, therefore, how one might attempt to modify the theory. Using Eq. (15.2) (compare Eq. (11.30)) we easily find that[4]

$$R_{K\to\mu+\nu_\mu} = \frac{g_K{}^2 m_\mu{}^2}{4\pi m_K{}^3}(m_K{}^2 - m_\mu{}^2)^2. \tag{15.3}$$

The measured lifetime of the K^+ is given as[5]

$$\tau(K^+) = 1.235 \times 10^{-8} \pm .006 \text{ sec.}$$

Unlike the pion, the K, since it is more massive, has a variety of decay modes; that is, $\pi^+\pi^0$, $\pi^+\pi^0\pi^0$, $\pi^0 e^+\nu$, $\pi^0\mu^+\nu$, etc., and the branching ratio into $\mu + \nu_\mu$ is given as 63.4 ± 0.590, which means that

$$|g_K|^2 m_K{}^2 = 2.03 \times 10^{-14}. \tag{15.4}$$

Thus

$$\frac{|g_K|^2}{|g_\pi|^2} = 7.4 \times 10^{-2}. \tag{15.5}$$

If we take the position that g_K and g_π measure the "intrinsic" strengths of these decays (that is, that the mass factors in the phase space are "accidental"), then we would draw the conclusion that this strangeness changing decay is sup-

---

3. This is now known to be false, but the T violation does not change anything essential in this discussion.
4. The same set of arguments used in Chapter 11 in connection with $\pi \longrightarrow e + \nu_e$ decay make it clear that $K \longrightarrow e + \nu_e$ is strongly suppressed relative to $K \longrightarrow \mu + \nu_\mu$. Indeed, theoretically, ignoring radiative corrections but assuming muon-electron symmetry,

$$\frac{R_{K\to e+\nu_e}}{R_{K\to\mu+\nu_\mu}} = \left(\frac{m_e}{m_\mu}\right)^2\left(\frac{m_K{}^2 - m_e{}^2}{m_K{}^2 - m_\mu{}^2}\right)^2 \simeq 2.5 \times 10^{-5}.$$

A branching ratio of $1.9 \pm 1.2 \times 10^{-5}\%$ is quoted for this mode (Rosenfeld et al., *Rev. Mod. Phys.*, **39**:1 (1967)), which is certainly consistent with the theoretical prediction. In a recent review of the *V-A* theory, R. E. Marshak (unpublished) has given a summary of the predictions of $\mu - e$ universality and their comparison with experiment:

pressed relative to the strangeness nonchanging decays by an order of magnitude, in violation of the naive idea that all the weak decays are governed by one universal constant. Such a breakdown of naive universality is also suggested by an analysis of the β-decay of the strange hyperons such as $\Lambda^0 \longrightarrow p + e^- + \bar{\nu}_e$. In general, the matrix element $\langle P|S_\mu|\Lambda^0\rangle$ is given by

$$\langle P|S_\mu|\Lambda^0\rangle = i\bar{u}(p)[\gamma_\mu f_V(q^2) + \sigma_{\mu\nu}q_\nu f_M(q^2) + iq_\mu f_S(q^2) + \gamma_5[\gamma_\mu g_A(q^2) - iq_\mu g_P(q^2) + i(\Lambda + P)_\mu g_E(q^2)]]u(\Lambda), \quad (15.6)$$

with

$$q_\mu = (\Lambda - P)_\mu. \quad (15.7)$$

There is no general principle, like G conjugation or time reversal invariance, that serves to eliminate form factors here, since Λ^0 and p are distinct particles. In analyzing the relatively meager data on Λ^0 β-decay, the assumption is made that all the terms, other than those involving $f_V(q^2)$ and $g_A(q^2)$, can be neglected since they are multiplied by the momenta P_μ and Λ_μ. Moreover, it is assumed that $f_V(q^2)$ and $g_A(q^2)$ can be correctly replaced by their values at $q^2 = 0$. A recent[6] and typical experimental determination of $f_V(0)$ gives for this quantity

$$f_V(0) = 0.24 \pm 0.02. \quad (15.8)$$

The most striking feature is that $f_V(0)$ for Λ^0 β-decay is an order of magnitude smaller than the $f_V(0) = 1$ of neutron β-decay, again suggesting that the strangeness-violating leptonic weak decays are strongly suppressed. The totality of experiments on these decays suggests that we are in the presence of a general principle; that is, the strangeness-changing weak interactions are intrinsically weaker than those of ordinary β-decay. Furthermore, experimentally, changes of strangeness by two units are even weaker than changes of strangeness by one

Test of Electron-Muon Universality

Decay Process	*Theory*	*Experiment*
$\frac{\Gamma(\pi^+ \longrightarrow e^+ + \nu_e)}{\Gamma(\pi^+ \longrightarrow \mu^+ + \nu_\mu)}$	1.23×10^{-4}	$(1.24 \pm 0.03) \times 10^{-4}$
$\frac{\Gamma(K^+ \longrightarrow e^+ + \nu_e)}{\Gamma(K^+ \longrightarrow \mu^+ + \nu_\mu)}$	2.47×10^{-5}	$(3 \pm 1.89) \times 10^{-5}$
$\frac{\Gamma(K^+ \longrightarrow \pi^0 + \mu^+ + \nu_\mu)}{\Gamma(K^+ \longrightarrow \pi^0 + e^+ + \nu_e)}$	0.69	0.703 ± 0.056
$\frac{\Gamma(\Lambda \longrightarrow p + e^- + \bar{\nu}_e)}{\Gamma(\Lambda \longrightarrow p + \mu^- + \bar{\nu}_\mu)}$	5.88	5.87 ± 0.75
$\frac{\Gamma(\Sigma^- \longrightarrow n + \mu^- + \bar{\nu}_\mu)}{\Gamma(\Sigma^- \longrightarrow n + e^- + \nu_e)}$	0.45	0.496 ± 0.26

5. A. H. Rosenfeld et al., *Rev. Mod. Phys.*, **39**:1 (1967).
6. N. Brene et al., *Phys. Rev.* **149**:1288 (1966).

unit and, indeed, such decays have not been definitely observed at all. To take a celebrated example: the nonleptonic weak decay, with $\Delta S = 2$, $\Xi^- \longrightarrow n + \pi^-$, has a branching ratio[7] of $<5 \times 10^{-3}\%$ compared with the $\Delta S = 1$ mode, $\Xi^- \longrightarrow \Lambda^0 + \pi^-$, which comprises essentially 100% of the Ξ^- decays. A more subtle manifestation of the same notion (the suppression of $|\Delta S| = 2$ weak decays compared with $|\Delta S| = 1$ decays) is found in the $K^0 - \bar{K}^0$ system. We outline the argument here, and below return to this system when we treat time reversal violations. The K^0 has $S = 1$ while the $\bar{K}^0$ has $S = -1$. They are the only known pair of metastable particles with the property that particle and antiparticle can communicate via a known interaction. Pairs like n, $\bar{n}$, e^+, e^-, etc., are rigidly separated by an absolute selection rule such as charge conservation or baryon number conservation. Both the decays $K^0 \longrightarrow \pi^+ + \pi^-$ and $\bar{K}^0 \longrightarrow \pi^+ + \pi^-$ take place weakly, and hence one may have the conversion[8]

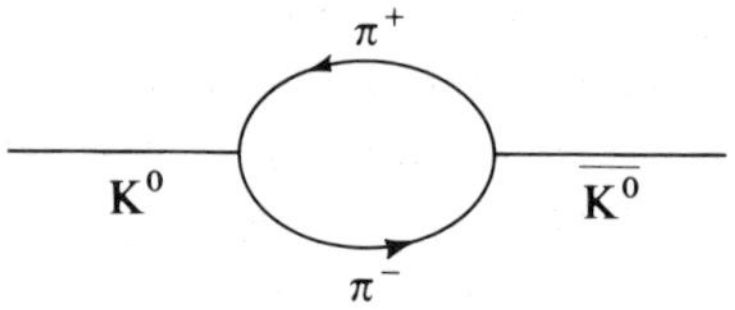

This conversion results in mixing the states and, in fact, assuming time-reversal or CP invariance, the states that decay exponentially have the form[9]

$$\begin{aligned} CP &= +1, & \frac{K^0 - \bar{K}^0}{\sqrt{2}} &\longleftrightarrow \gamma_S, \\ CP &= -1, & \frac{K^0 + \bar{K}^0}{\sqrt{2}} &\longleftrightarrow \gamma_L. \end{aligned} \tag{15.9}$$

We have indicated the CP eigenvalues of these states on the left. (Remember that the K^0 is a pseudoscalar.) Here γ_S and γ_L are the complex lifetimes of the decaying states, so that, if the time dependence is written in the form $e^{-\gamma t}$, then $\operatorname{Im}\gamma$ is the mass and $\operatorname{Re}\gamma/2$ is the lifetime of the respective linear combinations. Experimentally,[10]

$$\begin{aligned} \tau_s &= \frac{2}{\operatorname{Re}\gamma_S} = .87 \times 10^{-10} \pm .009 \text{ sec}, \\ \tau_L &= \frac{2}{\operatorname{Re}\gamma_L} = 5.68 \times 10^{-8} \pm .26 \text{ sec}, \end{aligned} \tag{15.10}$$

7. Rosenfeld et al., op. cit.
8. M. Gell-Mann and A. Pais, *Phys. Rev.* **97**:1387 (1955).
9. T. D. Lee, R. Oehme, and C. N. Yang, *Phys. Rev.* **106**:340 (1957). If K^0 is produced at $t = 0$, then, with the mixing,

$$\psi_{K^0}(t) = \frac{(K^0 - \bar{K}^0)}{\sqrt{2}} \frac{e^{-\gamma_S t}}{\sqrt{2}} + \frac{(K^0 + \bar{K}^0)}{\sqrt{2}} \frac{e^{-\gamma_L t}}{\sqrt{2}}.$$

so that the two linear combinations decay with well-separated lifetimes. If g is a constant characterizing the strength of the effective Hamiltonian governing the decay $K_S{}^0 \longrightarrow 2\pi$, which is the principal mode, then Re $\gamma_S \sim 0(g^2)$. But if the mixing is given only by two-step processes as in Figure 1, then also

$$|m_S - m_L| = |\text{Im}\ \gamma_S - \text{Im}\ \gamma_L| \sim 0(g^2). \tag{15.11}$$

In fact, in the usual units, $\hbar = c = 1$, experiment gives[11]

$$m_S - m_L = -(0.48 \pm .02)\frac{1}{\tau_S}, \tag{15.12}$$

confirming the two-step $\Delta S = 1$ character of the mixing. If there is an additional direct, one-step, $K^0 \longleftrightarrow \bar{K}^0$, $\Delta S = 2$ transition, this result shows that its effective Hamiltonian must be governed by a coupling constant $\sim 0(g^2)$, which is to say that $\Delta S = 2$ transitions are "forbidden." [12] In terms of the strangeness-changing current, S_μ, and the strangeness, S, we can build in the selection rule $\Delta S = \pm 1$, by demanding that S_μ and S have the commutation relation

$$[S_\mu, S]_- = \pm S_\mu. \tag{15.13}$$

From the charge character of decays like $\Lambda^0 \longrightarrow p + e^- + \nu_e$, which, in analogy with the Feynman-Gell-Mann theory of ordinary β-decay would be given by a Lagrangian of the form

$$\mathcal{L} = S_\mu{}^* L_\mu + S_\mu L_\mu{}^*, \tag{15.14}$$

where L_μ is the lepton current described in Chapter 7, it follows that S_μ changes ordinary charge by one unit, that is,

$$\Delta Q = 1,$$

so that

$$[S_\mu, Q]_- = S_\mu = \pm [S_\mu, S]_-. \tag{15.15}$$

There is now a large body of experimental evidence on leptonic decays of strange particles that indicates that in Eq. (15.15) the *plus* sign is correct and the minus sign is ruled out, which is expressed by the selection rule[13]

$$\Delta Q = \Delta S. \tag{15.16}$$

Typical consequences of this selection rule are

$$\Sigma^+ \longrightarrow n + e^+ + \nu_e, \qquad \text{forbidden,}$$

$$\Sigma^- \longrightarrow n + e^- + \bar{\nu}_e, \qquad \text{allowed.}$$

---

10. A. H. Rosenfeld et al., *Rev. Mod. Phys.*, **39**:1 (1967).
11. A. H. Rosenfeld et al., op. cit.
12. More quantitatively, $|\Delta m| \sim 10^{-5}$ eV, while the matrix element, H, dividing out phase space, for the $\Delta S = 1$ transition $K_S{}^0 \longrightarrow 2\pi$ is of order $H \sim \left(\frac{m_K}{\tau_S}\right)^{1/2} \sim 10^2$ eV. Thus $\Delta S = 2$ transitions are suppressed by a factor of 10^7.
13. Clearly this rule applies only to decays in which $\Delta S \neq 0$, since $\Delta Q \neq 0$ always.

Experimentally,[14]

$$\frac{\Sigma^+ \longrightarrow n + e^+ + \nu_e}{\Sigma^- \longrightarrow n + e^- + \bar{\nu}_e} < .12.$$

Moreover,

$$K^0 \longrightarrow \pi^- + e^+ + \nu_e, \qquad \text{allowed,}$$

$$\bar{K}^0 \longrightarrow \pi^- + e^+ + \nu_e, \qquad \text{forbidden,}$$

and, experimentally,[15]

$$\frac{\bar{K}^0 \longrightarrow \pi^- + e^+ + \nu_e}{K^0 \longrightarrow \pi^- + e^+ + \nu_e} < .5,$$

in conformity with the rule, which is consistent with all the data on strangeness-changing decays.[16]

In general,

$$Q = T_3 + \frac{(S + B)}{2}. \tag{15.17}$$

Thus

$$\Delta T_3 = \Delta(Q - S) + \frac{\Delta S}{2}. \tag{15.18}$$

Hence, given $\Delta Q = \Delta S$ and $\Delta S = \pm 1$, we have, for leptonic strangeness changing decays,

$$\Delta T_3 = \pm \tfrac{1}{2}. \tag{15.19}$$

The simplest way to arrange this from the point of view of the current S_μ is to demand that S_μ transforms under isotopic spin transformations like an isotopic spinor. In fact we can summarize the strangeness and isotopic spin requirements on S_μ by the demand that S_μ have the same transformation properties under these operations as the K^+ meson field.[17] We recall from the work of the last chapter that, in quark language, the field that transforms like the K^+ *creation* operator has the form $\bar{q}(\lambda_4 + i\lambda_5)q/2$; hence the selection rules that we want can be built into the theory, in the good SU_3 limit, if we demand that S_μ has the SU_3 transformation properties

$$S_\mu = S_{\mu 4} + iS_{\mu 5}, \tag{15.20}$$

where the subscripts 4 and 5 refer to the appropriate octet character of S_μ.

---

14. W. Willis et al., *Phys. Rev. Letters* **13**:291 (1964).
15. See, for example, Y. Cho et al., Proceedings of the 1966 Berkeley Conference on High Energy Physics.
16. See, for example, N. Cabibbo, Proceedings of the 1966 Berkeley Conference on High Energy Physics for a review.
17. Of course the K^+ is a pseudoscalar so that we do not refer to the space-time transformation properties of S_μ here.

Following Cabibbo,[18] we can summarize all of the insights we have so far obtained in the hypothesis that the full weak current W_μ can be written in the form[19]

$$W_\mu = \cos\theta\,(J_{\mu 1} + iJ_{\mu 2}) + \sin\theta\,(S_{\mu 4} + iS_{\mu 5}) + L_\mu. \tag{15.21}$$

Here

$$J_{\mu i} = V_{\mu i} + A_{\mu i}, \tag{15.22}$$

with V_μ and A_μ the strangeness nonchanging currents introduced previously and

$$S_{\mu i} = V_{\mu i}' + A_{\mu i}'. \tag{15.23}$$

We need both V_μ' and A_μ' for the strangeness changing decays since the process $K^+ \longrightarrow \mu^+ + \nu_\mu$ goes by pure A', while the process $K^0 \longrightarrow \pi^- + l^+ + \bar{\nu}_e$ goes by pure V' as K and π have the same parity.[20] In Eq. (15.23) we have made the simplest assumption, which is that V' and A' enter S_μ with the same intrinsic coupling strength, so that any differences in the effective couplings are due to renormalization effects in the matrix elements. The leptonic current L_μ is as before. The fact that the strangeness changing decays appear to be fundamentally weaker than the strangeness nonchanging decays is summarized by the introduction of the "Cabibbo angle," as indicated. We may make several remarks about this angle.

1. If we had simply

$$W_\mu = aJ_\mu + bS_\mu, \tag{15.24}$$

with no leptonic current, then the normalization condition

$$a^2 + b^2 = 1 \tag{15.25}$$

would be a trivial one, since we could always redefine the over-all coupling constant multiplying W_μ so that this was true. The fact that the coefficient of L_μ is taken to be unity relative to those of J_μ and S_μ makes the introduction of the Cabibbo angle nontrivial.

2. In general, one might have written

$$W_\mu = aJ_\mu + bS_\mu + L_\mu, \tag{15.26}$$

with

$$a^2 + b^2 \neq 1. \tag{15.27}$$

Following Gell-Mann[21] we shall now argue that the condition $a^2 + b^2 = 1$ is an expression (or, perhaps better, a new definition) of hadron-lepton "universality"

---

18. N. Cabibbo, *Phys. Rev. Letters* **10**:531, (1963). See also M. Gell-Mann and M. Lévy, *Nuovo Cimento* **16**:705 (1960).
19. As written, $J_{\mu i}$ and $S_{\mu i}$ might be members of *different* octets. As we shall emphasize below, most of the more interesting predictions of the Cabibbo theory flow from the assumption that they are actually members of the *same* octet.
20. That is to say, $\langle\pi|S_\mu|K\rangle = \langle\pi|V_\mu'|K\rangle$.
21. M. Gell-Mann, *Physics 1*, 63 (1964).

in the weak interactions. We recall from Chapter 12 that if we define a four-component lepton field

$$\psi = \begin{pmatrix} \psi_e \\ \psi_\mu \\ \psi_{\nu_e} \\ \psi_{\nu_\mu} \end{pmatrix} \tag{15.28}$$

and 4 × 4 "isotopic spins" (see Eq. (12.27)), we had, for (say) the vector leptonic "charges"

$$\mathbf{Q}_l = \int \psi^\dagger \frac{\boldsymbol{\tau}}{2} \psi \, d\mathbf{r} \tag{15.29}$$

the commutation relations

$$[Q_{li}, Q_{lj}]_- = i\epsilon_{ijk} Q_{lk}. \tag{15.30}$$

In particular, if we call

$$Q_{l+} = Q_{l1} + iQ_{l2}, \tag{15.31}$$

we have

$$[Q_{l+}, Q_{l-}]_- = 2Q_{l3} = N_{e^-} + N_{\mu^-} - N_{\nu_e} - N_{\nu_\mu}, \tag{15.32}$$

where the N's in Eq. (15.32) give the number of leptons minus the number of antileptons in each case. The particular structure of these commutation relations depends on the fact that there are two distinct species of neutrino, ν_μ associated with μ and ν_e associated with e. If we form the vector charges associated with the hadronic part of W_μ, that is,

$$H_\pm = \int d\mathbf{r} \, [a[V_{01} \pm iV_{02}] + b[V_{04} \pm iV_{05}]], \tag{15.33}$$

we can compute $[H_+, H_-]_-$ if we make the assumption that the currents in Eq. (15.33) are members of one single SU_3 octet. From this point of view we drop the notational distinction between J_μ and S_μ or between V_μ and V_μ', since we shall assume from now on that there is *one* octet of currents, $V_{\mu i}$, whose 1, 2 components give ordinary β-decay and whose 4, 5 components give the strangeness changing β-decays. This, in fact, was the assumption of Cabibbo.[22] Making use of the SU_3 commutation relations for

$$Q_i = \int d\mathbf{r} \, V_{0i}(\mathbf{r}, 0), \tag{15.34}$$

that is,

$$[Q_i, Q_j]_- = if_{ijk} Q_k, \tag{15.35}$$

we have

22. N. Cabibbo, op. cit.

$$[H_+, H_-]_- = 2a^2 Q_3 + b^2(Q_3 + \sqrt{3}Q_8) - 2abQ_6. \tag{15.36}$$

We can always *define*

$$H_3 = \tfrac{1}{2}[H_+, H_-]_-, \tag{15.37}$$

but the question is, does this H_3, along with H_+ and H_-, form a closed SU_2 algebra in analogy with the algebra formed by $Q_{l\pm}$ and Q_{l3}. In particular, we must have

$$[H_3, H_+]_- = H_+. \tag{15.38}$$

However, if we work out this commutator explicitly, we find instead of Eq. (15.38) the equation

$$[H_3, H_+]_- = (a^2 + b^2)H_+. \tag{15.39}$$

Thus to preserve the analogy with the commutation relations among the leptonic charges we are forced to the normalization

$$a^2 + b^2 = 1 \tag{15.40}$$

and hence to a Cabibbo *angle*.

3. While no convincing argument has been found that gives a priori the magnitude of the Cabibbo angle, it is important to emphasize that the magnitude of this angle as determined in several types of experiments[23] is essentially the same.

a. As discussed in Chapter 10 (Eq. (10.63)), one way of accounting for the apparent small discrepancy between the weak constant G, as measured in μ-decay and as measured in 0–0 nuclear transitions, such as O^{14}, is to assume that, a la Cabibbo, the hadronic, nonstrangeness changing constant is not $G/\sqrt{2}$ but rather $(G/\sqrt{2})\cos\theta$. From the rate of such transitions one obtains[23]

$$1 - \cos\theta = (2.2 \pm .2 \pm .5) \times 10^{-2}. \tag{15.41}$$

The first error is experimental and the second represents a measure of the theoretical uncertainty attached to computing radiative corrections in heavy nuclei. Eq. (15.41) gives

$$\sin\theta = .210 \pm .016 \tag{15.42}$$

b. In the Cabibbo theory the two decays $K^- \longrightarrow \mu^- + \nu_\mu$ and $\pi^- \longrightarrow \mu^- + \nu_\mu$ are given by the matrix elements

$$\frac{G}{\sqrt{2}}\langle 0|W_\mu|\mathbf{K}^-\rangle = \frac{i\sin\theta}{\sqrt{2E_K}}g_K(-m_K{}^2)p_{K\mu} \tag{15.43}$$

and

23. We follow here the review summary of N. Cabibbo, International Conference on High Energy Physics at Berkeley (1966).

$$\frac{G}{\sqrt{2}}\langle 0|W_\mu|\pi^-\rangle = \frac{i\cos\theta}{\sqrt{2E_\pi}}\,g_\pi(-m_\pi^2)p_{\pi\mu}, \tag{15.44}$$

respectively. Since only the axial vector contributes here, we have exploited the assumption that the intrinsic coupling of vector and axial vector is the same in W_μ. There is some ambiguity on how these formulas should be used to obtain, say, $\tan\theta$. In the good SU_3 limit we have the equation, with Q_i, the ith SU_3 generator (Eq. (13.62)),

$$Q_i|A_j\rangle = if_{ijk}|A_k\rangle. \tag{15.45}$$

Thus

$$(Q_6 + iQ_7)|A_1 - iA_2\rangle = -|A_4 - iA_5\rangle, \tag{15.46}$$

so that, using Eq. (13.30) and the fact that $Q_i|0\rangle = 0$,

$$\begin{aligned}\langle 0|A_{\mu 4} + iA_{\mu 5}|K^-\rangle &= -\langle 0|(A_{\mu 4} + iA_{\mu 5})(Q_6 + iQ_7)|\pi^-\rangle \\ &= -\langle 0|[A_{\mu 4} + iA_{\mu 5}, Q_6 + iQ_7]_-|\pi^-\rangle = -\langle 0|A_{\mu 1} + iA_{\mu 2}|\pi^-\rangle,\end{aligned} \tag{15.47}$$

which would indicate that, in the good SU_3 limit,

$$|g_K| = |g_\pi|. \tag{15.48}$$

In the absence of a detailed theory of the SU_3 symmetry breaking it is not clear how much this result will be changed by putting in the physical masses and breaking the symmetry. We might take the position, often adopted in the literature, that the dominant effect of the symmetry breaking is to replace the common octet mass by the actual physical masses, while even in the presence of symmetry breaking $|g_K| \sim |g_\pi|$. With this assumption the ratio of the $\pi \longrightarrow \mu + \nu$ and $K \longrightarrow \mu + \nu$ experimental rates can be used to fix θ. Indeed, using the relation

$$\frac{\Gamma(K^- \longrightarrow \mu^- + \nu)}{\Gamma(\pi^- \longrightarrow \mu^- + \nu)} = \tan^2\theta\,\frac{m_K}{m_\pi}\left(\frac{1 - \dfrac{m_\mu^2}{m_K^2}}{1 - \dfrac{m_\mu^2}{m_\pi^2}}\right)^2, \tag{15.49}$$

one finds[24]

$$\sin\theta \simeq 0.262, \tag{15.50}$$

in only fair agreement with the determination in (a). The discrepancy might be attributed to effects of the symmetry breaking on the g's. In fact, if θ is taken from method (a) above or method (c) below, which gives, as we shall see, the same value as (a), both of which may be freer of the symmetry breaking problem, one learns that

$$\frac{g_K}{g_\pi} \simeq 1.28, \tag{15.51}$$

---

24. T. D. Lee and C. S. Wu, "Decays of Charged K Mesons," *Annual Review of Nuclear Science*, Vol. 16, Chapter 8, 1966.

which is about the order of magnitude one might anticipate from a first-order SU_3 breaking interaction. The discrepancy between the Cabibbo angle as measured in vectorial and axial vectorial processes is sometimes attributed to distinct angles for these currents. It seems to us that this compromises the elegance of the model, so we prefer to attribute it to symmetry breaking.

c. The decay $K^- \longrightarrow \pi^0 + e^- + \bar{\nu}_e$ (or $K^0 \longrightarrow \pi^- + e^+ + \nu_e$) gives a rather direct determination of $\sin\theta$. Indeed (see Eq. (5.14)), since K and π have the same parity,

$$(2\pi)^3\sqrt{2E_\pi 2E_K}\langle\pi^0|W_\mu(0)|K^-\rangle = (2\pi)^3\sqrt{2E_\pi 2E_K}\,\sin\theta\langle\pi^0|V_{\mu 4}(0) + iV_{\mu 5}(0)|K^-\rangle$$
$$= \sin\theta\,[F_+(q^2)(p_\pi + p_K)_\mu + F_-(q^2)(p_K - p_\pi)_\mu], \quad (15.52)$$

with q the momentum transferred to the leptons:

$$q_\mu = (p_K - p_\pi)_\mu = (l + \nu_l)_\mu. \quad (15.53)$$

This matrix element is to be multiplied, in computing the decay amplitude, by L_μ, the lepton current. Since, to lowest order in G, the weak coupling constant, the fields in L_μ are free, and since L_μ contains terms of the form $\bar{u}_l\gamma_\mu(1 + \gamma_5)u_{\nu_e}$, the contribution to the matrix element of W_μ from $F_-(q^2)$ is multiplied, effectively, by m_l, the mass of the lepton. We can see this simply from the Dirac equation

$$(i\gamma l + m_e)u_l = 0. \quad (15.54)$$

Thus in the decay $K^- \longrightarrow \pi^0 + e^- + \bar{\nu}_e$, the contribution to the matrix element from F_- can be neglected. The simplest assumption one can make about F_+ is that, over the range of q^2 in the physical region for this decay, it is essentially constant. In fact, it is customary to represent F_+ in the parametric form

$$F_+(q^2) \simeq F_+(0)\left(1 - \frac{q^2}{M_+^2}\right), \quad (15.55)$$

where M_+ is a "mass." An expression of this form would arise in a dispersion relation treatment of the $K - \pi$ lepton vertex in terms of intermediate particles. One such particle, in fact the only known one of relatively low mass, with the correct quantum numbers,[25] is the K^{*+} at a mass 892.4 ± 0.8. It can enter in a dispersion diagram of the form

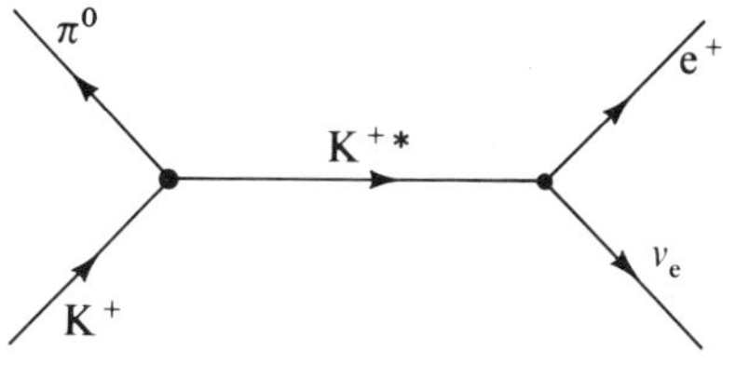

25. It must be a vector in Lorentz space, and it must carry strangeness and isotopic spin $\frac{1}{2}$ in order to maintain the $\Delta T = \frac{1}{2}$ rule.

The maximum q^2 in this decay is

$$q^2 = -(m_K - m_\pi)^2, \tag{15.56}$$

so that, from the diagram, or the dispersion relation

$$F_+(q^2) = \frac{F_+(0)}{1 + \dfrac{q^2}{m_{K^*}^2}} \simeq F_+(0)\left[1 - \frac{q^2}{m_{K^*}^2}\right]. \tag{15.57}$$

We can expand Eq. (15.57) in q^2, since

$$\frac{q^2}{m_{K^*}^2} \leq \frac{(m_K - m_\pi)^2}{m_{K^*}^2} \simeq .16. \tag{15.58}$$

We can now use SU_3 arguments to eliminate $F_-(q^2)$ from the problem, in general, and to determine $F_+(0)$. In the limit of good SU_3 the Cabibbo current, which generates the $K^- \longrightarrow \pi^0 + e^- + \bar{\nu}_e$ decay, $V_{\mu 4}(x) + iV_{\mu 5}(x)$, is conserved. We can see this directly from the commutation relation[26]

$$[Q_6 + iQ_7, V_{\mu 1}(x) + iV_{\mu 2}(x)]_- = -(V_{\mu 4}(x) + iV_{\mu 5}(x)) \tag{15.59}$$

if we take the four-divergence of Eq. (15.59) and use the conservation of the isotopic spin currents. The conservation of the strangeness-changing current is only possible in the limit in which the strange and nonstrange particles are degenerate in mass, since with H the strong Hamiltonian, the equation

$$[H, Q_i]_- = 0 \tag{15.60}$$

implies that if $|A\rangle$ is a state of zero strangeness, then $(Q_4 + iQ_5)|A\rangle$ is a state of finite strangeness and the same mass; that is, the strange and nonstrange particles are degenerate in mass. Thus the conservation of $V_{\mu 4} + iV_{\mu 5}$ implies that

$$F_-(q^2) = 0 \tag{15.61}$$

simply by multiplying Eq. (15.52) by q_μ and setting the result equal to zero. We can also use these SU_3 ideas to evaluate $F_+(0)$. In fact, arguing as in Chapter 10 (Eq. (10.77)), we have

$$F_+(0) = \langle \pi^0 | Q_4 + iQ_5 | K^- \rangle. \tag{15.62}$$

But, from Eq. (15.45),

$$(Q_4 + iQ_5)|A_4 - iA_5\rangle = |A_3\rangle.$$

Since

$$\pi^0 = |A_3\rangle,$$
$$K^- = \frac{|A_4 - iA_5\rangle}{\sqrt{2}}, \tag{15.63}$$

---

26. Recall that in the Cabibbo theory the SU_3 generators are given by

$$Q_i = \int V_{0i}(\mathbf{r}, 0)\, d\mathbf{r}.$$

27. M. Ademollo and R. Gatto, *Phys. Rev. Letters* **13**:264 (1964), and C. Bouchiat and P.

we have a well-known SU_3 result that states that

$$F_+(0) = \frac{1}{\sqrt{2}}. \tag{15.64}$$

This value is, of course, altered by the SU_3 symmetry-breaking interaction. However, we[27] can make use of the fact that $F_+(0)$ is related to the matrix element of Q_l, the SU_3 symmetry generator to show that, in this case, there are no *first-order* symmetry breaking effects. In essence, the argument goes as follows. If V is the symmetry breaking interaction, and $|A_i\rangle$ is a member of an SU_3 multiplet, then to first order in the symmetry breaking the physical state $|A_i\rangle_P$ is given by

$$|A_i\rangle_P = |A_i\rangle + \sum_n |n\rangle \frac{\langle n|V|A_i\rangle}{E_0 - E_n}, \tag{15.65}$$

where $|n\rangle$ is a member of a complete set of states *not* in the original multiplet. The symmetry breaking interaction V can have diagonal terms within a multiplet, but since it commutes, presumably, with Y, T_3, and T, it can only have off-diagonal terms connecting states in different multiplets.[28] The diagonal terms do not enter the perturbation sum. Now

$$\langle A_j|_P Q_l|A_i\rangle_P = \langle A_j|Q_l|A_i\rangle + \sum_n \frac{\langle A_j|Q_l|n\rangle\langle n|V|A_i\rangle}{E_0 - E_n} + \sum_n \frac{\langle A_j|V|n\rangle\langle n|Q_l|A_i\rangle}{E_0 - E_n}. \tag{15.66}$$

However [see Eq. (15.45)], Q_l acting on any $|A_i\rangle$ gives back states in the same multiplet, and since

$$\langle n|A_j\rangle = 0, \tag{15.67}$$

the first-order symmetry breaking effects on matrix elements of Q_l vanish. Hence we would expect that the determination of θ from the decay $K^- \longrightarrow \pi^0 + e^- + \bar{\nu}_e$ should agree more closely with the determination from ordinary β-decay than the determination from the $K \longrightarrow \mu + \nu_\mu$: $\pi \longrightarrow \mu + \nu_\mu$ ratio, which involves only the axial vector for which there is no comparable low-energy theorem. Using a recent experimental determination of the rate[29] for $K^+ \longrightarrow \pi^0 + e^+ + \nu_e$,

$$\Gamma(K^+ \longrightarrow \pi^0 + e^+ + \nu_e) = (3.61 \pm 0.20)10^6 \text{ sec}^{-1}.$$

(The theoretical discussion for this process is identical, so far as the rate is concerned, to the one we have given for the charge conjugate process.) We find,[29] with $M_+ = \infty$,

$$\sin\theta = 0.222 \pm 0.006, \tag{15.68}$$

Meyer, *Nuovo Cimento* **34**:1122 (1964). The treatment that we follow here is that of J. S. Bell, *Theory of Weak Interactions*, Les Houches Summer School Lectures (1965). See also R. E. Behrends and A. Sirlin, *Phys. Rev. Letters* **4**:186 (1960).

28. It can connect states like η^0 and x^0 or ω and ϕ.
29. N. Cabibbo, Berkeley Conference, op. cit.

and if $M_+ = 890\text{ MeV} \simeq m_K{}^*$, then

$$\sin\theta = .21 \tag{15.69}$$

Both of these results agree, excellently, with the value of $\sin\theta$ found from ordinary β-decay; Eq. (15.42).

d. The Cabibbo theory also makes, in the good SU_3 limit, a set of predictions for hyperon β-decay. The discussion here closely parallels the one given in Chapter 13 for the matrix elements of the electromagnetic currents. We introduce, once again, the 3×3 matrices $\bar{B}$, B, and λ_i of Chapter 13. We can now write the general octet matrix elements in the form (suppressing Lorentz indices) for the vector part,

$$\begin{aligned}\langle\bar{B}|W|B\rangle = \cos\theta\Big[F_V \operatorname{Tr}\Big[\frac{(\lambda_1+i\lambda_2)}{2}[B,\bar{B}]_-\Big] \\ + D_V \operatorname{Tr}\Big[\Big(\frac{\lambda_1+i\lambda_2}{2}\Big)[B,\bar{B}]_+\Big]\Big] \\ + \sin\theta\Big[F_V \operatorname{Tr}\Big[\frac{(\lambda_4+i\lambda_5)}{2}[B,\bar{B}]_-\Big] \\ + D_V \operatorname{Tr}\Big[\frac{(\lambda_4+i\lambda_5)}{2}[B,\bar{B}]_+\Big]\Big].\end{aligned} \tag{15.70}$$

There is a similar form for the axial part with D_V, F_V replaced by D_A, F_A, two new coupling constants (form factors) which are, in general, distinct from D_V and F_V. The fact that the *same* D's and F's apply to both the strangeness changing and nonchanging parts of the matrix element is a consequence of Cabibbo's assumption that the currents J_μ and S_μ are members of the *same* octet. Without this assumption the strangeness changing and nonchanging decays would not be simply related through the Cabibbo angle. These traces are simple to work out and we can express the result in the long formula[30]

$$\begin{aligned}&\cos\theta\Big[F(\sqrt{2}\bar{\Sigma}^0\Sigma^- - \sqrt{2}\bar{\Sigma}^+\Sigma^0 - \bar{\Xi}^0\Xi^- + \bar{P}N) \\ &\quad + D\Big(\frac{2}{\sqrt{6}}\bar{\Lambda}^0\Sigma^- + \frac{2}{\sqrt{6}}\bar{\Sigma}^+\Lambda^0 + \bar{\Xi}^0\Xi^- + \bar{P}N\Big)\Big] \\ &\quad + \sin\theta\Big[F\Big(\frac{3}{\sqrt{6}}\bar{\Lambda}^0\Xi^- + \frac{1}{\sqrt{2}}\bar{\Sigma}^0\Xi^- + \bar{\Sigma}^+\Xi^0 - \bar{N}\Sigma^- - \frac{3}{\sqrt{6}}\bar{P}\Lambda^0 - \frac{1}{\sqrt{2}}\bar{P}\Sigma^0\Big) \\ &\quad + D\Big(-\frac{1}{\sqrt{6}}\bar{\Lambda}^0\Xi^- + \frac{1}{\sqrt{2}}\bar{\Sigma}^0\Xi^- + \bar{\Sigma}^+\Xi^0 + \bar{N}\Sigma^- + \frac{1}{\sqrt{2}}\bar{P}\Sigma^0 - \frac{1}{\sqrt{6}}\bar{P}\Lambda^0\Big)\Big].\end{aligned} \tag{15.71}$$

30. We drop the A, V subscript on D and F.

From this equation one may now read off the matrix elements for all the energetically possible decays. We record the results in the table below.

Table 1

Decay	*Matrix Element*
1. $p \longrightarrow n + e^- + \bar{\nu}_e$	$(D + F)\cos\theta$
2. $\Sigma^\pm \longrightarrow \Lambda^0 + e^\pm + \nu_e$	$\sqrt{2}/3D\cos\theta$
3. $\Lambda^\circ \longrightarrow p + e^- + \bar{\nu}_e$	$-\left(3/\sqrt{6}F + \frac{1}{\sqrt{6}}D\right)\sin\theta$
4. $\Sigma^- \longrightarrow n + e^- + \bar{\nu}_e$	$(D - F)\sin\theta$
5. $\Xi^- \longrightarrow \Lambda^0 + e^- + \bar{\nu}_e$	$\left(3/\sqrt{6}F - \frac{1}{\sqrt{6}}D\right)\sin\theta$
6. $\Sigma^0 \longrightarrow p + e^- + \bar{\nu}_e$	$\left(-\frac{1}{\sqrt{2}}F + \frac{1}{\sqrt{2}}D\right)\sin\theta$
7. $\Sigma^+ \longrightarrow \Sigma^0 + e^+ + \nu_e$	$-\sqrt{2}F\cos\theta$
8. $\Xi^- \longrightarrow \Sigma^0 + e^- + \bar{\nu}_e$	$\left(\frac{1}{\sqrt{2}}F + \frac{1}{\sqrt{2}}D\right)\sin\theta$
9. $\Xi^- \longrightarrow \Xi^0 + e^- + \bar{\nu}_e$	$(D - F)\cos\theta$

There are several remarks that can be made about this table. The F's and D's are really functions of q^2, the momentum transferred to the leptons. For the q^2 relevant to these decays it is, probably, a good approximation to replace the form factors by their value at $q^2 = 0$. This means, in particular, that

$$F_V = 1, \qquad D_V = 0. \tag{15.72}$$

To see the latter result we argue as follows: at $q^2 = 0$ the matrix elements for the vector part of the leptonic decays are given, in the Cabibbo theory, by the matrix element of the associated charge or SU_3 generator. For the $\Delta S = 0$ decays this charge is the isotopic spin $T_\pm$. Now

$$[T_\pm, T^2]_- = 0, \tag{15.73}$$

so that

$$0 = \langle T'T_3'|[T_\pm, T^2]_-|TT_3\rangle = (T'(T'+1) - T(T+1))\langle T'T_3'|T_\pm|TT_3\rangle. \tag{15.74}$$

Thus we have the selection rule for allowed vector β-transitions:

$$\Delta T = 0. \tag{15.75}$$

This selection rule holds for the vector part of the $\Delta S = 0$, β-decays so long as $q^2 \simeq 0$. Since the decay $\Sigma^- \longrightarrow \Lambda^0 + e^- + \bar{\nu}_e$ is both $\Delta T \neq 0$ and pure D, we conclude that

$$D_V = 0. \tag{15.76}$$

But, from the nonrenormalization of the vector charge,

$$D_V + F_V = 1. \tag{15.77}$$

Thus

$$F_V = 1. \tag{15.78}$$

By the Ademollo-Gatto theorem[31] we expect that these results are not changed by first-order terms in the SU_3 symmetry breaking, but they will not hold when $q^2 \neq 0$.

No comparable theorems exist for D_A and F_A although, as we shall see, it is possible to compute F_A/D_A in some theoretical models. One may use the limited experimental data to find D_A/F_A and θ,[32] with the results

$$\theta = .240 \pm .014,$$
$$\frac{F_A}{D_A} \simeq .50, \tag{15.79}$$
$$|F_A + D_A| = |F_A(0)| = 1.14^{+0.23}_{-0.33},$$

which gives reasonable agreement with the other values of θ.[33]

That these four experimental determinations of θ agree, within the kind of errors encompassed by the theory, lends plausibility to the basic assumptions of the Cabibbo model and suggests making an attempt to explain, dynamically, the F_A/D_A ratio. (An explanation of θ itself has so far eluded theorists.) The simplest calculation of F_A/D_A is based on the quark model.[34] Nonrelativistically[35] the quark axial vector operator effective in, say, the transition $\Sigma^- \longrightarrow \Lambda^0 + e^- + \bar{\nu}_e$, is

$$\mathbf{A} = \sum_{i=1}^{3} \tau_i^+ \boldsymbol{\sigma}_i \cos\theta \tag{15.80}$$

with the sums over the quarks; in the same limit, the vector is given by

$$V_0 = \sum_{i=1}^{3} \tau_i^+ \cos\theta. \tag{15.81}$$

The transition elements can be computed directly using the baryon wave functions (Eq. (14.7)). The results can be related to D_A and F_A by using the table. In his computation Dalitz[34] allows for the fact that V and A may not enter with

31. Op. cit.
32. See, for example, W. Willis et al. *Phys. Rev. Letters* **13**:291 (1964), N. Brene et al. *Phys. Letters* **11**:344 (1964), and especially N. Brene et al., *Phys. Rev.* **149**:1288 (1964).
33. In giving the last number we have set $F_V(0) = 1$ as a consequence of the conserved vector current in the good SU_3 limit. What is actually measured from Λ^0 β-decay is

$$\frac{F_A(0)}{F_V(0)} = -1.14^{+0.23}_{-0.33}.$$

The D_A/F_A ratio is still not fixed with complete certainty by experiment. We have adopted the numbers given in Cabibbo's Berkeley review summary (op. cit.). To give an impression

the same coefficient in the total current by writing the effective interaction in the form $V + RA$. Thus

$$\langle P|V_0|N\rangle = \cos\theta, \qquad \langle P|A_3|N\rangle = 5/3\, R\cos\theta, \tag{15.82}$$

so that

$$\frac{F_A(0)}{F_V(0)} = \frac{5}{3}R, \tag{15.83}$$

which comparing with experiment gives

$$|R| \simeq \frac{1}{\sqrt{2}}. \tag{15.84}$$

Moreover,

$$D_A + F_A = 5/3\, R\cos\theta. \tag{15.85}$$

Likewise, with no V_0 contribution,

$$\langle\Lambda^0|A_3|\Sigma^-\rangle = \sqrt{\frac{2}{3}}\, R\cos\theta, \tag{15.86}$$

so that, from the table

$$D_A = R\cos\theta. \tag{15.87}$$

Thus

$$\frac{D_A}{D_A + F_A} = \frac{3}{5}. \tag{15.88}$$

This can be compared to the experimental number (Eq. (15.79)),

$$\frac{1}{1 + \dfrac{(F_A)_{\text{exp}}}{(D_A)_{\text{exp}}}} \simeq \frac{1}{1 + \frac{1}{2}} = \frac{2}{3}, \tag{15.89}$$

which is at least close enough to Eq. (15.88) to be suggestive.

A more profound, or at least, more complicated, discussion of the D/F ratio can be given in terms of a generalization of *PCAC*, the Goldberger-Treiman

---

of the existing uncertainties we may quote another set of values from the recent literature:

$$|D_A| = 0.766 \pm 0.037,$$
$$|F_A| = 0.415 \pm 0.035,$$
$$\theta = 0.245 \pm 0.010,$$

from C. Carlson, *Phys. Rev.* **152**:1433 (1966). Cabibbo's numbers, have, in any case, the virtue that they are simple to compute with.

34. See, for example, R. H. Dalitz, *Les Houches Lectures*, op. cit.
35. Remember that for $\Delta S = 0$ transitions the axial quark current is given by

$$A_\mu = i\bar{q}\gamma_5\gamma_\mu\left(\frac{\tau_1 + i\tau_2}{2}\right)q.$$

relation, and the Adler-Weisberger sum rules to SU_3. In good SU_3 all of the vector currents, $i = 1, \ldots, 8$, are, in the Cabibbo theory, conserved, since they are the currents associated with the SU_3 generators. This is consistent with the degeneracy among the masses in the SU_3 octets. It is natural to generalize the strict *PCAC* defined in Chapter 13 for the strangeness nonchanging A_μ to the full octet of currents in the Cabibbo theory by assuming that

$$\partial_\mu A_{\mu i} = \phi_i(x) C_i, \tag{15.90}$$

where the C_i are constants (to be discussed) and the ϕ_i are the field operators corresponding to the pseudoscalar octet. In the quark current model, at least with free quarks, one has the relation

$$\partial_\mu \bar{q}\left(\gamma_\mu \gamma_5 \frac{\lambda_i}{2}\right) q = 2m_q \bar{q} \frac{\lambda_i}{2} \gamma_5 q. \tag{15.91}$$

This is a kind of *PCAC* if one identifies the combination $\bar{q}(\lambda_i/2)\gamma_5 q$ with the pseudoscalar octet. Of course, there is no guarantee that this equation is maintained in the presence of quark interactions and, in any case, one cannot take the combination $\bar{q}(\lambda_i/2)\gamma_5 q$ literally as the pseudoscalar field, since it has the wrong commutation relations. We will exhibit a more sophisticated model below in which *PCAC*, in the strict sense, is built in. We can repeat the steps in Chapter 11 that lead to the G.T. relation, in outline. Let

$$\partial_\mu A_{\mu i} = O_i^5 \tag{15.92}$$

and let $|\bar{B}\rangle$ and $|B\rangle$ be the octet baryon states. Then, at zero momentum transfer,[36]

$$\langle \bar{B}|O_j^5|B\rangle = -i2M\left[F_A \operatorname{Tr}\left[\frac{\lambda_i}{2}[\gamma_5 B, \bar{B}]_-\right] + D_A \operatorname{Tr}\left[\frac{\lambda_i}{2}[\gamma_5 B, \bar{B}]_+\right]\right], \tag{15.93}$$

where M is the common octet mass. In fact, we are only interested in the components of this equation which change charge, since all the weak decays involve charge transfer. For the strangeness changing decays, Eq. (15.93) becomes

$$\langle \bar{B}|O_4^5 + iO_5^5|B\rangle = -i2M\left[F_A \operatorname{Tr}\left[\frac{\lambda_4 + i\lambda_5}{2}[\gamma_5 B, \bar{B}]_-\right]\right.$$
$$\left. + D_A \operatorname{Tr}\left[\frac{\lambda_4 + i\lambda_5}{2}[\gamma_5 B, \bar{B}]_+\right]\right]. \tag{15.94}$$

---

36. In the literature one often finds an equivalent representation of these matrix elements written in the form

$$\langle A_i(\mathbf{p})|A_{\mu k}(0)|A_j(\mathbf{p})\rangle = g_A[(1-\alpha)f_{ijk} + \alpha d_{ijk}]i\bar{u}(p)\gamma_\mu\gamma_5 u(p),$$

where the $|A_i\rangle$ are members of the baryon octet and $A_{\mu k}(0)$ is the axial current, $\mathbf{p}$ is the three-momentum of A_i, and the f_{ijk} and d_{ijk} are the SU_3 constants defined by Eq. (13.6) and Eq. (13.8). The connection between the parameters in the expression above and the ones that we have been using is given by the relations

$$g_A = D_A + F_A$$

and

To obtain the generalized *G.-T.* relation we assume that $O_4{}^5 + iO_5{}^5$ acts, at least at small q^2, like the K^+ field, which is to say that $\langle \bar{B}|O_4{}^5 + iO_5{}^5|B\rangle$ has a pole at $q^2 = -m_K{}^2$, whose residue is related to the coupling constants for the emission of K's by the various baryons. Repeating the steps of Chapter 11 we are led to the relation

$$\frac{G}{\sqrt{2}}\left[2M\left[F_A \operatorname{Tr}\left[\frac{\lambda_4 + i\lambda_5}{2}[\gamma_5 B, \bar{B}]_-\right] + D_A \operatorname{Tr}\left[\frac{\lambda_4 + i\lambda_5}{2}[\gamma_5 B, \bar{B}]_+\right]\right]\right]$$
$$\simeq g_K(-m_K{}^2)\left[f \operatorname{Tr}\left[\frac{\lambda_4 + i\lambda_5}{2}[\gamma_5 B, \bar{B}]_-\right] + d \operatorname{Tr}\left[\frac{\lambda_4 + i\lambda_5}{2}[\gamma_5 B, \bar{B}]_+\right]\right]. \tag{15.95}$$

The quantities on the left side of Eq. (15.95) have all been defined previously. On the right side g_K is as defined in Eq. (15.2). The constants f and d are the effective strong baryon-baryon-K-meson coupling constants. These couplings might arise from a fundamental, SU_3 invariant, Yukawa Lagrangian of the form (M is the 3×3 meson matrix)

$$\mathfrak{L}_{mBB} = f \operatorname{Tr}\left[M[\gamma_5 B, \bar{B}]_-\right] + d \operatorname{Tr}\left[M[\gamma_5 B, \bar{B}]_+\right]. \tag{15.96}$$

One can draw an important consequence from Eq. (15.95). If one takes the ratio of the terms involving F_A and D_A, on the left, to the terms involving f and d, on the right, the result is a constant that is independent of any of the parameters of SU_3. This is only possible if

$$\frac{F_A}{D_A} = \frac{f}{d}, \tag{15.97}$$

which gives a remarkable connection between the strong and weak coupling constants. A detailed discussion of how f/d is determined experimentally is beyond the scope of this book[37] since it involves strong interaction dynamics in an essential way. The conclusion appears to be that Eq. (15.97) is in reasonable agreement with the data. Equation (15.95) leads to several *G.T.* relations, depending on the choice of particles. If we stay with the p, Λ, K^+ system we have, as a special case,

$$g_K(-m_K{}^2) \simeq \frac{G}{\sqrt{2}}(M_P + M_\Lambda)\frac{(3F_A + D_A)}{(3f + d)}. \tag{15.98}$$

$$\alpha = \frac{\dfrac{D_A}{F_A}}{1 + \dfrac{D_A}{F_A}} = \frac{D_A}{g_A},$$

while

$$1 - \alpha = \frac{F_A}{g_A}.$$

37. See, for example, *Elementary Particle Physics* by S. Gasiorowicz, Wiley, 1966, Chapter 25 and references cited therein.

In the same notation,

$$g_\pi(-m_\pi^2) \simeq \frac{G}{\sqrt{2}} 2M_P \frac{(F_A + D_A)}{(f + d)}, \tag{15.99}$$

where

$$|F_A + D_A| = |g_A(0)| \tag{15.100}$$

and

$$f + d = \sqrt{2} f_\pi, \tag{15.101}$$

where f_π is the pion-nucleon coupling constant. Following the tradition in this subject we have made an attempt at allowing for the symmetry breaking by replacing $2M$ by $M_{\Lambda^0} + M_P$ in Eq. (15.98). Thus

$$\frac{g_K}{g_\pi} \simeq \left(\frac{M_P + M_\Lambda}{2M_P}\right) \frac{\left(\dfrac{3 + \dfrac{D_A}{F_A}}{1 + D_A/F_A}\right)}{\left(\dfrac{3 + \dfrac{d}{f}}{1 + \dfrac{d}{f}}\right)}. \tag{15.102}$$

In the limit of good SU_3 this gives

$$\frac{g_K}{g_\pi} \simeq 1 \tag{15.103}$$

and, if we put in the physical masses, but equate D_A/F_A and d/f, we obtain

$$\frac{g_K}{g_\pi} \simeq 1.09. \tag{15.104}$$

In fact, experimentally,[38]

$$\frac{d}{f} \simeq 2.7. \tag{15.105}$$

Thus

$$\frac{g_K}{g_\pi} \simeq 1.18, \tag{15.106}$$

to be compared to the Cabibbo angle value of 1.28. The general agreement of these numbers lends credence to the theory.[39]

---

38. C. Jarlskog and H. Pilkuhn, *Phys. Letters* **20**:428 (1966), B. R. Martin, *Phys. Rev.* **138**:B1136 (1965).
39. In the $\Delta S = 1$ applications of *PCAC* to the *G.T.* relation, the extrapolations between $q^2 = 0$ and $q^2 = -m_K^2$, which are necessary if SU_3 is broken (so that $m_\pi \neq m_K$), make the approximations even more dubious than in the $\Delta S = 0$ case and hence the general agreement even more remarkable.
40. W. I. Weisberger, *Phys. Rev.* **143**:1302 (1966), D. Amati, C. Bouchiat, and J. Nuyts, *Phys.*

PCAC can be used along with the sorts of arguments given in Chapter 12 to derive Adler-Weisberger relations for the $\Delta S = 1$ transitions.[40] The axial equal-time charge commutation relation

$$[Q_{5i}(t), Q_{5j}(t)]_- = if_{ijk}Q_k(t) \tag{15.107}$$

yields, as a special case,

$$[Q_{54} + iQ_{55}, Q_{54} - iQ_{55}]_- = \sqrt{3}Q_8 + Q_3 = \tfrac{3}{2}Y + T_3. \tag{15.108}$$

This commutation relation can be sandwiched in between nucleon states of spin s' and s. For the neutron,[41]

$$\begin{aligned}\langle N|_{s'}[Q_{54} + iQ_{55}, Q_{54} - iQ_{55}]_-|N\rangle_s &= 1 \\ &= \sum_{s'} |\langle \Sigma^-|_{s'}A_{04}(0) - iA_{05}(0)|N\rangle_s|^2 + R_N(\mathbf{p}, \mathbf{p}),\end{aligned} \tag{15.109}$$

reflecting the fact that the Σ^- is the *only* one-particle state that can be obtained from N by operating with $A_{04} - iA_{05}$. The coupling constant for Σ^- decays is given in terms of F_A and D_A by Table 1 and in the limit $v \longrightarrow 1$ (see Chapter 12). $R_N(P, P)$ is related to $K - N$ scattering. Indeed,[42]

$$1 = (D_A - F_A)^2 + \frac{2g_K^2}{\pi G^2}\int \frac{dq_0}{q_0}\,[\sigma_{K^-N}(q_0) - \sigma_{K^+N}(q_0)], \tag{15.110}$$

where the $\sigma_{K^\pm N}$ are the total $K^\pm$-neutron cross sections. We can use the experimental value of g_K taken from the $K \longrightarrow \mu + \nu_\mu$ decay or, invoking SU_3 and the *G.T.* relation, we can write

$$g_K^2 \simeq g_\pi^2 \simeq 2G^2M_P^2\frac{(F_A + D_A)^2}{(f + d)^2}. \tag{15.111}$$

Thus we have

$$1 = (D_A - F_A)^2 + \frac{2M_P^2 g_A^2}{\pi g_\pi^2}\int \frac{dq_0}{q_0}\,[\sigma_{K^-N}(q_0) - \sigma_{K^+N}(q_0)]. \tag{15.112}$$

On the other hand, taking the proton matrix elements gives

$$2 = \sum_{s'} |\langle \Lambda^0|_{s'}(A_{04} - iA_{05})|P\rangle_s|^2 + \sum_{s'} |\langle \Sigma^0|_{s'}A_{04} - iA_{05}|P\rangle_s|^2 + R_P(\mathbf{p}, \mathbf{p}). \tag{15.113}$$

Again using Table 1 and the *G.T.* relation we can write

$$1 = F_A^2 + \frac{1}{3}D_A^2 + \frac{M_P^2 g_A^2}{\pi g_\pi^2}\int \frac{dq_0}{q_0}\,[\sigma_{K^-P}(q_0) - \sigma_{K^+P}(q_0)]. \tag{15.114}$$

Letters **19**:59 (1965), C. A. Levinson and I. J. Muzinich, *Phys. Rev. Letters* **15**:715 (1965), L. K. Pandit and J. Schechter, *Phys. Letters* **19**:56 (1965).

41. Refer to Chapter 12 for details.
42. Our definition of f_π is related to Weisberger's (op. cit.) by

$$(f_\pi)_{us} = \frac{G}{\sqrt{2}}(f_\pi)_{\text{W.W.}}.$$

To orient ourselves we can set $M_P = 0$, which might correspond to the limit of a conserved axial current and solve the resulting equations for F_A and D_A. There are four solutions,

F_A	D_A
$\frac{1}{2}$	$\frac{3}{2}$
-1	0
1	0
$-\frac{1}{2}$	$-\frac{3}{2}$

of which the first produces the most promising D/F ratio; ~3. The authors cited above have evaluated the corrections due to the integrals over the cross sections. This evaluation is nontrivial and the interested reader should consult the papers. The quantity computed by them is

$$\alpha = \frac{D_A}{D_A + F_A} = \frac{D_A/F_A}{1 + D_A/F_A}, \tag{15.115}$$

with the results

$\alpha = 0.75 \pm .10$	Weisberger
$\alpha = 0.73$	Amati et al.
$\alpha = 0.63$	C. A. Levinson and I. J. Muzinich

These results are in general agreement with the experimental numbers. Weisberger also solves for $|g_A|$ and finds

$$|g_A| = 1.28 \pm 0.10,$$

again in accord with experiment, within the errors.

All of this is encouraging but none of it sheds much light on the basic question of why the strangeness changing decays are intrinsically weaker than the $\Delta S = 0$ decays: the answer is not known.

Nonleptonic decays. We now turn to the nonleptonic decays and, at the end of the chapter, come back to the question of the existence of field theoretic models that fulfill the conditions used in the above discussions—*PCAC* and the commutation algebra.

The fundamental problem posed by the nonleptonic decays of strange particles can be seen by considering the simplest selection rule that is apparently obeyed in these decays, the $\Delta T = \frac{1}{2}$ isotopic spin selection rule. For the leptonic decays this rule is very simply built into the theory by assuming certain transformation properties for the currents. Typically, for the $\Delta S = 1$ currents,

$$[J_{\mu 4} + iJ_{\mu 5}, T_3]_- = \tfrac{1}{2}[J_{\mu 4} + iJ_{\mu 5}]. \tag{15.116}$$

(See the table of f_{ijk} in Chapter 13.) Thus matrix elements of the form $\langle T'T_3'|J_{\mu 4} + iJ_{\mu 5}|TT_3\rangle$ vanish unless $\Delta T_3 = \frac{1}{2}$. Now the nonleptonic decays also appear to obey a $\Delta T = \frac{1}{2}$ rule, which we can summarize by saying that the effective Hamiltonian for nonleptonic decays, such as $\Sigma^- \longrightarrow n + \pi^-$, $K^0 \longrightarrow \pi^+ + \pi^-$, transforms in isotopic space like a spinor. Before discussing the theoretical questions, let us invoke some experimental data in support of the rule. To derive consequences of the rule we use the "spurion" technique, which consists in imagining that a typical reaction of the form

$$A \xrightarrow{H_{WK}^{\Delta S=1}} B + C \tag{15.117}$$

is written

$$S + A \longrightarrow B + C, \tag{15.118}$$

where S is a fictitious object of isospin $\frac{1}{2}$, a "spurion," that simulates the role of $H_{WK}^{\Delta S=1}$ in isospace. For example, in isotopic space, the decay $\Lambda^0 \longrightarrow n + \pi$ is treated by writing $\psi_{1/2}^{\text{in}}$ for the initial state (which, counting the spurion, is $T = \frac{1}{2}$). The final states, which must have zero electric charge, can be written

$$\begin{aligned} |\pi^- P\rangle &= \sqrt{\tfrac{1}{3}}\psi_{3/2}{}^{\text{out}} - \sqrt{\tfrac{2}{3}}\psi_{1/2}{}^{\text{out}}, \\ |N\pi^0\rangle &= \sqrt{\tfrac{2}{3}}\psi_{3/2}{}^{\text{out}} + \sqrt{\tfrac{1}{3}}\psi_{1/2}{}^{\text{out}}. \end{aligned} \tag{15.119}$$

Having counted the spurion in the initial state, isotopic spin is now, in effect, conserved. Thus we have from Eq. (15.119) a prediction of the $\Delta T = \frac{1}{2}$ rule,

$$\frac{R(\Lambda^0 \longrightarrow \pi^- + p)}{R(\Lambda^0 \longrightarrow \pi^0 + n)} = \frac{2}{1} \tag{15.120}$$

to be compared with the experimental numbers[43]

$$\frac{R(\Lambda^0 \longrightarrow \pi^- + p)_{\text{exp}}}{R(\Lambda^0 \longrightarrow \pi^0 + n)_{\text{exp}}} = \frac{66.4}{33.6} \pm 1.1. \tag{15.121}$$

Another application of the same idea is to the decay of the short-lived K^0 meson into $\pi^+ + \pi^-$ and $\pi^0 + \pi^0$. Because of Bose statistics the two π mesons can only be in a $T = 0$ or $T = 2$ S-state.[44] (The kaon has spin zero.) If we accept the $\Delta T = \frac{1}{2}$ rule, then the $T = 2$ state is ruled out. Now we can write the $T = 0$ state in the form

$$\pi_1 \cdot \pi_2 = \pi_1{}^+\pi_2{}^- + \pi_1{}^-\pi_2{}^+ + \pi^0\pi^0, \tag{15.122}$$

which means that

$$\frac{R(K_S{}^0 \longrightarrow \pi^+ + \pi^-)}{R(K_S{}^0 \longrightarrow \pi^0 + \pi^0)} = \frac{2}{1}, \tag{15.123}$$

to be compared with the experimental number[43]

$$\frac{R(K_S{}^0 \longrightarrow \pi^+ + \pi^-)_{\text{exp}}}{R(K_S{}^0 \longrightarrow \pi^0 + \pi^0)_{\text{exp}}} = \frac{69.3}{30.7} \pm 1.2. \tag{15.124}$$

---

43. A. H. Rosenfeld et al., *Rev. Mod. Phys.* **39**:1 (1967).
44. The $T = 1$ state is given by an antisymmetric function in T space, $\pi_1 \times \pi_2$ which, by Bose statistics, would require an antisymmetric function in space; consequently, it is not an S-state.

This rule should not hold better than isospin conservation itself, but it is clear from both examples (and there are others) that it gives quite decent comparison with experiment. Now comes the dilemma, and it is entirely typical of this subject. In the current-current theory these decays are described by products of currents of the form $J_\mu S_\mu^*$, where J_μ is the $\Delta S = 0$ current with $\Delta T = 1$ and S_μ is the $\Delta S = 1$ current with $\Delta T = \frac{1}{2}$. But this product will contain both $T = \frac{3}{2}$ and $T = \frac{1}{2}$ components. To give a concrete example, a coupling of the form $(\overline{N}P)(\overline{P}\Lambda)$, which would occur in the theory, leads to both $\Delta T = \frac{1}{2}$ and $\Delta T = \frac{3}{2}$. We can get rid of the $\Delta T = \frac{3}{2}$ part only at the expense of introducing *neutral* currents; for example, the combination[45]

$$(\overline{N}P)(\overline{P}\Lambda) + \tfrac{1}{2}(\overline{P}P - \overline{N}N)\overline{N}\Lambda \tag{15.125}$$

transforms like an isospinor. But these neutral currents do not, experimentally, appear to couple to leptons. Thus the elegance of the $W_\mu W_\mu^*$ theory is lost. Hence to ensure the $\Delta T = \frac{1}{2}$ rule for nonleptonic decays one must either provide a dynamical mechanism for enhancing those graphs that lead to $\Delta T = \frac{1}{2}$ or one must introduce neutral currents that couple asymmetrically to hadrons and leptons.

This problem reappears in the discussion of the SU_3 properties of the Hamiltonian for nonleptonic weak decays. If we start from the Hamiltonian

$$H_{WK} = \frac{G}{\sqrt{2}}\,[W_\mu W_\mu^* + \text{h.c.}] \tag{15.126}$$

and assume the Cabibbo theory, then for the nonleptonics this Hamiltonian will have the form

$$H_{WK}^{\Delta S=1} = \frac{G}{\sqrt{2}} \sin\theta\cos\theta \sum_{ij} [J_{\mu i}J_{\mu j}^* + J_{\mu i}^*J_{\mu j}], \tag{15.127}$$

where the sum extends over suitable, but not all, SU_3 indices.[46] In general, we have the multiplication table for SU_3 representations

$$8 \times 8 = 1 + 8_A + 8_S + 10 + 10^* + 27. \tag{15.128}$$

However, since Eq. (15.127) is symmetric in i and j, some of the representations in Eq. (15.128) do not contribute to $H_{WK}^{\Delta S=1}$. In particular, both 10 and 10* are ruled out. To see this we construct, say, the 10, by combining two octets $T_\beta{}^\alpha$ and $T_\beta'{}^\alpha$. Thus

---

45. Dirac matrices are suppressed.
46. So far as the SU_3 properties are concerned we can ignore the distinction between $J_{\mu i}$ and $J_{\mu i}^*$, since terms of the form $i\bar{q}\gamma\lambda_i q$ are Hermitian. In any case, $8 = 8^*$.
47. In the $T_\beta{}^\alpha$ language these ways correspond to taking suitable components of the symmetric tensor

$$T_\gamma{}^\alpha = T'_\beta{}^\alpha T''_\gamma{}^\beta + T'_\gamma{}^\beta T''_\beta{}^\alpha - \tfrac{2}{3}\delta_\gamma{}^\alpha T'_\lambda{}^\beta T''_\beta{}^\lambda,$$

where, in this case, $T' = T''$.

$$D_{\lambda\beta\delta} = \epsilon_{\lambda\alpha\gamma}[T_\beta{}^\alpha T_\delta{}'^\gamma + T_\delta{}^\alpha T_\beta{}'^\gamma] + \epsilon_{\beta\alpha\gamma}[T_\delta{}^\alpha T_\lambda{}'^\gamma + T_\lambda{}^\alpha T_\delta{}'^\gamma] + \epsilon_{\delta\alpha\gamma}[T_\lambda{}^\alpha T_\beta{}'^\gamma + T_\beta{}^\alpha T_\lambda{}'^\gamma]. \quad (15.129)$$

If we now set $T_\beta{}^\alpha = T_\beta{}'^\alpha$, identifying the two octets, from the antisymmetry of $\epsilon_{\lambda\alpha\gamma}$ we have

$$D_{\lambda\beta\delta} = 0. \quad (15.130)$$

The same discussion works for the 10*. For the 8's there are two possibilities, which we have distinguished by the subscripts 8_A and 8_S, whose significance we will now make clear. When we combine $J_{\mu i}$ and $J_{\mu j}{}^*$ the resultant combinations will contain only products involving the SU_3 components $J_{\mu 1}J_{\mu 4}{}^*$, $J_{\mu 1}J_{\mu 5}{}^*$, and so on. For 8's there are two ways[47] of obtaining these particular linear combinations,

$$\begin{aligned} &d_{6ij}(J_{\mu i}J_{\mu j}{}^* + \text{h.c.}), \\ &d_{7ij}(J_{\mu i}J_{\mu j}{}^* + \text{h.c.}). \end{aligned} \quad (15.131)$$

Here we have used the table of d_{ijk} in Chapter 13, where we have shown the only nonvanishing d_{ijk}, involving 1, 2 combined with 4, 5, have 6 or 7 as the third index. We can form 8_A by replacing d_{ijk} by f_{ijk},[48] but this is ruled out in the $H_{WK}^{\Delta S=1}$ of the Cabibbo theory because of its SU_3 index symmetry. We can now use the presumed *CP* invariance of $H_{WK}^{\Delta S=1}$ to rule out the d_{7ij} octet coupling. In the $T_\beta{}^\alpha$ language $H_{WK}^{\Delta S=1}$ has the form Tr $(\lambda_i TT)$, where $i = 6$ or 7. We can find the *CP* transformation properties of this coupling by the following simple, albeit indirect, argument.[49] Since *CP* is assumed to be conserved we can argue, using either the vector or axial vector currents: the conclusion must be the same. This means that under *CP* the SU_3 tensor T must transform like the divergence of the axial vector current, which, in turn, transforms like M, the meson octet matrix. But in Chapter 4 (Eq. (14.15)) we have already settled the question of how M transforms under $\mathcal{C}$:

$$\mathcal{C}M\mathcal{C}^{-1} = M^T. \quad (15.132)$$

Thus, since this is a *pseudo*scalar octet,

$$(\mathcal{C}P)M(\mathcal{C}P)^{-1} = -M^T. \quad (15.133)$$

Thus in $H_{WK}^{\Delta S=1}$

$$(\mathcal{C}P)T(\mathcal{C}P)^{-1} = -T^T. \quad (15.134)$$

---

48. These combinations correspond in the $T_\beta{}^\alpha$ language to taking suitable components of the antisymmetric tensor

$$T_\gamma{}^\alpha = T'_\beta{}^\alpha T''_\gamma{}^\beta - T'_\alpha{}^\beta T''_\beta{}^\gamma$$

with $T' = T''$.

49. For other arguments leading to the same answer, see J. S. Bell, *Les Houches Lectures* (1965), op. cit., R. H. Dalitz, *Varenna Lectures* (1964), op. cit., and S. Gasiorowicz *Elementary Particle Physics*, Wiley (1966), or the reader can simply apply the indicated transformations directly to the quark currents and see what happens.

Hence to assure the *CP* invariance of $H_{WK}^{\Delta S=1}$ we must have[50]

$$\mathrm{Tr}\,(\lambda_i TT) = \mathrm{Tr}\,(\lambda_i T^T T^T) = \mathrm{Tr}\,(\lambda_i{}^T TT). \tag{15.135}$$

But

$$\lambda_6{}^T = \lambda_6 \tag{15.136}$$

and

$$\lambda_7{}^T = -\lambda_7, \tag{15.137}$$

which rules out the d_{7ij} coupling. This still leaves the 27-fold representation,[51] which contains isotopic spin-$\frac{3}{2}$ components, which in $H_{WK}^{\Delta S=1}$ would lead to a violation of the $\Delta T = \frac{1}{2}$ rule.[52] (The octet terms, since they transform like the K^+, lead to the $\Delta T = \frac{1}{2}$ rule.) Since the rule appears to agree well with experiment and since the octet character of $H_{WK}^{\Delta S=1}$ leads to additional predictions that also appear to agree with experiment, one is led to the conclusion that some mechanism must be at work suppressing the 27. There is no unanimity of opinion in the literature[53] as to how this is to be arranged, and the problem is one of the most active ones in the field. In what follows we shall, for simplicity, assume that $H_{WK}^{\Delta S=1}$ has been fixed so that somehow the 27 is suppressed. We shall study the consequences.

The first consequence is, of course, the $\Delta T = \frac{1}{2}$ rule for the nonleptonics. As mentioned above, this rule cannot hold any better than isotopic spin conservation itself. Historically[54] the rule had its origins in an apparent anomaly in K decay. The experimental ratio

$$R_{\exp} = \frac{\Gamma(K^+ \longrightarrow \pi^+ + \pi^0)}{\Gamma(K_S{}^0 \longrightarrow \pi^+ + \pi^-)} \sim \frac{1}{500} \tag{15.138}$$

appears to augur for a selection rule, since the phase space for the two modes is nearly identical. In fact a *strict* $\Delta T = \frac{1}{2}$ rule would reduce this ratio to zero, since $\pi^+\pi^0$ cannot be in a $T = 0$ state, the only relevant one allowed by the rule. (The $T = 1$ state has the wrong symmetry.) At first look, the situation appears

---

50. $\mathrm{Tr}\,(A) = \mathrm{Tr}\,(A^T)$.
51. Since $J_{\mu i}J_{\mu j}{}^*$ only involves the indices 1, 2, 4, 5, we cannot make an SU_3 scalar out of them. Such a scalar could not, in any case, contribute to H_{WK}, since the weak decays all have $\Delta T \neq 0$ and $\Delta Y \neq 0$. The reader will note that $1 + 8 + 27 = 36$, which is the number of components in a symmetric tensor $T_\beta{}^\alpha$, where α and β both run from 1, . . . , 8.
52. For completeness we give the structure of 27-fold SU_3 tensor in the $T_\beta{}^\alpha$ language:

$$\begin{aligned} T_{\beta\delta}{}^{\alpha\gamma} &= T'_\beta{}^\alpha T''_\delta{}^\gamma + T'_\beta{}^\gamma T''_\delta{}^\alpha + T'_\delta{}^\alpha T''_\beta{}^\gamma + T'_\delta{}^\gamma T''_\beta{}^\alpha \\ &\quad - \tfrac{1}{5}\{\delta_\beta{}^\alpha(T'_\delta{}^\lambda T''_\lambda{}^\gamma + T'_\lambda{}^\gamma T''_\delta{}^\lambda) + \delta_\beta{}^\alpha(T'_\delta{}^\lambda T''_\lambda{}^\alpha + T'_\lambda{}^\alpha T''_\delta{}^\lambda) \\ &\quad + \delta_\delta{}^\alpha(T'_\beta{}^\lambda T''_\lambda{}^\gamma + T'_\lambda{}^\gamma T''_\beta{}^\lambda) + \delta_\delta{}^\gamma(T'_\beta{}^\lambda T''_\lambda{}^\alpha + T'_\lambda{}^\alpha T''_\beta{}^\lambda)\} \\ &\quad + \tfrac{1}{10}(\delta_\beta{}^\alpha\delta_\delta{}^\gamma + \delta_\delta{}^\alpha\delta_\beta{}^\gamma)(T'_\mu{}^\lambda T''_\lambda{}^\mu), \end{aligned}$$

which has the correct symmetry to be contained in the decomposition of the currents in $H_{WK}^{\Delta S=1}$.

very good since we do not expect the rule to hold exactly anyway. However, one might have expected, since the corrections are electromagnetic, that typically[55]

$$R_{\text{exp}} \sim \left(\frac{\alpha}{\pi}\right)^2 \simeq 10^{-5}, \tag{15.139}$$

which, if one takes this numerical estimate very seriously, appears to be a paradox if this number is compared to Eq. (15.136). Insofar as this is a paradox then the $H_{WK}^{\Delta S=1}$ that appears in the Cabibbo theory, assuming octet dominance, provides an answer. We shall use a method of argument due to Okubo,[56] which yields an identity that we will also use later. Consider four 3 × 3 matrices A, B, C, D and form the sum

$$\sum_P (-1)^P A_\lambda{}^\alpha B_\mu{}^\beta C_\nu{}^\gamma S_\sigma{}^\delta, \tag{15.140}$$

where the sum is over the permutations of the lower indices and the factor $(-1)^P$ is ± 1, depending on whether the permutation is even or odd. This sum, in fact, vanishes, since the indices only run from 1 to 3, which means that there will always be terms with two like indices; 1, 1 or 2, 2, and so on. The remaining two terms will cancel in pairs because of the factor $(-1)^P$. As a special case we can choose

$$\lambda = \beta, \qquad \mu = \gamma, \qquad \nu = \delta, \qquad \alpha = \sigma, \tag{15.141}$$

which, assuming that A, B, C, D are *traceless*, leads to the identity

$$\begin{aligned} &\operatorname{Tr}(ABCD) + \operatorname{Tr}(ABDC) + \operatorname{Tr}(ACBD) + \operatorname{Tr}(ACDB) \\ &\quad + \operatorname{Tr}(ADBC) + \operatorname{Tr}(ADCB) = \operatorname{Tr}(AB)\operatorname{Tr}(CD) + \operatorname{Tr}(AC)\operatorname{Tr}(BD) \\ &\quad + \operatorname{Tr}(AD)\operatorname{Tr}(BC). \end{aligned} \tag{15.142}$$

As a first application of this identity we can consider the matrix element for

$$M \longrightarrow M + M,$$

---

53. For a few sample tries, see the following: S. Oneda et al. *Phys. Rev.* **119**:482 (1960); A. Salam and J. C. Ward, *Phys. Rev. Letters* **5**:390 (1960); S. Coleman and S. L. Glashow, *Phys. Rev.* **134**:B681 (1964); R. Dashen et al., "Eight-Fold Way," op. cit., p. 254; R. F. Dashen and S. Frautchi, *Phys. Rev.* **137**:B1331 (1964).

    The SU_3 commutator algebra has led to some success with this problem. For this approach, see H. Sugawara, *Phys. Rev. Letters* **15**:870 and 997 (1965), and M. Suzuki, *Phys. Rev. Letters* **15**:986 (1965).
54. M. Gell-Mann and A. Pais, *Proc. Intl. Conf. on High Energy Physics* at Glasgow, Pergamon Press, London, 1955, p. 324.
55. We toss the factor of $1/\pi$ in for good measure, since radiative corrections usually produce such factors.
56. S. Okubo, *Phys. Letters* **8**:362 (1964). The procedure we use here follows a discussion by R. H. Dalitz, lectures given at the 1964 summer school "Enrico Fermi" at Varenna.

that is, a meson in a given octet going into two mesons in the *same* octet via $H_{WK}^{\Delta S=1}$, as in $K \longrightarrow \pi + \pi$. In the limit of perfect SU_3, since $H_{WK}^{\Delta S=1}$ transforms like λ_6, the only combinations that can be formed[57] are $\mathrm{Tr}\,(\lambda_6 M)\,\mathrm{Tr}\,(M^2)$ and $\mathrm{Tr}\,(M^3\lambda_6)$. But since M and λ_6 are traceless 3×3 matrices we can use Eq. (15.142) to conclude that

$$6\,\mathrm{Tr}\,(M^3\lambda_6) = 3\,\mathrm{Tr}\,(M^2)\,\mathrm{Tr}\,(M\lambda_6), \tag{15.143}$$

so that there is really only *one* combination. However,

$$\mathrm{Tr}\,(\lambda_6 M) = K^0 + \overline{K}^0, \tag{15.144}$$

which is CP odd and cannot be connected to the $\pi^+\pi^-$ or $\pi^0\pi^0$ states, which are CP even; so that $K_S{}^0 \longrightarrow \pi + \pi$ is *forbidden* in the good SU_3 limit in the Cabibbo theory.[58] Before discussing the significance of this result we shall put the proof in a slightly different way that may make it clearer. In effect, in this decay $H_{WK}^{S\Delta=1}$ is replaced by an octet spurion S, whose transformation property under CP is

$$(\mathcal{C}P)S(\mathcal{C}P)^{-1} = S^T. \tag{15.145}$$

But this spurion cannot be coupled in an SU_3 CP invariant way to M^3, where M is the meson octet, since $\mathrm{Tr}\,(M^3S)$ has the wrong transformation under CP.[59] As a consequence

$$\Gamma(K_S{}^0 \longrightarrow \pi + \pi) \simeq 0(\lambda^2), \tag{15.146}$$

where λ measures the SU_3-breaking coupling strength. In light of this, and the $\Delta T = \frac{1}{2}$ rule,

$$\frac{\Gamma(K^+ \longrightarrow \pi^+ + \pi^0)}{\Gamma(K_S{}^0 \longrightarrow \pi^+ + \pi^-)} \sim \frac{\left(\frac{\alpha}{\pi}\right)^2}{\lambda^2} \sim 10^{-3}, \tag{15.147}$$

which is more in line with the experimental result and appears to remove the paradox.

As a final application of octet dominance in nonleptonics we give a partial derivation of the celebrated Lee-Sugawara sum rule.[60] It relates the amplitudes for the three nonleptonic decays

$$\begin{aligned} &\Xi^- \longrightarrow \Lambda^0 + \pi^-, && a(\Xi_-{}^-), \\ &\Lambda^0 \longrightarrow p + \pi^-, && a(\Lambda_-{}^0), \\ &\Sigma^+ \longrightarrow p + \pi^0, && a(\Sigma_0{}^+), \end{aligned} \tag{15.148}$$

---

57. In general, one might also have derivatives of the M's but in the good SU_3 limit these either vanish or reduce to the form above. See R. H. Dalitz, *Varenna Lectures*, 1964, op. cit.
58. This result was first obtained by N. Cabibbo, *Phys. Rev. Letters* **12**:62 (1964).
59. $(cP)M(cP)^{-1} = -M^T$.
60. H. Sugawara, *Prog. Theor. Physics*, **31**:213 (1964), and B. W. Lee, *Phys. Rev. Letters* **12**:83 (1964). For a modern current algebra treatment of the rule, see H. Sugawara, *Phys. Rev. Letters* **15**:870 and 997(E) (1965), and M. Suzuki, *Phys. Rev. Letters* **15**:986 (1965).

as follows:

$$2a(\Xi_-^-) = a(\Lambda_-^0) + \sqrt{3}a(\Sigma_0^+). \tag{15.149}$$

It appears to hold experimentally,[61] at least to within a few percent.

As we have indicated above we will only give a "partial" derivation of this result. The meaning of "partial" will become clear shortly. All of the decays indicated can occur in two orbital angular momentum states for the π-hyperon system; S and P, since the initial hyperon has spin $\frac{1}{2}$. The amplitude for the decay $B \longrightarrow B + M$ will therefore take the form[62] (parity is *not* conserved!)

$$\text{Tr}\,[\bar{B}(a_0 + a_5\gamma_5)BMS], \tag{15.150}$$

where a_0 and a_5 are complex numbers and S is the octet spurion. From the point of view of SU_3 there are nine possible invariant terms,[63] that is, with the M_i complex numbers, $i = 1, \ldots, 9$:

$$\begin{aligned} M_1\,\text{Tr}\,(\bar{B}SMB) &+ M_2\,\text{Tr}\,(\bar{B}MBS) + M_3\,\text{Tr}\,(\bar{B}BSM) \\ &+ M_4\,\text{Tr}\,(\bar{B}SBM) + M_5\,\text{Tr}\,(\bar{B}MSB) + M_6\,\text{Tr}\,(\bar{B}BMS) \\ &+ M_7\,\text{Tr}\,(\bar{B}M)\,\text{Tr}\,(BS) + M_8\,\text{Tr}\,(\bar{B}S)\,\text{Tr}\,(BM) \\ &+ M_9\,\text{Tr}\,(\bar{B}B)\,\text{Tr}\,(MS). \end{aligned} \tag{15.151}$$

We can now use Eq. (15.142) to eliminate one of these forms; there are only eight independent M_i. However, in the Cabibbo theory,

$$S = \lambda_6. \tag{15.152}$$

Thus, we can use *CP* invariance to reduce the number of *M*'s still further. First we consider parity. Under P (M is a pseudoscalar octet)

$$\begin{aligned} \bar{B}BM &\longrightarrow -\bar{B}BM, \\ \bar{B}\gamma_5BM &\longrightarrow \bar{B}\gamma_5BM, \end{aligned} \tag{15.153}$$

while under C

$$\begin{aligned} M &\longrightarrow M^T, \\ B &\longrightarrow \bar{B}^T c, \\ \bar{B} &\longrightarrow (c^{-1}B)^T, \end{aligned} \tag{15.154}$$

where c is the 4×4 antisymmetric matrix defined so that

$$c\gamma_\mu{}^T c^{-1} = -\gamma_\mu. \tag{15.155}$$

---

61. M. L. Stevenson et al., *Physics Letters* **9**:349 (1964) and especially the review summary by J. P. Berge, International Conference on High Energy Physics at Berkeley (1966).
62. We give the other SU_3 invariant possibilities below. There are other forms involving momenta but the reader can verify (using the Dirac equation) that they do not lead to anything new.
63. We follow the notation of R. H. Dalitz, *Varenna Lectures* (1964), op. cit. We use the fact that all these 3×3 matrices, including S, are traceless, to eliminate combinations involving a single trace.

With these results we can evaluate the CP characteristics of the nine terms in Eq. (15.151). We do a typical one, say $\mathrm{Tr}\,(\bar{B}SMB)$, and then write down the answers for the rest. Thus

$$\mathrm{Tr}\,(\bar{B}SMB) \xrightarrow{CP} \mp\,\mathrm{Tr}\,(B^T SM^T \bar{B}^T) = \mp\,\mathrm{Tr}\,(\bar{B}MSB). \tag{15.156}$$

Here we have assumed that S is the *symmetric* spurion, or λ_6, which corresponds to the CP invariance of $H_{WK}^{\Delta S=1}$. The $\mp$ signs in Eq. (15.156) arise from the parity transformation and go along with terms of the form $\bar{B}B$ and $\bar{B}\gamma_5 B$, respectively. Completing the list we have

$$\begin{aligned}
\mathrm{Tr}\,(\bar{B}SMB) &\longleftrightarrow \mp\,\mathrm{Tr}\,(\bar{B}MSB),\\
\mathrm{Tr}\,(\bar{B}BSM) &\longleftrightarrow \mp\,\mathrm{Tr}\,(\bar{B}BMS),\\
\mathrm{Tr}\,(\bar{B}M)\,\mathrm{Tr}\,(BS) &\longleftrightarrow \mp\,\mathrm{Tr}\,(BM)\,\mathrm{Tr}\,(\bar{B}S),\\
\mathrm{Tr}\,(\bar{B}MBS) &\longleftrightarrow \mp\,\mathrm{Tr}\,(\bar{B}MBS),\\
\mathrm{Tr}\,(\bar{B}SBM) &\longleftrightarrow \mp\,\mathrm{Tr}\,(\bar{B}SBM),\\
\mathrm{Tr}\,(\bar{B}B)\,\mathrm{Tr}\,(MS) &\longleftrightarrow \mp\,\mathrm{Tr}\,(\bar{B}B)\,\mathrm{Tr}\,(MS).
\end{aligned} \tag{15.157}$$

At this point the discussion divides. For the Dirac scalar, or odd parity, part of the $B \longrightarrow B + \pi$ coupling, the $\bar{B}BM$ part,[64] CP invariance gives the following conditions on the M_i:

$$M_2 = M_4 = M_9 = 0 \tag{15.158}$$

and

$$M_1 = -M_5, \qquad M_3 = -M_6, \qquad M_7 = -M_8. \tag{15.159}$$

For the $\bar{B}\gamma_5 BM$ terms there are no constraint conditions on M_2, M_4, and M_9, and this argument cannot be used to derive Eq. (15.149). However, for the $\bar{B}BM$ terms we can use these conditions to derive the Lee-Sugarawa sum rule, as follows. In addition to the three processes listed in Eq. (15.148) there are four others:

$$\begin{aligned}
\Sigma^- &\longrightarrow n + \pi^-, \qquad && a(\Sigma_-{}^-),\\
\Lambda^0 &\longrightarrow n + \pi^0, && a(\Lambda_0{}^0),\\
\Sigma^+ &\longrightarrow p + \pi^0, && a(\Sigma_0{}^+),\\
\Xi^0 &\longrightarrow \Lambda^0 + \pi^0, && a(\Xi_0{}^0).
\end{aligned} \tag{15.160}$$

But we can use the $\Delta T = \frac{1}{2}$ rule to connect some of them. We have already in Eq. (15.120) shown that

64. In the literature this term is usually called the "parity violating" part. Of course, by itself, it does not violate parity. It only does so if the rest of the Hamiltonian contains terms with the opposite transformation property under P, which is the case.
65. These follow directly from evaluating the traces in Eq. (15.151) since the $\Delta T = \frac{1}{2}$ rule is built into the SU_3 structure of these amplitudes.
66. This argument is due essentially to M. Gell-Mann, *Phys. Rev. Letters* **12**:199 (1964).

$$\sqrt{2}a(\Lambda_0{}^0) = a(\Lambda_-{}^0), \tag{15.161}$$

and we can also derive two additional relations[65]

$$\begin{aligned}\sqrt{2}a(\Xi_0{}^0) &= a(\Xi_-{}^-),\\ \sqrt{2}a(\Sigma_0{}^+) &= -a(\Sigma_-{}^-) + a(\Sigma_+{}^+),\end{aligned} \tag{15.162}$$

which are confirmed experimentally. This means that, *without* invoking CP invariance, there are four independent amplitudes, given just the $\Delta T = \frac{1}{2}$ rule or octet dominance. For the $\bar{B}BM$ terms, CP invariance reduces the number of possible parameters available to fit the amplitudes to only *three*, which means that there must be a relation among the four amplitudes. Indeed, evaluating the traces one is led to Eq. (15.149).[66] This derivation *fails* for the even parity, $\bar{B}\gamma_5 BM$, terms, although experimentally Eq. (15.149) holds for these as well. Many physicists have worked on generalizing the treatment so that it covers both cases. All of these attempts involve additional assumptions of one kind or another and the reader is invited to study the literature.

We conclude this long chapter with a discussion of three topics that are at the center of present interest: (1) current algebra as applied to low-energy physics; (2) Lagrangians; and (3) CP violation in nonleptonic K^0 decay.

1. *Current algebra.* The first item on this agenda involves relations between matrix elements of the form $\langle B\pi|_{\text{out}}H(0)|A\rangle$ and $\langle B|H(0)|A\rangle$, where $|A\rangle$ is a state that contains no pions and $|B\rangle$ is a state that may contain pions. For example, we may look for relations among the matrix elements $\langle\pi^+\pi^-e^+\nu_e|H_{WK}|K^+\rangle$, $\langle\pi^0\mu^+\nu_\mu|H_{WK}(0)|K^+\rangle$, and $\langle\mu^+\nu_\mu|H_{WK}|K^+\rangle$, corresponding to K_{l4}, K_{l3}, K_{l2} decays, where the numerical subscript refers to the number of particles in the final state.[67] All of the work on this subject proceeds from a variation of the "reduction formula" used in connection with pion-nucleon scattering in Chapter 12, which we now derive. We introduce a standard notation. Let

$$A(x)\overleftrightarrow{\partial}_\mu B(x) = A(x)\frac{\partial}{\partial x_\mu}B(x) - \frac{\partial}{\partial x_\mu}A(x)B(x), \tag{15.163}$$

and hence we can write, as usual,

$$\langle B\pi|_{\text{out}}\,H(0)|A\rangle = \lim_{t\to+\infty}\left(-i\int d\mathbf{r}\,\langle B|f^*(x)\overleftrightarrow{\partial}_t\,\pi(x)H(0)|A\rangle\right), \tag{15.164}$$

where

$$f^*(x) = \frac{e^{-i(qx)}}{(2\pi)^{3/2}\sqrt{2q_0}} \tag{15.165}$$

---

67. This problem was first treated in the literature by C. Callan and S. B. Treiman, *Phys. Rev. Letters* **15**:153 (1966). Important earlier papers on *PCAC* and soft pion emission are those of Y. Nambu and D. Lurie, *Phys. Rev.* **125**:1429 (1962), and Y. Nambu and E. Shrauner, *Phys. Rev.* **128**:862 (1962). These matters are nicely reviewed by N. Cabibbo, Berkeley Conference on High Energy Physics (1966).

and

$$(-\nabla^2 + \partial_t^2 + m_\pi^2)f^*(x) = 0. \tag{15.166}$$

Since it is f^* that occurs in Eq. (15.164) and, since $|A\rangle$ does not contain pions, we can write

$$\begin{aligned}\langle B\pi|_{\text{out}}H(0)|A\rangle &= \lim_{t\to+\infty}\left(-i\int d\mathbf{r}\, f^*(x)\,\overleftrightarrow{\partial_t}\,\langle B|[\pi(x), H(0)]_-|A\rangle\right)\\ &= \lim_{t\to+\infty}\left(-i\int d\mathbf{r}\, f^*(x)\,\overleftrightarrow{\partial_t}\,\langle B|\theta(t)[\pi(x), H(0)]_-|A\rangle\right),\end{aligned} \tag{15.167}$$

where we can introduce the commutator into Eq. (15.167), since acting to the right $f^*\,\overleftrightarrow{\partial_t}\,\pi$ annihilates pions, of which there are none, in the initial state. We have inserted the $\theta(t)$ function in Eq. (15.167), since we are always working in the region $t > 0$ so that

$$\theta(t) = 1. \tag{15.168}$$

Next, using the fact that $\theta(t)$ vanishes for negative t, we can write

$$\begin{aligned}\langle B\pi|_{\text{out}}H(0)|A\rangle &= -i\int d\mathbf{r}\,dt\,\frac{\partial}{\partial t}\,[f^*(x)\,\overleftrightarrow{\partial_t}\,\langle B|\theta(t)[\pi(x), H(0)]_-|A\rangle]\\ &= -i\int d\mathbf{r}\,dt\,f^*(x)(-\Box^2 + m_\pi^2)[\theta(t)\langle B|[\pi(x), H(0)]_-|A\rangle],\end{aligned} \tag{15.169}$$

where we have used Eq. (15.166) and have assumed, as is customary, that when we integrate by parts, over *space*, the surface terms vanish at infinity. By partial integration onto f^* we finally write Eq. (15.169) as

$$\langle B\pi|_{\text{out}}H(0)|A\rangle = -i(q^2 + m_\pi^2)\int d^4x\, f^*(x)\theta(t)\langle B|[\pi(x), H(0)]_-|A\rangle. \tag{15.170}$$

We will make use of this basic result shortly. Before doing so, let us consider the object[68]

$$T_\mu = \int d^4x\, e^{-i(qx)}\langle B|[\mathbf{A}_\mu(x), H(0)]_-|A\rangle\theta(t), \tag{15.171}$$

where $\mathbf{A}_\mu$ is the $\Delta S = 0$ axial vector weak current. We now evaluate $q_\mu T_\mu$ by exploiting the identity

$$q_\mu e^{-i(qx)} = i\frac{\partial}{\partial x_\mu}\,e^{-i(qx)}, \tag{15.172}$$

and, integrating by parts and again setting surface terms equal to zero, we have[69]

---

68. We follow the logic and notation of N. Cabibbo, Berkeley Conference Proceedings 1966, op. cit.

69. Remember that $\mathbf{A}_\mu \dfrac{\partial}{\partial x_\mu}\theta(t) = \mathbf{A}_0\dot{\theta}(t) = \mathbf{A}_0\,\delta(t)$.

$$q_\mu T_\mu = i \int d^4x \frac{\partial}{\partial x_\mu} (e^{-i(qx)}) \langle B | [\mathbf{A}_\mu(x), H(0)]_- | A \rangle \theta(t)$$

$$= -i \int d^4x \, e^{-i(qx)} \langle B | [\partial_\mu \mathbf{A}_\mu(x), H(0)]_- | A \rangle \theta(t)$$

$$-i \int d^4x \, e^{-i(qx)} \langle B | [\mathbf{A}_0(x), H(0)]_- | A \rangle \, \delta(t). \quad (15.173)$$

There is no reason for T_μ to have a singular behavior at $q = 0$ so that, in the limit as $q_\mu \longrightarrow 0$, the left side of Eq. (15.173) vanishes. We now use *PCAC*, in the strong sense, to evaluate the right side of Eq. (15.173). Thus we assume (see Eq. (12.103) and the discussion preceding it) that

$$\partial_\mu A_\mu = m_\pi^2 \frac{g_\pi(-m_\pi^2)}{G} \sqrt{2}(2\pi)^{3/2} \pi^+, \quad (15.174)$$

where as usual $g_\pi(-m_\pi^2)$ is defined by the equation

$$\frac{G}{\sqrt{2}} \langle 0 | A_\mu(0) | \mathbf{p}_\pi \rangle = i \frac{p_{\pi\mu}}{\sqrt{2E_\pi}} g_\pi(-m_\pi^2). \quad (15.175)$$

Thus[70]

$$\lim_{q_\mu \to 0} m_\pi^2 g_\pi(-m_\pi^2) \frac{\sqrt{2}}{G} \frac{(2\pi)^{3/2}}{\sqrt{2}} \int d^4x \, e^{-i(qx)} \langle B | [\pi(x), H(0)]_- | A \rangle \theta(t)$$

$$= -\langle B | [\mathbf{Q}_5, H(0)]_- | A \rangle \quad (15.176)$$

where we have used the $\delta(t)$ to do the time integration in the second term on the right side of Eq. (15.173) and the definition

$$\mathbf{Q}_5(\mathbf{r}, 0) = \int \mathbf{A}_0(\mathbf{r}, 0) \, d\mathbf{r}. \quad (15.177)$$

Therefore, from Eq. (15.170),

$$g_\pi(-m_\pi^2) \frac{\sqrt{2}}{G} \frac{(2\pi)^3}{\sqrt{2}} \lim_{q_\mu \to 0} \sqrt{2q_0} \langle B\pi |_{\text{out}} H(0) | A \rangle$$

$$= i \langle B | [\mathbf{Q}_5, H(0)]_- | A \rangle, \quad (15.178)$$

which gives a useful relation between low-energy pionic emission processes and matrix elements involving one less pion. We now can use Eq. (15.178) to derive the so-called Callan-Treiman formula,[71] which connects K_{l3} and K_{l2} decay. To this end let

---

70. The factor of $1/\sqrt{2}$ here comes from the relation $\pi^+ = (\pi_1 + i\pi_2)/\sqrt{2}$.
71. C. Callan and S. B. Treiman, op. cit., and V. S. Mathur et al., *Phys. Rev. Letters* **16**:371 (1966).

$$
\begin{aligned}
|A\rangle &= |K^+\rangle, \\
|B\rangle &= |0\rangle, \qquad \text{the vacuum,} \\
H(0) &= J_{\mu 4}(0) - iJ_{\mu 5}(0).
\end{aligned} \tag{15.179}
$$

Since we are considering the decay $K^+ \longrightarrow \pi^0 + l^+ + \nu_e$, the relevant component of $\mathbf{Q}_5$ to use in Eq. is Q_{53} corresponding to *neutral* pion emission. Now,[72] from Eq. (13.30),

$$[Q_{53}, (V_{\mu 4} - iV_{\mu 5}) + (A_{\mu 4} - iA_{\mu 5})]_- = -\tfrac{1}{2}[A_{\mu 4} - iA_{\mu 5}] - \tfrac{1}{2}[V_{\mu 4} - iV_{\mu 5}] \tag{15.180}$$

so that

$$
\begin{aligned}
g_\pi(-m_\pi^2)\frac{\sqrt{2}}{G}\frac{(2\pi)^3}{\sqrt{2}} \lim_{q_\mu \to 0} \sqrt{2q_0}\langle q\pi^0|V_{\mu 4} - iV_{\mu 5}|p_{K^+}\rangle \\
= -\frac{i}{2}\langle 0|A_{\mu 4} - iA_{\mu 5}|K^+\rangle \\
= \frac{1}{2}\frac{\sqrt{2}}{G}\frac{p_{K\mu}}{\sqrt{2E_K}}g_K(-m_K^2).
\end{aligned} \tag{15.181}
$$

Thus

$$\lim_{q\pi \to 0}(2\pi)^3\sqrt{2E_\pi}\sqrt{2E_K}\langle q_\pi|V_{\mu 4} - iV_{\mu 5}|K^+\rangle = \frac{1}{\sqrt{2}}\frac{g_K(-m_K^2)}{g_\pi(-m_\pi^2)}p_{K\mu}. \tag{15.182}$$

Therefore from Eq. (15.52) we have, in the limit,

$$f_+(q_\pi^2 = 0) + f_-(q_\pi^2 = 0) = \frac{1}{\sqrt{2}}\frac{g_K(-m_K^2)}{g_\pi(-m_\pi^2)}, \tag{15.183}$$

which is the Callan-Treiman relation connecting K_{l3} and K_{l2} decays. The form factors $f_\pm$ that occur in Eq. (15.183) are evaluated for unphysical pions of zero rest mass. If we assume, as in the other applications of current algebras, that we can use the measured values of $f_\pm$ as obtained in $K_{\mu 3}$ decays without worrying unduly about the extrapolation to the physical pion mass, then we can compare Eq. (15.183) to experiment. Before commenting on this we note that in good SU_3 Eq. (15.183) is satisfied exactly, since

$$
\begin{aligned}
f_+(0) &= \frac{1}{\sqrt{2}}, \\
f_-(0) &= 0,
\end{aligned} \tag{15.184}
$$

and

---

72. The commutations are taken at equal times.
73. G. H. Trilling, report to the International Conference on Weak Interactions, Argonne National Laboratory (1965).
74. C. Callan and S. B. Treiman, op. cit., S. Weinberg, *Phys. Rev. Letters* **17**:336 (1966), and C. Bouchiat and P. Meyer, *Phys. Letters* **22**:198 (1966).

$$g_K = g_\pi. \tag{15.185}$$

In deriving Eq. (15.183) we have implicitly broken SU_3, in that $m_K \neq m_\pi$. There is at least one set of experimental data[73] on K_{l3} decays that gives reasonably good agreement with Eq. (15.183):

$$\frac{f_-}{f_+} = 0.46 \pm 0.27, \qquad |f_+| \simeq \frac{1}{\sqrt{2}}, \tag{15.186}$$

so that the left side of Eq. (15.183) is, cancelling the $1/\sqrt{2}$,

$$1 + 0.46 \pm 0.24$$

as compared to the right side, which is 1.28. The experimental situation is not completely settled but the Callan-Treiman relation gives an interesting connection involving the SU_3 symmetry breaking as it affects the vector and axial vector currents.

The same "zero energy" theorem (Eq. (15.178)) has been applied to K_{l4} decay,[74] where a connection is derived among the three form factors that occur in the matrix element $\langle \pi^+\pi^- | J_{\mu 4} - iJ_{\mu 5} | K^+ \rangle$ and the $f_\pm$ and g_π defined above. There are some important technical questions that arise in exploiting these connections and we refer the reader to the literature[75] for their discussion and for a comparison of these predictions to experiment, which again seem to lead to reasonable results. The technique has also been used[76] to connect $K \longrightarrow 3\pi$ to $K \longrightarrow 2\pi$ to $K \longrightarrow \pi$ to $K \longrightarrow$ vacuum. An interesting feature of this work is that, insofar as the zero energy theorems are applicable, one obtains some understanding of the $\Delta T = \frac{1}{2}$ rule in these decays, since the $K \longrightarrow$ vacuum matrix element only involves the $T = \frac{1}{2}$ part of H_{WK}. Thus, no $T = \frac{3}{2}$ contributions are possible for the processes involving several pions so long as these are related by the low-energy theorem to the leptonic K decay.

2. *Lagrangians.* We may now turn to the second topic on the agenda: Lagrangians. As we have stressed throughout the book, a Lagrangian gives the most convenient expression of the symmetries of a dynamical system. In the early days of field theory it was optimistically assumed that the equations of motion derived from these Lagrangians might be solved in, say, successive approximations, and that the results could be straightforwardly compared to experiment. This program has been realized to a certain extent in quantum electro-

75. See N. Cabibbo, Berkeley Conference 1966, op. cit.
76. M. Suzuki, *Phys. Rev.* **144**:1154 (1966); Y. Hara and Y. Nambu, *Phys. Rev. Letters* **16**:875 (1966); S. K. Bose and S. N. Biswas, *Phys. Rev. Letters* **16**:340 (1966); D. K. Elias and J. C. Taylor, *Nuovo Cimento* **44A**:518 (1966); H. D. Abarbanel, *Phys. Rev.* **153**:1547 (1967).

dynamics, but elsewhere it has become snared in a quagmire of intractable mathematics. Hence a change in attitude toward the Lagrangian seems to have materialized in the last few years. We might summarize the new attitude in the following way. Starting from the experimental data one attempts to intuit the invariance group structure, SU_2, SU_3, etc., that *any* Lagrangian must have if it is to describe, decently, the symmetries of the world, or, in any case, some part of the world. The next step is to construct model Lagrangians that exhibit this structure. Associated with each of these Lagrangians is a set of conserved currents, and charges, generated by the continuous transformations of the symmetry group. This would be almost tautological and uninteresting except that the matrix elements of these currents and charges themselves describe physically observable processes that take place among real partical states that are the ap-approximate eigenstates of the Hamiltonian associated with the original Lagrangian. These currents and charges, since they have matrix elements that are directly observable, have essentially replaced the equations of motion of the fields as the quantities of physical interest. In the actual physical case, the strong couplings that determine the gross properties of the particle eigenstates show the maximum symmetry. Hence one is led to interesting relations among the weaker processes, which break these symmetries, because these processes are generated, at least in some low-energy limits, by the group generators that express the strong symmetries. Of course, the symmetry breaking also shows up directly in the weak processes. For example, the weak axial vector current A_μ is not conserved because the group $SU_2 \otimes SU_2$ is not an exact invariance of the strong interactions, a fact which, at least in many models, can be traced back to the nonvanishing of the pion mass. In general one would like to construct the full Lagrangian so as to break these partial symmetries as "simply" as possible. For example one might demand *PCAC*, in the strong sense, as an additional constraint on the exact Lagrangian with a nonvanishing pion mass. Apart from the problems of how many of these constraints one is to impose and what forms they should take, there is the additional and no doubt related question of how many particles, and which ones, should go into the original Lagrangian. Clearly the naive quark model provides the most economical answer to this question, since a Lagrangian with three quarks of identical mass, invariant under SU_3, has as much symmetry as experiment entitles us to believe is present in the Lagrangian, and it involves the minimum number of fields needed to express this symmetry. However, the problem of how these quarks are to be bound to give the observed particles is unsolved, to say nothing in the matter of the existence of the quarks themselves. It might be possible in principle to base a physical theory on "particles" put into a Lagrangian that are not realized in nature as observable states, but it would certainly be a radical departure from past procedures.

Leaving aside the quarks we might summarize the Lagrangian situation, as it appears to be at present, somewhat as follows. If one insists on restricting oneself to a theory of pions and nucleons alone, and if, furthermore, one imposes the condition that the pion-nucleon couplings are of the form[77] $\bar{B}BM$, Yukawa type, then no Lagrangian exists that exhibits both $SU_2 \otimes SU_2$ symmetry—one that leads to the Gell-Mann commutator algebra—and has *PCAC* in the strong sense. The pseudoscalar pion-nucleon coupling and the pseudovector pion-nucleon coupling fail in one of these respects or the other. We may then attempt to remedy this situation in two ways, both of which have been explored in the literature.

1. Add new, even unobserved, particles to the pion-nucleon Lagrangian, but still insist on bilinear Yukawa couplings of mesons to nucleons.

2. Stay within the pion-nucleon system but allow non-Yukawa-like couplings.

The most celebrated example of the first type is the so-called σ-model of Gell-Mann and Lévy.[78] The σ is a particle, perhaps nonexistent, with all of the quantum numbers of the vacuum—no spin, isotopic spin, charge, or hypercharge, and even parity. If such an object had a mass greater than two pions it could decay into two pions in an S-state and, indeed, it might show up as an S-wave pion-pion resonance. In fact, there are occasional reports of such resonances in the literature[79] at various masses. The success of the "zero energy" theorems for processes involving more than one pion makes it unlikely that such a resonance can exist at low energies (say 400 Mev total energy) since the extrapolations that are used in these calculations would break down. However, it may well be that such mesons abound at large masses. In any event, one may study the Lagrangian that describes the interactions of pions, nucleons, and hypothetical σ's. It consists of four pieces, which, following Lévy,[80] we write as follows:

$$\begin{aligned}
\mathcal{L}_N &= -\bar{N}\gamma_\mu\partial_\mu N, \\
\mathcal{L}_M &= -\tfrac{1}{2}\left[\boldsymbol{\phi}_{,\mu}{}^2 + \sigma_{,\mu}{}^2 + \mu_0{}^2[\boldsymbol{\phi}^2 + \sigma^2]\right], \\
\mathcal{L}_{\text{int}} &= g\bar{N}[i\gamma_5\boldsymbol{\tau}\cdot\boldsymbol{\phi} + \sigma]N, \\
\mathcal{L}_{MM} &= -\lambda_0\left[\boldsymbol{\phi}^2 + \sigma^2 - \frac{1}{4f_0{}^2}\right]^2 - \frac{\mu_0{}^2}{2f_0}\,\sigma.
\end{aligned} \tag{15.187}$$

Here $\mathcal{L}_N$ and $\mathcal{L}_M$ are the free nucleon, pion, and σ Lagrangians. It may appear as if we have committed ourselves to equal "bare" pion and σ masses. However, if one translates the σ field by the displacement,[81]

77. Dirac matrices and Lorentz indices are suppressed.
78. M. Gell-Mann and M. Lévy, *Nuovo Cimento* **16**:705 (1960), and M. Lévy (to be published). The σ was first introduced by J. Schwinger, *Ann. Phys.* **2**:407 (1957), and first used in the weak interaction context by the authors above.
79. See A. H. Rosenfeld et al., Berkeley High Energy Physics Conference (1966), for a review.
80. M. Lévy, op. cit.
81. M. Gell-Mann and M. Lévy, op. cit.

$$\sigma' = \sigma + \frac{1}{2f_0}, \tag{15.188}$$

one can convince oneself that the σ mass is effectively $\mu_0^2 + (2\lambda_0/f_0)$ and is hence arbitrary. $\mathfrak{L}_{\text{int}}$ is the π-nucleon and σ-nucleon coupling written in Yukawa form, taking into account the fact that σ is a scalar and π a pseudoscalar. Finally, $\mathfrak{L}_{MM}$ is a meson-meson coupling required to make the theory renormalizable. Using Eq. (15.188) it too can be rewritten in a somewhat less peculiar fashion.[82] The interest in this Lagrangian arises from the fact that it is a realization of a theory with both *PCAC*, in the strong sense, and the $SU_2 \otimes SU_2$ part of the Gell-Mann commutator algebra. We can generate the full weak current $V_\mu + A_\mu$ by the combined transformations[82]

$$\begin{aligned} N &\longrightarrow \left[1 + i\frac{\boldsymbol{\tau}}{2}\cdot \boldsymbol{\Lambda}(1+\gamma_5)\right]N, \\ \boldsymbol{\phi} &\longrightarrow \boldsymbol{\phi} - \boldsymbol{\Lambda}\sigma - \boldsymbol{\Lambda}\times\boldsymbol{\phi}, \\ \sigma &\longrightarrow \sigma + \boldsymbol{\Lambda}\cdot\boldsymbol{\phi}, \end{aligned} \tag{15.189}$$

with $\boldsymbol{\Lambda} \ll 1$. For constant $\boldsymbol{\Lambda}$ these transformations leave the total Lagrangian invariant, providing that $\mu_0 = 0$. In fact Eq. (15.187) was so arranged that all of the terms are invariant except $(\mu_0^2/2f_0)\sigma$, which makes the symmetry breakdown explicit and simple. For $\boldsymbol{\Lambda} = \boldsymbol{\Lambda}(x)$ these transformations generate the familiar isotopic vector current, which we do not need to write down again. It is, of course, conserved. They also generate

$$\mathbf{A}_\mu = \bar{N}\frac{\boldsymbol{\tau}}{2}\gamma_\mu\gamma_5 N + i(\sigma\partial_\mu\boldsymbol{\phi} - \boldsymbol{\phi}\partial_\mu\sigma) \tag{15.190}$$

and using the Lagrangian prescription of Chapter 2 (Eq. (2.14)) to compute $\partial_\mu A_\mu$, we find at once that

$$\partial_\mu \mathbf{A}_\mu = -\frac{\mu_0^2}{2f_0}\boldsymbol{\phi}, \tag{15.191}$$

or *PCAC*, in the strong sense. This set of transformations generates the group $SU_2 \otimes SU_2$, so if one forms the charges Q_i and Q_{5i} they satisfy the Gell-Mann commutator algebra, Eq. (12.35).[83] Recently Lévy[84] has generalized the model by introducing an octet of σ mesons coupled in an SU_3-invariant fashion to the baryon octet:

$$\mathfrak{L}_{\text{int}} = g\,\text{Tr}\,[\bar{B}[i\gamma_5 M + S]B], \tag{15.192}$$

---

82. M. Gell-Mann and M. Lévy, op. cit.
83. The reader who is sceptical of this abstract statement may check the commutation relations explicitly. He will be aided by the fact that there are no derivative couplings in $\mathfrak{L}$ so that $\dot{M}$ is the canonical field to any meson M. Thus, for example, at equal times,

$$[\sigma\phi_1 - \phi_1\dot{\sigma},\, \sigma\dot{\pi}_2 - \dot{\pi}_2\dot{\sigma}] = i\,\delta(\mathbf{r}-\mathbf{r}')[\phi_1\dot{\phi}_2 - \phi_2\dot{\phi}_1].$$

84. M. Lévy (to be published).

and a suitably generalized SU_3-invariant meson-meson coupling. In this theory one has *PCAC* in the strong sense for the entire octet as well as a set of charges that obey the full $SU_3 \otimes SU_3$ commutator algebra.

In the original paper of Gell-Mann and Lévy[85] Lagrangians of the second type (nonlinear, non-Yukawa coupling) were also studied. At the time this paper was written (1959) renormalizability of the field theory resulting from the Lagrangian equations was considered a positive virtue, so that the nonlinear Lagrangian, despite its symmetry, was viewed with a certain unease. With the new attitude, renormalizability, or its lack, is not taken very seriously, since one simply uses the Lagrangian to generate operators whose matrix elements are taken to lowest order. It is unlikely that one will be able to get away with this procedure indefinitely, but its success in low-energy processes has been remarkably good.

Recently, Weinberg[86] revived interest in nonlinear Lagrangians by exhibiting one that, in the $m_\pi \longrightarrow 0$ limit, is invariant under $SU_2 \otimes SU_2$ (that is, the Gell-Mann commutator algebra), and from which one can obtain[87] a simple derivation of the Adler-Weisberger result for $-g_A(0)/f_V(0)$. We give here a version of this Lagrangian due to Schwinger.[88] It takes the exotic-looking form

$$\begin{aligned}\mathfrak{L} = -\frac{1}{2}\left[\frac{\boldsymbol{\phi}_{,\mu}{}^2}{\left[1+\left(\frac{f_0}{m_\pi}\right)^2\boldsymbol{\phi}^2\right]^2}\right] - \frac{1}{2}\left(\frac{m_\pi{}^2}{f_0}\right)^2 \log\left[1+\left(\frac{f_0}{m_\pi}\right)^2\boldsymbol{\phi}^2\right] \\ - \bar{N}[\partial_\mu\gamma_\mu + M]N + \frac{1}{\left[1+\left(\frac{f_0}{m_\pi}\right)^2\boldsymbol{\phi}^2\right]}\left\{\frac{f}{m_\pi}\, i\bar{N}\gamma_\mu\gamma_5\boldsymbol{\tau}N\cdot\partial_\mu\boldsymbol{\phi}\right. \\ \left. - \left(\frac{f_0}{m_\pi}\right)^2 \bar{N}\gamma_\mu\boldsymbol{\tau}N\cdot\boldsymbol{\phi}\times\partial_\mu\boldsymbol{\phi}\right\}. \qquad (15.193)\end{aligned}$$

In the limit $m_\pi \longrightarrow 0$ it is invariant ($\boldsymbol{\Lambda} \ll 1$) under the transformations

$$\begin{aligned} N &\longrightarrow \left[1 + i\left(\frac{f_0}{m_\pi}\right)^2 \boldsymbol{\tau}\cdot\boldsymbol{\phi}\times\boldsymbol{\Lambda}\right]N, \\ \boldsymbol{\phi} &\longrightarrow \boldsymbol{\phi} + \boldsymbol{\Lambda} + \left(\frac{f_0}{m_\pi}\right)^2 [2\boldsymbol{\phi}\boldsymbol{\Lambda}\cdot\boldsymbol{\phi} - \boldsymbol{\Lambda}\phi^2], \end{aligned} \qquad (15.194)$$

which generate $SU_2 \otimes SU_2$. With $m_\pi \neq 0$ and $\boldsymbol{\Lambda} = \boldsymbol{\Lambda}(x)$ one is led to an A_μ that satisfies *PCAC* in the strong sense and is identical for $\boldsymbol{\phi} \ll 1$ to the A_μ given at the end of Chapter 12 (Eq. (12.144)). We have seen in Chapter 12 how this A_μ

85. Op. cit.
86. S. Weinberg, *Phys. Rev. Letters* **18**:185 (1967).
87. See Chapter 12, where we exhibited the approximation to Eq. (15.193) which obtains if $\boldsymbol{\phi}$ is taken to be infinitesimal.
88. J. Schwinger, *Phys. Rev. Letters* **18**:923 (1967), and *Physics Letters* **24B**:473 (1967); also J. Wess and B. Zumino, *Phys. Rev.*, **163**:1727 (1967).

gives a formula for $-g_A(0)/f_V(0)$ that is equivalent to the Adler-Weisberger sum rule. This Lagrangian has been extended to include other particles and resonances with very interesting consequences, which the reader may study in the growing literature.

3. *The violation of CP*. As the last item on the agenda and the final topic in the book, we discuss the question of *CP* conservation or time reversal invariance. Until the summer of 1964 all of elementary particle physics appeared to be consistent with the absolute conservation of *CP*. Just prior to 1957[89] Lüders proved that any local, Lorentz-invariant field theory with a vacuum state was invariant under *CPT* and, soon after, Jost[90] showed that *CPT* invariance was essentially equivalent to the analytic continuation of the Lorentz group to complex transformations. Hence it is difficult to imagine how a Lorentz invariant theory could fail to be *CPT* invariant as well and, indeed, no one has succeeded in finding such a theory. Thus, if *CP* is violated and *CPT* commutes with the Lagrangian, then either T must be violated or there must be something radically wrong with our entire approach to relativistic field theory. As of this writing there is only one system in physics where *CP* violation has definitely been observed[91] and that is the neutral K meson system. Interestingly enough, this system, as we shall see, also provides the best experimental verification of *CPT* conservation so that this *CP* violation is also, no doubt, a violation of T invariance.

The K^0 is the only observed metastable particle that can transform into its (distinct) antiparticle via known interactions.[92] It is clear that only a neutral meson can perform such a transformation without violating baryon and charge conservation. The only known quantum number that distinguishes such a meson from its antiparticle is the strangeness or hypercharge, and hence the meson must be strange if it is to be distinct from its antiparticle. Running through the list we are led,[93] uniquely, to the $K^0 - \bar{K}^0$ complex. As we have indicated earlier in the chapter both K^0 and $\bar{K}^0$ can decay, by $H_{WK}^{\Delta S=1}$, into two pions: $\pi^+ + \pi^-$ or $\pi^0 + \pi^0$. There are several additional common modes but this one involves the fewest particles and will serve to make the argument. According to figure 1 (p. 266), K^0 and $\bar{K}^0$ can communicate with each other by a two-step weak process in which $\Delta S = 1$ at each step. There is no evidence for a direct $\Delta S = 2$ connection and we will not consider such couplings here. This means that if a K^0 meson

---

89. G. Lüders, *Ann. Phys. N.Y.* **2**:1 (1957), J. Schwinger, *Phys. Rev.* **91**:720 (1953), and W. Pauli, *Niels Bohr and the Development of Physics*, McGraw-Hill, New York (1955).
90. R. Jost, *Helv. Phys. Acta* **30**:409 (1957).
91. J. H. Christenson, J. W. Cronin, V. L. Fitch, and R. Turlay, *Phys. Rev. Letters* **13**:138 (1964).
92. In principle the excited K^0's (the K^{0*}) can also communicate with their antiparticles. But since these objects decay by strong couplings, so that their lifetimes are about 10^{-21} sec or less, this effect may not be observable.
93. The first physicists to exploit this aspect of the K^0 system were M. Gell-Mann and A. Pais, *Phys. Rev.* **97**:1387 (1955). See also A. Pais and O. Piccioni, *Phys. Rev.* **100**:1487 (1955).

is produced, at $t = 0$, by a strong interaction such as $\pi^- + p \longrightarrow \Lambda^0 + K^0$, it will begin at once to transform itself into a $\overline{K}^0$. The particle observed at $t > 0$ will be a linear combination of K^0 and $\overline{K}^0$. We can represent the arbitrary combination of K^0 and $\overline{K}^0$ at time t as a complex vector in a two-dimensional space in which, say, the upper component refers to K^0 and the lower component refers to $\overline{K}^0$. Thus, in an obvious notation, the mixed state at time t can be written[94]

$$\psi(t) = \begin{pmatrix} a(t) \\ b(t) \end{pmatrix}. \tag{15.195}$$

If the K^0 were stable, then $\psi(t)$ would be obtained from a unitary transformation, e^{iMt}, of the 2×2 vector $\psi(0)$ so that, for infinitesimal t,

$$d\psi = -iM\psi\, dt. \tag{15.196}$$

In the actual case, since the state decays, we can write

$$\dot{\psi} = -(\Gamma + iM)\psi, \tag{15.197}$$

where Γ and M are Hermitian 2×2 matrices. (*Any* matrix, A, has a decomposition into a Hermitian and an anti-Hermitian part: $A = (A + A^\dagger)/2 + (A - A^\dagger)/2$.) The fact that the Schrödinger equation in this case involves a non-Hermitian "Hamiltonian" reflects the fact that this equation just describes the decaying state, not the total wave function of the system, which would include the decay products as well. This approximation is essentially that of Wigner and Weisskopf.[95] For reasons that will shortly become clear M is known as the "mass matrix" and Γ is known as the "lifetime matrix." For simplicity of notation let us define four *complex* parameters by the equation

$$\Gamma + iM = \begin{pmatrix} m & p^2 \\ q^2 & \overline{m} \end{pmatrix}, \tag{15.198}$$

where, in the absence of any special symmetry requirements, m, $\overline{m}$, p^2, and q^2 are all *distinct* complex numbers. This 2×2 matrix problem can be diagonalized by assuming that there exist two states $\psi_\pm(t)$ that have the time dependence $e^{-\gamma_\pm t}$ and solving for $\psi_\pm(0)$ and $\gamma_\pm$ in terms of the four complex parameters. We find at once that

---

94. In general, this t should be the proper time of the K^0. (For a discussion of this point and a very elegant general treatment of *CP* violations in K decays, see J. S. Bell, *Les Houches Lectures* (1965).) We work in a frame in which the K^0 is at rest.
95. V. F. Weisskopf and E. P. Wigner *Z. Physik* **63**:54 (1930) and **65**:18 (1930). T. D. Lee, R. Oehme, and C. N. Yang, *Phys. Rev.* **105**:1671 (1957), made the fundamental analysis of the time dependence of the K^0 system for the situation in which C, P, T, or *CPT* might be violated. For an elementary review of the Wigner-Weisskopf approximation and several of its applications to particle physics, see J. Bernstein, *Cargèse Lectures.* Gordon and Breach, 1966.

$$\gamma_{\pm} = \frac{m+\overline{m}}{2} \pm \tfrac{1}{2}\sqrt{(m-\overline{m})^2 + 4p^2q^2},$$
$$\frac{a_{\pm}}{b_{\pm}} = \frac{p^2}{\gamma_{\pm} - m} = \frac{\gamma_{\pm} - \overline{m}}{q^2}. \tag{15.199}$$

Before using these results in the general case, let us see what they imply if *CP* is good. In this case it is natural to call, since the K^0 is a pseudoscalar,

$$-|\overline{K}^0\rangle = CP|K^0\rangle, \tag{15.200}$$

thus defining the phase of the *CP* transformation. Hence, for good *CP*, if $\begin{pmatrix} a \\ b \end{pmatrix}$ is a solution to the equation

$$\gamma \begin{pmatrix} a \\ b \end{pmatrix} = (\Gamma + iM) \begin{pmatrix} a \\ b \end{pmatrix}, \tag{15.201}$$

then so is[96] $\begin{pmatrix} b \\ a \end{pmatrix}$, since interchanging the roles of a and b corresponds to *CP* conjugation. Thus from Eq. (15.199), if *CP* is good then

$$p^2 = q^2, \qquad m = \overline{m}, \tag{15.202}$$

or

$$\gamma_{\pm} = m \pm p^2 \tag{15.203}$$

and

$$\frac{a_{\pm}}{b_{\pm}} = \pm 1. \tag{15.204}$$

This result means that, for good *CP*, the states that decay exponentially are also eigenstates of *CP*. In fact, if we define two normalized states

$$|K_S{}^0\rangle = \frac{|K^0\rangle - |\overline{K}^0\rangle}{\sqrt{2}},$$
$$|K_L{}^0\rangle = \frac{|K^0\rangle + |\overline{K}^0\rangle}{\sqrt{2}}, \tag{15.205}$$

then these states, according to Eq. (15.204), decay exponentially and have *CP* eigenvalues ± 1. It is clear that if we turn off the interaction connecting K^0 to $\overline{K}^0$—that is, the weak coupling—then the matrix $\Gamma + iM \longrightarrow iM_0$, where

$$M_0 = \begin{pmatrix} m_0 & 0 \\ 0 & m_0 \end{pmatrix}. \tag{15.206}$$

Here m_0 is the common *real* mass associated with the $|K^0\rangle$ and $|\overline{K}^0\rangle$ states which, under these circumstances, would be stable. As we shall see shortly, $|K^0\rangle$ and

96. Clearly $\begin{pmatrix} -b \\ -a \end{pmatrix}$, which *is* the *CP* conjugate, is also a solution.

$|\bar{K}^0\rangle$ have a common mass under the weaker assumption of CPT invariance. Thus, calling M the "mass matrix" in general is appropriate. With the weak interaction turned on, M is no longer diagonal and its matrix elements are no longer real. Moreover the Schrödinger equation is no longer Hermitian and Γ, the "decay matrix," reflects the instability of the $|K^0\rangle$ and $|\bar{K}^0\rangle$. In the general case, clearly[97] Im $\gamma_{\substack{S\\L}}$ is the mass of $|K_S{}^0\rangle$ and $|K_L{}^0\rangle$, respectively, while Re $(\gamma_{\substack{S\\L}}/2)$ are the reciprocal lifetimes of these two states. A priori there is no reason to expect that γ_S and γ_L should have any particular numerical relation to each other. However, as indicated earlier in the chapter (Eq. (15.10)), experimentally

$$\text{Re}\,\gamma_S \gg \text{Re}\,\gamma_L,$$

and in fact the CP invariance, which is only violated by a relatively small amount in this system, gives us an understanding of why this is true. It is easy to see that two pions with zero total charge and orbital angular momentum l have, under CP, the eigenvalue

$$CP|2\pi\rangle = (-1)^{2l}|2\pi\rangle = |2\pi\rangle. \tag{15.207}$$

Thus the 2π final state is only accessible to $|K_S{}^0\rangle$, which is CP even. On the other hand, $|K_L{}^0\rangle$ can decay into 3π's, $\pi\mu\nu$, etc., but into *no* two-body final state. Hence phase space considerations imply that Re $\gamma_S >$ Re γ_L. To make such arguments quantitative one needs an expression for the lifetimes in terms of the perturbation that is responsible for the instability of the system. One way of obtaining this connection is through the conservation of probability, or unitarity.[98] We can write the total transition rate out of the state $\begin{pmatrix} a \\ b \end{pmatrix}$ as

$$2\pi \sum_f \rho_f |a\langle f|H|K^0\rangle + b\langle f|H|\bar{K}^0\rangle|^2$$
$$= 2\pi[\sum_f [|a|^2|\langle f|H|K^0\rangle|^2\rho_f + |b|^2|\langle f|H|\bar{K}^0\rangle|^2\rho_f$$
$$+ [a^*b\rho_f\langle f|H|K^0\rangle^*\langle f|H|\bar{K}^0\rangle + ab^*\rho_f\langle f|H|K^0\rangle\langle f|H|\bar{K}^0\rangle^*]]. \tag{15.208}$$

(We do not assume CP here so that a and b are arbitrary complex numbers.) Here $\langle f|$ is an arbitrary final state (such as 2π, 3π, etc.), H is the Hermitian decay Hamiltonian, and ρ_f is the density of states appropriate to $\langle f|$. From the conservation of probability this must equal

$$-\frac{d}{dt}[|a|^2 + |b|^2] = -2\,\text{Re}\,(a^*\dot{a} + b^*\dot{b}). \tag{15.209}$$

But the equation of motion Eq. (15.197) says that

$$\begin{pmatrix} \dot{a} \\ \dot{b} \end{pmatrix} = -\begin{pmatrix} m & p^2 \\ q^2 & \overline{m} \end{pmatrix}\begin{pmatrix} a \\ b \end{pmatrix}, \tag{15.210}$$

---

97. We shall henceforth arbitrarily call $\gamma_- = \gamma_S$, $\gamma_+ = \gamma_L$.
98. We follow the discussion of J. S. Bell, *Les Houches Lectures*, op. cit.

so that

$$\dot{a} = -(ma + p^2 b),$$
$$\dot{b} = -(q^2 a + \overline{m} b). \quad (15.211)$$

Thus

$$\mathrm{Re}\,(a^*\dot{a} + b^*\dot{b})$$
$$= -\mathrm{Re}\,(m|a|^2 + \overline{m}|b|^2 + p^2 a^* b + q^2 b^* a)$$
$$= -\mathrm{Re}\,(m)|a|^2 - \mathrm{Re}\,(\overline{m})|b|^2 - \mathrm{Re}\,(p^2 a^* b + q^2 b^* a). \quad (15.212)$$

Thus, from Eq. (15.208),

$$\mathrm{Re}\,(m) = \pi \sum_f \rho_f |\langle f|H|K^0\rangle|^2,$$
$$\mathrm{Re}\,(\overline{m}) = \pi \sum_f \rho_f |\langle f|H|\overline{K}^0\rangle|^2, \quad (15.213)$$
$$\frac{(q^2)^* + p^2}{2} = \pi \sum_f \rho_f \langle K^0|H|f\rangle\langle f|H|\overline{K}^0\rangle,$$

where we have used the hermiticity of H in obtaining the equations in this form. These equations show that 2 Re (m) and 2 Re $(\overline{m})$ have the usual perturbation theory forms for the decay rates, while p^2 and q^2 express the effects of the mixing on the lifetimes. From the form of Eq. (15.213) we can rediscover the restrictions imposed by CP, T, and CPT invariance on the parts of m, $\overline{m}$, p^2, and q^2 given by Eq. (15.213). From CP we have at once (remembering that the sum over the states, $CP|f\rangle$, is identical in Eq. (15.213) to the sum over the states, $|f\rangle$, since these are both eigenstates of the strong Hamiltonian that commutes[99] with CP)

$$\mathrm{Re}\,(m) = \mathrm{Re}\,(\overline{m}),$$
$$p^2 + (q^2)^* = (p^2)^* + q^2.$$

If we apply CPT to Eq. (15.213) we can also conclude, assuming CPT invariance, that

$$\mathrm{Re}\,(m) = \mathrm{Re}\,(\overline{m}).$$

In order to find the imaginary parts of m and $\overline{m}$—the masses, and the mass shifts caused by the weak coupling—we need to invoke a dynamical model such as perturbation theory. It is easy to show[100] that the time dependent perturbation theory gives, in general,

---

99. The direct evidence for CP conservation in strong interactions comes from pp scattering, polarization measurements, and detailed balance, which relates the reactions $A + B \longrightarrow C + D$ to $C + D \longrightarrow A + B$. (See, for example, L. Rosen and J. E. Brolley, Jr., *Phys. Rev. Letters* **2**:98 (1959), and D. Bodansky et al., *Phys. Rev. Letters* **2**:101 (1959).) The data show that the CP violating amplitude cannot be more than a percent or so of the CP conserving amplitude. We come back to detailed balance below.

100. See, for example, T. D. Lee, R. Oehme, and C. N. Yang, op. cit.

$$\text{Im}\,(m) = m_0 + \sum_f \frac{\langle K^0|H|f\rangle\langle f|H|K^0\rangle}{m_0 - E_f} + \langle K^0|H|K^0\rangle,$$
$$\text{Im}\,(\bar{m}_0) = \bar{m}_0 + \sum_f \frac{\langle \bar{K}^0|H|f\rangle\langle f|H|\bar{K}^0\rangle}{\bar{m}_0 - E_f} + \langle \bar{K}^0|H|\bar{K}^0\rangle, \tag{15.214}$$

where, if H_0 is the strong Hamiltonian,

$$m_0 = \langle K^0|H_0|K^0\rangle,$$
$$\bar{m}_0 = \langle \bar{K}^0|H_0|\bar{K}^0\rangle. \tag{15.215}$$

Thus, if *CPT* is good, then from Eqs. (15.213), (15.214), and (15.215) we can conclude that

$$m = \bar{m}. \tag{15.216}$$

This is a consequence of *CPT* that can be checked. It is possible to measure $\text{Im}\,\gamma_L - \text{Im}\,\gamma_S$, using the following idea. If a K^0 is produced at $t = 0$ then the state at $t > 0$ has the form[101]

$$\psi(t) = \frac{1}{\sqrt{2}}|K_S{}^0\rangle e^{-\gamma_S t} + \frac{1}{\sqrt{2}}|K_L{}^0\rangle e^{-\gamma_L t}, \tag{15.217}$$

where all of these quantities have been defined above. (We suppose good *CP* in writing the coefficients in this form. $|K_S{}^0\rangle$ and $|K_L{}^0\rangle$ are always assumed normalized.) This means that any decay amplitude for this state to go into, say, $\pi\mu\nu$, will have a complicated, nonexponential time dependence. Indeed, if one computes the probability (by multiplying Eq. (15.217) by the final state in question and squaring) there will be an interference term of the form $\cos((\text{Im}\,\gamma_L - \text{Im}\,\gamma_S)t)$ and it is the measurement of such interferences that gives the value quoted in Eq. (15.12) for $m_L - m_S$. We can rewrite it as

$$\left|\frac{m_L - m_S}{m_0}\right| \simeq 6.5 \times 10^{-15}, \tag{15.218}$$

where m_0 is the mass of the K^0 as measured in the strong interactions. Since, from Eq. (15.199),

$$\gamma_S - \gamma_L = \sqrt{(m - \bar{m})^2 + 4p^2q^2} \tag{15.219}$$

and since the experimental mass difference is consistent with what would be expected from a second-order weak coupling, we can conclude that[102]

---

101. U. Camerini and C. T. Murphy in Appendix A of the review talk of N. Cabibbo, Berkeley Conference (1966), op. cit., give the most recent experimental data and discuss how the interpretation of the experiment is modified if *CP* is violated.
102. It can be shown (see, for example, J. S. Bell and J. Steinberger, Oxford Conference Proceedings (1965), p. 195 and ff.), essentially from the implications of unitarity, that, in the general case,

$$|m - \bar{m}| \leq 2|m_L - m_S|.$$

$$\frac{|m - \bar{m}|}{m_0} \simeq 10^{-14}, \tag{15.220}$$

which is the strongest evidence in particle physics for *CPT* invariance.

Until 1964, all of the evidence on K^0 decays, and on weak interactions in general, was consistent with *CP* conservation. In the summer of 1964[103] it was discovered and, in later independent experiments,[104] confirmed, that the long-lived K^0 has a 2π decay mode.[105] This 2π state is, as we have seen, *CP* even. On the other hand, among its other decays the long-lived K^0 has a $3\pi^0$ mode with a branching ratio[106] of $23.5 \pm 2.19\%$. It is easy to see that a $3\pi^0$ state with total angular momentum zero must have *odd CP*, essentially because each of the π^0's has odd intrinsic parity. Thus *CP* must be violated in K^0 decay since the long-lived K^0 decays into both a *CP* even and a *CP* odd final state. Before discussing the possible theoretical significance of this fact let us carry out the rest of the computation of the eigenstates that decay exponentially for the case in which *CPT* is good but *CP* is violated. Thus, from Eq. (15.199),

$$\begin{aligned} \gamma_L &= m + pq, \\ \gamma_S &= m - pq, \\ \frac{a_L}{b_L} &= \frac{p}{q}, \\ \frac{a_S}{b_S} &= -\frac{p}{q}. \end{aligned} \tag{15.221}$$

Thus, normalized,[107]

$$\begin{aligned} |K_S{}^0\rangle &= \frac{1}{\sqrt{|p|^2 + |q|^2}}\,[p|K^0\rangle - q|\bar{K}^0\rangle], \\ |K_L{}^0\rangle &= \frac{1}{\sqrt{|p|^2 + |q|^2}}\,[p|K^0\rangle + q|\bar{K}^0\rangle]. \end{aligned} \tag{15.222}$$

We note that

$$\langle K_S{}^0|K_L{}^0\rangle = \frac{|p|^2 - |q|^2}{|p|^2 + |q|^2}, \tag{15.223}$$

so that the long- and short-lived states are not orthogonal in the *CP* violating case, which is intuitively plausible since they then have common decay modes. It is possible to use unitarity to set an upper bound on this inner product as follows.[108] Written in terms of $|K_S{}^0\rangle$ and $|K_L{}^0\rangle$ the general form of the neutral K^0 state at time t is (X and Y are arbitrary complex numbers)

---

103. J. H. Christenson et al., op. cit.
104. A. Abashian et al., *Phys. Rev. Letters* **13**:243 (1964), W. Galbraith et al., *Phys. Rev. Letters* **14**:383 (1965), and X. de Bouard et al., *Phys. Letters* **15**:58 (1965).
105. The first experiments detected the $\pi^+ + \pi^-$ mode.
106. A. H. Rosenfeld et al., *Rev. Mod. Phys.* **39**:1 (1967).

$$\psi(t) = Xe^{-\gamma_L t}|K_L{}^0\rangle + Ye^{-\gamma_S t}|K_S{}^0\rangle, \tag{15.224}$$

whose norm, $N(t)$, is

$$N(t) = |X|^2 e^{-\Gamma_L t} + |Y|^2 e^{-\Gamma_S t} + X^* Y e^{-(\gamma_L{}^* + \gamma_S)t}\langle K_L{}^0|K_S{}^0\rangle + Y^* X e^{-(\gamma_S{}^* + \gamma_L)t}\langle K_S{}^0|K_L{}^0\rangle \tag{15.225}$$

where the lifetimes are given by

$$\begin{aligned} \Gamma_L &= 2\,\mathrm{Re}\,(\gamma_L), \\ \Gamma_S &= 2\,\mathrm{Re}\,(\gamma_S). \end{aligned} \tag{15.226}$$

Thus at $t = 0$

$$\frac{dN}{dt} = |X|^2\Gamma_L + |Y|^2\Gamma_S + X^*Y(\gamma_L{}^* + \gamma_S)\langle K_L{}^0|K_S{}^0\rangle + Y^*X(\gamma_S{}^* + \gamma_L)\langle K_S{}^0|K_L{}^0\rangle, \tag{15.227}$$

and this must equal, for arbitrary X and Y, the total transition rate

$$2\pi \sum_f \rho_f |X\langle f|H|K_L{}^0\rangle + Y\langle f|H|K_S{}^0\rangle|^2. \tag{15.228}$$

Thus

$$\begin{aligned} \Gamma_S &= 2\pi \sum_f \rho_f |\langle f|H|K_S{}^0\rangle|^2, \\ \Gamma_L &= 2\pi \sum_f \rho_f |\langle f|H|K_L{}^0\rangle|^2, \\ (\gamma_L{}^* + \gamma_S)\langle K_L{}^0|K_S{}^0\rangle &= 2\pi \sum_f \rho_f \langle f|H|K_L{}^0\rangle^* \langle f|H|K_S{}^0\rangle. \end{aligned} \tag{15.229}$$

Using the last equation and the Schwarz inequality,

$$|\gamma_L{}^* + \gamma_S|\,|\langle K_L{}^0|K_S{}^0\rangle| \le (\Gamma_L\Gamma_S)^{1/2}. \tag{15.230}$$

Since

$$|\gamma_L{}^* + \gamma_S| = \left|\frac{(\Gamma_L + \Gamma_S)}{2} + i(m_S - m_L)\right|, \tag{15.231}$$

we can put in the experimental numbers and conclude that[109]

$$|\langle K_L{}^0|K_S{}^0\rangle| \lesssim 0.06. \tag{15.232}$$

This means that

$$|p| \sim |q| \tag{15.233}$$

no matter how badly CP is violated. Hence, ironically, the K^0 system, which so far is the only one where CP violation has definitely been seen, is not an especially good detector.

---

107. This choice of phase coincides with the previous definitions of $|K_s{}^0\rangle$ and $|K_L{}^0\rangle$ in the good CP limit.
108. See, for example, J. S. Bell and J. Steinberger, Oxford Conference, op. cit.
109. On the basis of additional experimental information on K^0 branching ratios this number can be reduced to 0.01; J. S. Bell and J. Steinberger, op. cit.

We are now in a position to discuss the data.[110]

We introduce the following notation; let

$$A(K^0 \longrightarrow \pi^+ + \pi^-) = c,$$
$$A(\bar{K}^0 \longrightarrow \pi^+ + \pi^-) = \bar{c},$$
$$A(K^0 \longrightarrow \pi^0 + \pi^0) = d,$$
$$A(\bar{K}^0 \longrightarrow \pi^0 + \pi^0) = \bar{d}, \tag{15.234}$$
$$r = \frac{q}{p} = 1 - \epsilon,$$
$$\frac{\bar{c}}{c} = \epsilon' - 1.$$

With good CP[111]

$$c = -\bar{c},$$
$$d = -\bar{d}, \tag{15.235}$$
$$r = 1.$$

With this notation and Eq. (15.222) we can define

$$\eta_{+-} = \frac{A(K_L{}^0 \longrightarrow \pi^+ + \pi^-)}{A(K_S{}^0 \longrightarrow \pi^+ + \pi^-)} = \frac{1 + r\dfrac{\bar{c}}{c}}{1 - r\dfrac{\bar{c}}{c}}$$
$$= \frac{\epsilon + \epsilon' - \epsilon\epsilon'}{2 - \epsilon - \epsilon' + \epsilon\epsilon'}. \tag{15.236}$$

Given that $|\epsilon| \ll 1$, $|\epsilon'| \ll 1$ we may write

$$\eta_{+-} \simeq \frac{\epsilon + \epsilon'}{2}. \tag{15.237}$$

The original experiments cited above measured $|\eta_{+-}|$. A compilation of the various experiments gives[112]

$$|\eta_{+-}| = 1.94 \pm 0.09 \times 10^{-3}, \tag{15.238}$$

suggesting that $|\epsilon|$ and $|\epsilon'|$ are both $\ll 1$. The phase of η_{+-} is not yet completely well known but the latest results[113] give $78° \pm 15°$ for it. We can get more insight into this result if we analyze the 2π states into their isotopic spin compo-

---

110. The general analysis in the form given here is due to T. T. Wu and C. N. Yang, *Phys. Rev. Letters* **13**:501 (1964), and L. Wolfenstein, CERN preprint TH, 583, July 13 (1965).

111. Remember in deriving these results that with our phase convention,

$$CP|K^0\rangle = -|\bar{K}^0\rangle.$$

112. The parameter $|\eta_{+-}|$ is known experimentally, to be energy independent; that is, independent of the kinetic energy of the incident K^0's up to energies of 10 BeV. See X. de Bouard et al., *Phys. Letters* **15**:58 (1965), and W. Galbraith et al., *Phys. Rev. Letters* **14**:383 (1965). This observation rules out the "long range force" theories of J. Bernstein, N. Cabibbo, and T. D. Lee, *Phys. Letters* **12**:146 (1964), and J. S. Bell and J. K. Perring,

nents, $T = 0$ and $T = 2$. ($T = 1$ is ruled out for 2π's in an S-state by Bose statistics.) It can be shown,[114] assuming *CPT*, that

$$\begin{aligned} c &= A(K^0 \longrightarrow \pi^+ + \pi^-) = \sqrt{\tfrac{2}{3}}\, A_0 e^{i\delta_0} + \sqrt{\tfrac{1}{3}}\, A_2 e^{i\delta_2}, \\ \bar{c} &= A(\overline{K}^0 \longrightarrow \pi^+ + \pi^-) = -\sqrt{\tfrac{2}{3}}\, A_0 e^{i\delta_0} - \sqrt{\tfrac{1}{3}}\, A_2{}^* e^{i\delta_2}, \end{aligned} \tag{15.239}$$

where the subscripts 0 and 2 refer to the isotopic spins. The δ_i are the strong interaction pion-pion phase shifts in the isotopic state denoted by the subscript. Thus

$$\epsilon' = 1 + \frac{\bar{c}}{c} \simeq -\sqrt{2}\, i e^{i(\delta_2 - \delta_0)} \frac{\operatorname{Im}(A_2)}{A_0}, \tag{15.240}$$

where A_0 has been chosen, arbitrarily, to be real. If *CP* were good, then A_2 would also be real. Thus a nonvanishing ϵ' implies a *CP* violating effect in the "decay matrix" since the decay matrix gives the transition amplitudes. Moreover, $\epsilon' \neq 0$ also would imply that the $\Delta T = \frac{1}{2}$ rule is violated in the decay since, if H transforms like an isospinor, the $T = 2$ final state could not be reached from the $T = \frac{1}{2}$, K^0 state. We return to these matters shortly.

We shall now discuss the parameter ϵ:

$$\begin{aligned} \epsilon &= 1 - \frac{q}{p} = \frac{p - q}{p} = \frac{p^2 - q^2}{p(p+q)} \simeq \frac{p^2 - q^2}{2qp} = \frac{p^2 - q^2}{\gamma_L - \gamma_S} \\ &= \frac{p^2 - q^2}{\dfrac{(\Gamma_L - \Gamma_S)}{2} + i(m_L - m_S)} = \frac{\Delta\Gamma + i\,\Delta M}{\dfrac{(\Gamma_L - \Gamma_S)}{2} + i(m_L - m_S)}. \end{aligned} \tag{15.241}$$

In writing this equation we have used the fact that $p \sim q$ and Eq. (15.199). In the last step we have split p^2 and q^2 into their real and imaginary parts and have called the differences $\Delta\Gamma$ and ΔM, respectively. These parameters, which would be zero if *CP* were good, measure the *CP* violating effects on the off-diagonal elements of the mass and lifetime matrix elements. We can now see that any theory that makes the *CP* violation a *mass-matrix* effect (we give examples below) will predict that the phase of η_{+-} is about 45°. Thus, let

$$\Delta\Gamma \simeq 0, \qquad \epsilon' \simeq 0. \tag{15.242}$$

Hence

$$\eta_{+-} \simeq \frac{1}{2} \frac{i\,\Delta M}{\dfrac{(\Gamma_L - \Gamma_S)}{2} + i(m_L - m_S)}. \tag{15.243}$$

Phys. Rev. Letters **13**:348 (1964). The long range force, of cosmic origin, was imagined to act asymmetrically on K^0 and $\overline{K}^0$, thus producing an asymmetry in the mass matrix. This asymmetry was energy dependent for all forces transmitted by bosons with spin greater than zero. However, as we shall see below, new results on the mode, $K_L{}^0 \longrightarrow \pi^0 + \pi^0$, appear to rule out *all* theories in which *CP* violation occurs *exclusively* in the mass matrix.

113. C. Rubbia and J. Steinberger, *Phys. Letters* **24B**:531 (1967), and M. Bott-Bodenhanson et al., *Phys. Letters* **42B**:432 (1967).
114. See, for example, J. S. Bell and J. Steinberger, op. cit.

Thus

$$|\tan(\arg \eta_{+-})| \simeq \left|\frac{\dfrac{(\Gamma_L - \Gamma_S)}{2}}{m_L - m_S}\right| \simeq 1 \tag{15.244}$$

since

$$m_L - m_S \simeq \Gamma_S/2 \tag{15.245}$$

and

$$\Gamma_S \gg \Gamma_L. \tag{15.246}$$

We can now discern two limiting cases:

$$1. \begin{cases} \epsilon' \simeq 0, \\ \eta_{+-} \simeq \epsilon/2. \end{cases}$$

In this case the *CP* violation would be primarily in the mass matrix and the phase of η_{+-} would be about 45°.

$$2. \begin{cases} \epsilon \simeq 0, \\ \eta_{+-} \simeq \epsilon'/2. \end{cases}$$

In this case the *CP* violation would be primarily in the decay matrix and the $\Delta T = \frac{1}{2}$ rule would also be violated. The phase of η_{+-} would be fixed by the δ_2, δ_0 phase shift difference and could be as large as 90°; see Eq. (15.240) if $\delta_2 \sim \delta_0$.

Since the phase of η_{+-} is large, Case 1 appears ruled out. In fact we shall now present a second piece of evidence which, if it holds up, definitely rules out Case 1 and several of the theories that predicted Case 1. This involves the measurement of $|\eta_{00}|$, where

$$\eta_{00} = \frac{A(K_L{}^0 \longrightarrow \pi^0 + \pi^0)}{A(K_S{}^0 \longrightarrow \pi^0 + \pi^0)}. \tag{15.247}$$

Before giving the experimental results we rewrite η_{00} as follows: let

$$\frac{\bar{d}}{d} = -1 + 2\epsilon''. \tag{15.248}$$

Thus

$$\eta_{00} = \frac{1 + r\dfrac{d}{\bar{d}}}{1 - r\dfrac{\bar{d}}{d}} = -\frac{\epsilon - 2\epsilon'' + 2\epsilon\epsilon''}{2 - \epsilon + 2\epsilon'' - 2\epsilon\epsilon''} \simeq \epsilon'' - \frac{\epsilon}{2}. \tag{15.249}$$

But we can also expand the $\pi^0\pi^0$ states in their isotopic projections. Thus

$$\begin{aligned} d &= A(K^0 \longrightarrow \pi^0 + \pi^0) = \sqrt{\tfrac{1}{3}}\, A_0 e^{i\delta_0} - \sqrt{\tfrac{2}{3}}\, A_2 e^{i\delta_2}, \\ \bar{d} &= A(\bar{K}^0 \longrightarrow \pi^0 + \pi^0) = -\sqrt{\tfrac{1}{3}}\, A_0 e^{i\delta_0} + \sqrt{\tfrac{2}{3}}\, A_2{}^* e^{i\delta_2}, \end{aligned} \tag{15.250}$$

which gives from Eqs. (15.248) and (15.240)

$$\epsilon'' \simeq \epsilon'. \tag{15.251}$$

We can again distinguish two cases:

$$1.\quad \begin{cases} |\epsilon'| \simeq 0, \\ |\eta_{+-}| \simeq |\eta_{00}|, \end{cases}$$

$$2.\quad \begin{cases} |\epsilon| \simeq 0, \\ |2\eta_{+-}| \simeq |\eta_{00}|. \end{cases}$$

In fact two recent experiments[115] give

$$|\eta_{00}| = (4.3^{+1.1}_{-.08}) \times 10^{-3} \qquad \text{(Gaillard et al.),}$$

$$|\eta_{00}| = (4.17 \pm 0.3) \times 10^{-3} \qquad \text{(Cronin et al.).}$$

These results are consistent with Case 2 and clearly inconsistent with Case 1, although they do not rule out some intermediate possibilities.[116] The phase of η_{00} is not known experimentally.

---

115. M. Gaillard et al., *Phys. Rev. Letters* **18**:20 (1967); J. W. Cronin et al., *Phys. Rev. Letters* **18**:25 (1967). A theoretical analysis anticipating large $\Delta T = \frac{3}{2}$ components in the *CP* violating part of *H* was given by T. N. Truong, *Phys. Rev. Letters* **13**:358 (1964), and T. T. Wu and C. N. Yang, *Phys. Rev. Letters* **13**:380 (1964). The value of $|\eta_{00}|$, quoted above, for Cronin et al. is an unpublished value given at the Rochester Conference in High Energy Physics in August 1967. It supersedes the published value of Cronin et al. which was $(4.9 \pm 0.5) \times 10^{-3}$.

116. Some very recent experiments suggest that this intermediate situation may in fact obtain. The experiments, by D. Dorfan, M. Schwartz, and S. Wojcicki (unpublished) and by J. Steinberger and J. Sunderland (unpublished), concern the ratio of the decay rates for $K_L{}^0 \longrightarrow \pi^\pm + \mu^\mp + \nu$ and $K_L{}^0 \longrightarrow \pi^\pm + e^\mp + \nu$. These ratios would be unity if *CP* were good. Before giving the experimental numbers it is important to note that *CPT* does not guarantee equal partial decay rates in general. In this case, the state $K_L{}^0$ is not an eigenstate of *CPT* since p and q, the quantities that fix the ratio of K^0 and $\overline{K}^0$ in this state, are not real numbers. Hence no conclusion can be drawn about this ratio from *CPT* invariance. The two groups find the following results ($N_e{}^\pm$ is the number of leptons e with the given charge $\pm$).

$$\frac{N_\mu{}^+}{N_\mu{}^-} = 1.0080 \pm 0.0026 \qquad \text{(Dorfan et al.),}$$

$$\frac{N_{e^+} - N_{e^-}}{N_{e^+} + N_{e^-}} = +(2.16 \pm 0.4) \times 10^{-3} \qquad \text{(Steinberger et al.).}$$

In the first place these results show that *CP* is violated in a new process involving the K^0 (namely, a leptonic decay), as opposed to a nonleptonic decay. Moreover, assuming that the $\Delta S = \Delta Q$ rule is valid, one can show that

$$\frac{N^+ - N^-}{N^+ + N^-} = \text{Re}\ \epsilon,$$

so that these experiments give a direct measurement of Re ϵ, which turns out to be

$$2\ \text{Re}\ \epsilon = (2.0 \pm 0.65) \times 10^{-3} \qquad \text{(Dorfan et al.),}$$
$$2\ \text{Re}\ \epsilon = (1.08 \pm 0.20) \times 10^{-3} \qquad \text{(Steinberger et al.).}$$

If the $\Delta S = \Delta Q$ rule is not valid, which could be true on a small scale, then the analysis involves a parameter that measures this violation and is correspondingly complicated. But in any case it would appear as if the *CP* violation occurs both in the mass matrix and the decay matrix, although one is still in the dark as to the actual interaction that produces the violation.

Hence we may summarize the apparent experimental situation by saying that there seems to be a substantial *CP* violation in the decay matrix and that this violation also violates the $\Delta T = \frac{1}{2}$ rule. This would rule out a class of theories in which the *CP* violation was assigned to a new superweak[117] interaction. This interaction, conjectured to be perhaps 10^{-7} of the strength of the normal weak coupling, would have modified the off-diagonal terms of the mass matrix since it would have allowed a direct, first-order $K^0 - \bar{K}^0$ coupling with $\Delta S = 2$. The new experiment also seems to rule out the possibility that the Wigner-Weisskopf exponential approximation is breaking down in the $K^0 - \bar{K}^0$ system and that what is being studied is a possible nonexponential tail of $K_S{}^0$ mesons rather than a *CP* violating mode of $K_L{}^0$. This is ruled out because the $K_S{}^0$ decays into 2π's *do* obey the $\Delta T = \frac{1}{2}$ rule. Indeed, it is well known that, experimentally,[118] the ratio of $K_S{}^0$ going into $\pi^+ + \pi^-$ and $\pi^0 + \pi^0$ is, sensibly $\frac{2}{1}$ which, as Eqs. (15.239) and (15.250) show, is a consequence of the $\Delta T = \frac{1}{2}$ rule. The *CP* violating decays show, at the moment very nearly the opposite ratio and hence cannot be coming from the $K_S{}^0$ state.

We may ask what is left to explain the *CP* violation? At present there appears to be no completely satisfactory model, although the literature is replete with dead and dying theories. In broad outline there seem to be two viable lines.

1. There may exist a new weak interaction, which violates both *CP* and the $\Delta T = \frac{1}{2}$ rule. Such a weak interaction should also have consequences elsewhere, in other decays, but, at this time[119] there is no experimental evidence of *CP* violation in any other weak process[120] and, moreover, no one has found a simple, convincing general principle that leads to selection of a given model from the wide class of models that meet these general requirements.

2. There may be a violation of *CP* in the electromagnetic interactions of hadrons.[121] In this theory the total electromagnetic current, $J_\mu{}^\gamma$, is given by the sum of conserved currents,

$$J_\mu{}^\gamma = J_\mu + K_\mu, \tag{15.252}$$

where, if C is charge conjugation,

---

117. L. Wolfenstein, *Phys. Rev. Letters* **13**:562 (1964), and T. D. Lee and L. Wolfenstein, *Phys. Rev.* **138**:B1490 (1965).
118. A. H. Rosenfeld, *Rev. Mod. Phys.*, **39**:1 (1967). See Eq. (15.124).
119. September 1967. For some sample theories with *CP* violating weak Hamiltonian, see R. G. Sachs, *Phys. Rev. Letters* **13**:286 (1964), and N. Cabibbo, *Phys. Letters* **12**:137 (1964). For a lucid review of many of these possibilities, see J. Prentki, Proceedings of the Oxford Conference (1965), op. cit., p 47.
120. A typical experiment is that of K. A. Young et al., *Phys. Rev. Letters* **18**:806 (1967). These authors study $K_L{}^0 \longrightarrow \mu^+ + \pi^- + \nu_\mu$. As we have seen, this decay is characterized by the two form factors $f_\pm$ that arise in the decomposition of $\langle\pi^-|J_\mu(0)|K^0\rangle$. It is easy to show (see Eq. (5.56)) that if T is good we can always choose the phases in the time reversal transformation so that

$$z = \frac{f_-(q^2)}{f_+(q^2)}$$

$$CJ_\mu C^{-1} = -J_\mu \tag{15.253}$$

and

$$CK_\mu C^{-1} = K_\mu. \tag{15.254}$$

There is a certain amount of flexibility in choosing the new current K_μ—it may be an SU_3 scalar or an SU_3 vector or a mixture. In any case Q_K, the new charge, would have to have vanishing matrix elements between the known particles, since particle and antiparticle are known experimentally to have equal and opposite charges.[122] Despite the fact that the conventional electromagnetic theory of hadronic currents, with *CP* invariance, is one of the most beautiful and most successful theories in physics, some authors were led to propose a radical change in its structure because of two observations, which, if the proposal turns out to be wrong, may simply be irrelevant coincidences.

a. Numerically,

$$|\eta_{+-}| \simeq 2 \times 10^{-3} \simeq \frac{\alpha}{\pi},$$

which is the typical order of magnitude of an electromagnetic correction to a weak process.

b. At the time the proposal was made there was no *direct* experimental evidence that *CP* was a valid symmetry for hadronic electromagnetic couplings. Before discussing this point, which contains a number of useful lessons about testing *CP*, we may also add a third point based on the recent observations of a large $|\eta_{00}|$.

c. A large $|\eta_{00}|$ implies that the decay matrix for the K^0 decays must have a sizable $\Delta T = \frac{3}{2}$ component. Such a component in the decay matrix could arise naturally from the interference of a *CP*-violating isovector electromagnetic current with a *CP*-conserving weak Hamiltonian that obeys $\Delta T = \frac{1}{2}$.

At the first look it is almost impossible to believe that the *CP* invariance of the electromagnetic hadronic current is not well established experimentally. That this is so, despite twenty years of experimentation on electron scattering by hadrons, photoproduction of mesons, and the like, is a consequence of our basic inability to extract the exact consequences of a field theory in which

---

is a real number. Moreover, in general, a nonzero Im z implies that the T violating quantity $\boldsymbol{\sigma}_\mu \cdot (\mathbf{p}_\pi \times \mathbf{p}_\mu)$, the "transverse muon polarization," should have a nonzero expectation value. By measuring this correlation, the above authors conclude that Im $z = -0.014 \pm 0.066$ and Re $z = -1.3$, which is consistent with no T violation in this mode and contradicts at least one of the weak interaction models cited above (N. Cabibbo, op. cit.), according to which arg z is predicted to be $\pi/2$.

121. J. Bernstein, G. Feinberg, and T. D. Lee, *Phys. Rev.* **139**:B1650 (1965); see also S. Barshay, *Phys. Letters* **17**:78 (1965). In these papers an analysis was made of possible experiments to detect *CP* violation in electromagnetism. The list of possible experiments was expanded in N. Christ and T. D. Lee, *Phys. Rev.* **143**:1310 (1966) and **148**:1520 (1966), and G. Feinberg, *Phys. Rev.* **140**:1402 (1965).

122. T. D. Lee, *Phys. Rev.* **140**:B959 (1965), has given a model of such a current in the context of field theory.

strongly interacting particles play a role. In general, we are reduced to exploiting the symmetries of these theories and collecting our ignorance of the dynamics into unknown form factors that are fixed by experiment and left uncomputed in the theory. The *CP* invariance of the theory expresses itself in the symmetry properties of these form factors, and, as we shall see, even a theory that violates *CP* can imply the same, or nearly the same, symmetry properties of the form factors as in a *CP*-conserving theory, because of general requirements on the theory such as *CPT* invariance, Hermiticity, or the conservation of various currents. If experiment were to show a violation of one of these form factor symmetries, it would mean that several basic principles of relativistic field theory might break down simultaneously and it would not single out *CP* alone as the culprit. Leptonic electromagnetism is on a different footing. The leptons have no strong couplings. Everything about the dynamics is essentially well known and we cannot change it without doing violence to the all-but-perfect agreement that the theory has with experiment. Here, we are better off to leave well enough alone.

The simplest illustration of how *CP* invariance can be masked by more general symmetries is given by studying the matrix elements of observables, such as the current or the Hamiltonian, taken between physical particles on their mass shells. We present a few simple examples in outline; the reader will have no trouble supplying the proofs.

1. The matrix element of a conserved, Hermitian electromagnetic current, taken between single, spin-zero (or spin-$\frac{1}{2}$) particles will have the same structure in terms of *real* form factors, even if[123] $K_\mu \neq 0$, provided that the initial and final particles are the same. Hence, for example, none of the experiments done so far on electron scattering, which can be interpreted with a single virtual photon exchange between matrix elements of currents, can test *CP* invariance.

2. The Hermiticity of the electromagnetic Hamiltonian is enough to imply detailed balance between the rates for $\gamma + A \longrightarrow B + C$ and $B + C \longrightarrow \gamma + A$, providing that this reaction is correctly given by first-order perturbation theory in the electromagnetic field, since $\langle f|H_\gamma|i\rangle = \langle i|H_\gamma|f\rangle^*$.

---

123. That is, whether or not *CP* is good.

124. This asymmetry has been studied experimentally; see, for example, Baltay et al., *Phys. Rev. Letters* **16**:1224 (1966), A. Chops et al., Berkeley Conference 1966, op. cit.; and while the matter is not yet completely settled it appears as if there is little or no asymmetry, which may mean that electromagnetism conserves *CP* or it may be a statement about $\pi\pi$ couplings.

125. See N. Cabibbo, *Phys. Rev. Letters* **14**:965 (1965), where this general question is carefully reviewed.

126. A recent experiment (D. Berley et al., *Phys. Rev.* **142**:893 (1966)) gives

$$\frac{R(\eta^0 \longrightarrow \pi^0 + e^+ + e^-)}{R(\eta^0 \longrightarrow \pi^+ + \pi^- + \pi^0)} < 0.07.$$

127. It is possible that the study of the neutron-electric dipole moment may shed some light. This moment is proportional to $\langle \boldsymbol{\sigma}\cdot\mathbf{r}\rangle_N$, where $\boldsymbol{\sigma}$ is the neutron spin and $\mathbf{r}$ some typical neutron distance. Its existence would violate *both* *CP* and *P* and would represent an

On a more subtle level we can consider the decay $\eta^0 \longrightarrow \pi^+ + \pi^- + \pi^0$, which is, presumably, a virtual electromagnetic transition since it violates G parity conjugation. Then the next point follows.

3. As a consequence of CPT invariance the energy distribution of π^+ and π^- (the number at a given energy) must be symmetrical between π^+ and π^-, providing that final state interactions among the pions can be neglected.[124]

On a still more subtle level, invoking the approximate SU_3 invariance of the strong couplings,[125] we have point 4.

4. If the CP-violating electromagnetic current, K_μ has a well-defined SU_3 character—that is, is a member of an octet—then from Hermiticity and current conservation it follows that, in the good SU_3 limit, the matrix element of this current between members of an octet has the same form as if CP were conserved. Thus, as a special case, in good SU_3, $\eta^0 \longrightarrow \pi^0 + e^+ + e^-$ is forbidden, to lowest order in α, since, if J_μ^γ is an octet then $\langle\eta^0|J_\mu^\gamma|\pi^0\rangle$ is proportional to $\langle\pi^0|J_\mu^\gamma|\pi^0\rangle$, which vanishes by Hermiticity, current conservation, and CPT.

To prove this we make a more general definition of an antiparticle; that is, we call

$$|\bar{A}\rangle = CPT|A\rangle. \tag{15.255}$$

In fact, with CP violated, we *must* make this definition, since the CP operator is no longer a constant of the motion. With this definition, particles and antiparticles also must have the same masses and lifetimes. From this point of view a "neutral" particle is one which is self-conjugate under CPT; all of its electromagnetic form factors are then identically zero. Thus, for example,

$$\langle\pi^0|J_\mu(0)|\pi^0\rangle = 0, \tag{15.256}$$

since both J_μ and K_μ have the same transformation under CPT (see Chapter 5). This new definition of an antiparticle does not change anything in the analysis above of the K^0 system.[126]

The moral of these examples is that nature appears to do its best to hide any potential CP violation beneath other symmetries. Despite the best efforts of both theorists and experimenters, no CP violation not involving the K^0 has turned up as yet, and the subject is still deeply shrouded in mystery and confusion.[127]

---

interference between the normal weak interaction that violates P and some interaction that violates CP. In the electromagnetic CP-violating theory it might be of order (G is the weak Fermi constant)

$$GM_N e \sim 10^{-20} e \text{ cm},$$

while in a CP violating weak interaction theory it might be about 10^{-3} times smaller, since, on the face of it, the CP-violating weak interaction is about a thousand times smaller than the CP conserving one. Some recent experimental results give

$(-2 \pm 3)e \times 10^{-22}$ cm (P. D. Miller, et al. *Phys. Rev. Letters* **19**:381 (1967).)

$(+2.4 \pm 3.9)e \times 10^{-22}$ cm (O. E. Overseth and R. Roth, *Phys. Rev. Letters* **19**:385 (1967).)

If these results persist they may favor a CP-violating weak interaction with a $\Delta T = \frac{3}{2}$ part (to account for the K^0-decay), although, since the computation of the dipole moment is, in all theories, fairly crude, one cannot, as yet, rule out definitively any of them.

On this somber note we close this chapter and the book. The reader may have the feeling, shared by the writer, that he has been left suspended in midair. But a physics book, unlike a novel, not only has no happy ending, but has no real ending at all.

Index